Nanosensors: Materials and Technologies

Nada F. Atta
Editor

Nanosensors:
Materials and Technologies

International Frequency Sensor Association Publishing

Editor
Nada F. Atta
Nanosensors: Materials and Technologies

ISBN 10: 84-616-5378-5
ISBN 13: 978-84-616-5378-2
BN-20130705-XX
BIC: TJF

Contents

Chapter 3
Properties and Applications of Modified Carbon Nanotubes

Chapter 4
Nanosensors Based on Surfactant Modified Electrodes

Chapter 5
Synthesis and Sensing Applications of Nano-structured Conducting
Polymers and Conducting Polymers-based Nanocomposites

Contributors

Nada F. Atta

Nada F. Atta currently is a Professor of Chemistry at the Department of Chemistry, Faculty of Science, University of Cairo, Egypt. She earned a PhD degree from the University of Cincinnati, Ohio, USA. Current research interests are: new and advanced materials with emphasis on nano-structured materials, imprinted sol-gel materials, molecular recognition, nano-particles modified surfaces for Catalysis and sensors applications.

Ahmed Galal

Ahmed Galal is a Professor of Chemistry at the Department of Chemistry, Faculty of Science, University of Cairo, Egypt. He is currently Visiting Professor, at the University of Kuwait. Research interests are in the areas of electrochemical sensors, nano-materials, conducting Polymers, corrosion and passivity of metals and environmental chemistry. Other research interest is the use of chemically converted graphene in electrocatalysis and surface coating.

Shimaa M. Ali

Shimaa M. Ali is currently Lecturer of Chemistry at the Department of Chemistry, Faculty of Science, University of Cairo, Egypt. She earned a PhD degree from the same university. Research interests are the used of different surfaces for electrocatalytic production of hydrogen and oxygen gases.

Ali M. Abdel-Mageed

Ali M. Abdel-Mageed is Assistant Lecturer at the Department of Chemistry, Faculty of Science, University of Cairo, Egypt. He earned an MSc degree from the same university.

Shereen M. Azab

Shereen Azab is currently a Researcher at the National Organization for Drug Control and Research of Egypt. She earned a PhD degree from the University of Cairo, Egypt.

Ekram H. El-Ads

Ekram El-Ads is Assistant Lecturer at the Department of Chemistry, Faculty of Science, University of Cairo, Egypt. She earned an MSc degree from the same university.

Hagar K. Hassan

Hagar Hassan is Assistant Lecturer at the Department of Chemistry, Faculty of Science, University of Cairo, Egypt. She earned an MSc degree from the same university.

Preface

Sensing is among the most vital necessities of life. The last few centuries have witnessed growing concern for threats from all types of pollutants and the rising number of newly synthesized chemicals that poses human health risks. On the other hand, quality control and quality assurances in many occasions find its need to continuing development in sensors. Therefore, from the security and human well-being point of view the necessity of newly developed sensors is of paramount interest to different sectors of our societies. In particular, fast developed domains of new and advanced materials namely nano-structured, hybrids, and conducting polymers have promised new applications and potentials for sensor technology.

Electrochemistry has developed concurrently with natural science in the first half of the nineteenth century and was further explored with further development of the theory of electricity. Electrochemical science merged independently thereafter while interconnecting different areas of chemistry, physics, biology and engineering. Among the important fields of application where electrochemistry was explored extensively are electrochemical sensors and biosensors among many others. The pH electrode using a glass membrane was introduced in the beginning of the last century by Cremer in 1906, and was followed by ion-selective electrode (ISE) for chloride ion in 1937 by Kholtoff and Sanders. Different materials were introduced thereafter such as in the case of lanthanum fluoride and ceramics for fluoride ions determination and oxygen analyses in 1960. Surface modified electrodes were introduced in the beginning of the 1970s that allowed the employment of several materials that spanned between monolayer(s) as in the case of thiols, different types of materials such as polymeric films, and biological entities such as enzymes.

The perception of employing sensors in all practical fields namely medicine, industrial operations, and food industry triggered the interest in their development and pursuing new discoveries in this important field.

The present book aims at providing the readers with some of the most recent development of new and advanced materials such as carbon nanotubes, graphene, sol-gel films, self-assembly layers in presence of surface active agents, nano-particles, and conducting polymers in the surface structuring for sensing applications. The emphasis of the presentations will be devoted to the difference in properties and its relation to the mechanism of detection and specificity. Miniaturization on the other hand, is of unique importance for sensors applications. The chapters of this book present the usage of robust, small, sensitive and reliable sensors that take advantage of the growing interest in nano-structures. Different chemical species are taken as good example of the determination of different chemical substances industrially, medically and environmentally. A separate chapter in this book will be devoted to molecular recognition using surface templating.

The present book will find a large audience of specialists and scientists or engineers working in the area of sensors and its technological applications. The manuscript will also be useful for researchers working in the field of electrochemical and biosensors since it presents a collection of achievements in different areas of sensors applications. While the book is not intended to explore intensively on the type of materials used in sensor technology, however it will meet the interest of researchers in the field of materials it presents the different applications of the newly discovered advanced materials including carbon-based materials, nano-structures and hybrids in sensors applications.

Chapter 1 is devoted to modern applications of molecularly imprinted materials. Molecular imprinting has been widely studied and applied as an innovative tool for various technological and scientific fields. Molecularly imprinted materials have been applied as trapping elements for separating and extraction of molecules, antibody analogues and receptors, molecular catalysts, as recognition elements for sensory applications, and as membranes for separation of molecules and different processes which are diffusion dependent. Moreover, molecularly imprinted technique has been ameliorated for the chiral recognition and separation of stereo-isomers of the same compound. On the other hand, it introduced a very reasonable solution for one of the most challenging problems in molecular recognition by imprinting big molecules like enzymes, hormones, and other big protein molecules. This category of huge molecules is very difficult to be recognized by ordinary techniques. In addition, molecularly imprinted materials offer very high degree of selectivity and specificity, so they may be named as artificial antibodies.

Chapter 2 presents Graphene as electrochemical sensor and biosensor and describes its synthesis, characterization and applications. Graphene, the mother of all carbon materials, this interesting and

14

promising material that has a wide range of applications was prepared for the first time in 2004 by Andre Geim and his colleague, Konstantin Novoselov, and it was the main reason for their receiving a Nobel Prize in 2010. Since 2004 till now a lot of papers concerning using graphene in many applications have been published in many international journals of various interests. The scope of this chapter is to highlight the previous literatures that concerned with using graphene in the field of electrochemical sensing and biosensing up to date. Along this chapter using of graphene and the modified graphene as electrochemical sensors for neurotransmitters, drug ingredients, some biologically important substances and some hazardous materials are illustrated and discussed. Moreover, we will also focus on using graphene and the modified one as biosensor as well as field effect transistors based on graphene and using graphene as smart sensor.

Chapter 3 describes the properties and applications of modified carbon nanotubes. New carbon materials such as fullerenes, carbon nanotubes, and graphene have attracted tremendous research interest and have led to a Nobel Prize.Nanotubes are members of the fullerene structural family. Their name is derived from their long, hollow structure with the walls formed by one-atom-thick sheets of carbon, called graphene. It is the structure, topology and size of nanotubes that make their properties exciting compared to the parent, planar graphite-related structures. CNT modified electrode shows the properties of electrocatalytic activity and electroseparation. This is because CNT has unique electronic properties and can promote electron transfer reactions, which can be applied for the detection of analytes in a low concentration or in the complex matrix. Recently, nanoparticles have been used as modifier in chemically modified electrode for analysis of drug molecules, metals and ions. Many other applications are foreseen, among which nanoscopic gas sensing in which one property of the nanotube, sensitive to adsorbed molecules, is measured. It is remarkable that these beautiful molecules can be produced in such many different physical and chemical conditions (electric arc discharge, catalytic chemical

Chapter 4 is about the nanosensors based on surfactant modified electrodes. Surfactants have been widely used in chemistry and in particular affecting several electrochemical processes. The area of surface modified electrodes is of particular interest because of its application in sensors and biosensors for biologically and pharmaceutically important compounds and drugs. In some cases, synergistic preconcentration may occur for the studied compounds on the electrode surface, in other cases, surfactant may act as an antifouling and homogenizing agent. Moreover, there is a synergistic effect between the conducting substrate (conducting polymer, carbon paste, gold nanoparticles, self-assembly monolayer, etc.) and surfactant which enhances the use of surfactant modified electrodes as nanosensors with excellent reproducibility, high sensitivity, unique selectivity and exceptional stability. On the other hand, the use of surfactants can be applied for the analysis of drugs with a direct analytical procedure in aqueous, drug formulations and urine samples. This chapter will view the different sensory applications of surfactant modified electrodes in detail.

Chapter 5 presents the synthesis and sensing applications of nano-structured conducting polymers and conducting polymers-based nanocomposites. Because their chemical and physical properties may be tailored over a wide range of characteristics, the use of conducting polymers (CPs) is finding a permanent place in sophisticated electronic measuring devices such as sensors. Better selectivity and rapid measurements have been achieved by replacing classical sensor materials with CPs involving nanotechnology. Nanomaterials of CPs are found to have superior performance relative to conventional materials due to their much larger exposed surface area. In addition, Polymer-based nanocomposites, which incorporate advantages of both nanoparticles and polymers, have received increasing attention in both academia and industry. They present outstanding mechanical properties and compatibility owing to their polymer matrix, the unique physical and chemical properties caused by the unusually large surface area to volume ratios and high interfacial reactivity of the nanofillers. The composites provide an effective approach to overcome the bottleneck problems of nanoparticles in practice such as separation and reuse. Thus, polymers used in sensor devices either participate in sensing mechanisms or immobilize the component responsible for sensing the analyte. This chapter summarizes the preparation methods and sensing applications of CPs nanomaterials and polymer-based nanocomposites. Challenges and future directions are also discussed.

Nada F. Atta

16

Chapter 1

Modern Applications of Molecularly Imprinted Materials

Nada F. Atta and Ali M. Abdel-Mageed

1.1. Introduction

Imprinted materials have been used in the last two decades as smart recognition elements for a wide range of applications. The first use of imprinted materials dates back to 1949, when Dicky et al succeeded to imprint silicate materials by dye molecule [1, 2]. Nevertheless, these trials did not bring the current breakthrough in recognition characteristics and the wide application in different fields [3]. The creation of synthetic tailor-made receptors capable of recognizing desired molecular targets with high affinity and selectivity is a persistent long-term goal for researchers in the fields of chemical, biological, and pharmaceutical research. Compared to biomacromolecular receptors, these synthetic receptors promise simplified production and processing, less costs, and more robust receptor architectures. During recent decades, molecularly imprinted polymers (MIPs) have grown as widely considered mimics of natural molecular receptors suitable for a diversity of applications ranging from biomimetic sensors, to separations and biocatalysis [4].

To get started we will refer first to the general principle underlying imprinting process and possible mechanisms responsible for the binding of the template molecule to a specific material. Molecular imprinting process is based on creating of specific molecular cavities within polymer matrix. The imprinted materials are tailored in such a way that their internal structure is complementary with the template molecule. The ideal description of the outcome of this process is the creation of two structural fragments like a lock which fits only with its own key. These molecular cavities are created by the insertion of template molecule during the synthesis process leaving a three dimensional network with the same fitting structural details of that molecule. Following this synthesis of the template-imprint complex, the molecule should be extracted from the cross-linked host material leaving the molecular cavity [5-7].

Due to their smart and easily tailored characteristic, the applications of molecularly imprinted elements covered a wide range of basic scientific fields, including for example, high performance liquid chromatography [8], food analysis [9], capillary chromatography, solid phase extraction [10] and drug delivery techniques [11]. Among these applications molecular recognition stands as the most versatile use for the molecularly imprinted materials. Usually imprinted materials play the role of recognition element for the specific template molecule. Sensors based on imprinted materials could help to introduce a recognition mechanism of various molecular species with detection limits going down to few ppm. Recognition method is largely correlates to a large extent the enzyme recognition in human bodies. In our bodies, there are a large number of different protein molecules, and cells, without which we cannot survive, are able to work cooperatively in such a way that certain function are carried out very precisely and accurately. The binding characteristics between these molecules depend largely on the fine molecular details including the conformation of molecules, folding of proteins and their complex structure complementarities with each other. One example to manifest this mechanism is interaction between receptors existing on the surface of cell membranes and hormones responsible for activating these cells for a specific function. Following the interaction between a specific hormone and its counterpart receptor on cell membranes (chemical binding), a new complex is built with a new spatial conformation. After that a message of the hormone is transferred in terms of a conformational change which derives the cell to exert certain function. In molecular recognition, the molecules being imprinted can rebind to their molecular cavities with very high degree of selectivity and specificity, so that these materials may be named as artificial antibodies. The most crucial characteristic of the recognition system based on imprinted materials is the degree of selectivity for the template molecule which is governed by the appropriate selection of the carrier material and its ability for interaction and matrix growth around the molecule. Base on this concise introduction we will expand on explaining in much more details the mechanism and factors governing the molecular recognition process.

18

1.2. Imprinting Methodology and Recognition Mechanism

In this section we are much concerned about the basic imprinting methodologies and the mechanism of recognition of template molecules with the imprinted elements. First of all we will talk about the imprinting process in terms of chemical bonding between the template molecule and imprinted matrix. In general, there are two main strategies employed in the molecular imprinting processes (Fig. 1.1). Both strategies are mainly based on the type of interaction with building units of the matrix being imprinted. The first strategy is covalent imprinting which involves covalent interaction between the functional monomer units and template molecules being imprinted. On the other hand the second strategy is non-covalent imprinting which involves non-covalent interaction between the functional monomers and template molecules. In the following paragraphs the two types are discussed in details.

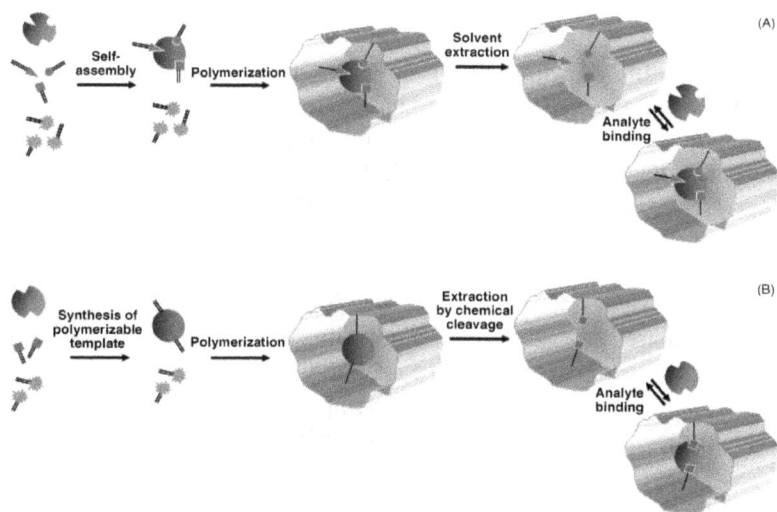

Fig. 1.1. General idea underlying the molecular imprinting process: (A) Molecular assembly between template molecule and functional monomers; (B) polymerization of functional monomers around template molecule.

In principle there are three basic steps involved in synthesis of molecularly imprinted materials whether it uses covalent or non-covalent approach including i) Contact between the template molecule and the pre-selected functional monomer so that a pre-polymerization complex (adduct) is formed; ii) Involves polymerization and cross-linking of the monomer building units around the complex formed in the first step; iii) And finally extraction of the imprinted template molecules leaving well established molecular cavities in the polymeric material. In this aspect it should be stressed out here that the first step is very crucial since it determines stability of the complex structure and consequently recognition capacity and selectivity of template versus other structurally similar molecules. Based on that we can classify two basic imprinting approaches:

1. Covalent Imprinting involves a preliminary step in which a template-monomer unit is covalently formed, then these complex units copolymerize with additional building blocks forming components, with suitable amount of the cross linker, in a suitable solvent able to dissolve both the monomer and template molecules. Molecularly template material is finally obtained with the template being encapsulated inside its matrix [11]. Condensation reactions are the main processes involved in covalent imprinting approaches. It is found that moderate and reversible covalent bonds are formed under mild aqueous conditions while strong covalent bonds, notably ester bonds, are formed under fairly acidic or basic conditions. It should be noted that chemical change occurring due to the formation of chemical bonds between molecular templates and host material is reversed under the effect of specific chemical agents but with keeping the polymerized matrix unchanged (hydrolysis of ester bonds).

The mild covalent imprinting process is found to be more advantageous when compared to strong covalent imprinting. On one hand the imprinting process itself is much easier and needs less processing steps. Most importantly, the extraction of template and its rebinding to the created molecular cavities would be easier in the first case.

A group of different molecules with ability to be imprinted with this approach includes glyceric acid [12], derivatives of mannose [13], galactose and fructose [14], sialic acid [13], castasterone [17] L-dopa [15], and some nucleosides [18]. Also, amino acid derivatives were successfully imprinted by condensation with carbonyl compounds (mainly aldehydes) [17, 19] which is an example of Schiff bases. Moreover, ketals and/or acetals, formed between a diol and a carbonyl compound, have been employed in molecular imprinting protocol. Mono- and di-ketone template molecules have been extensively studied by Shea et al [20-22]. Damen et al [23-25] moved further by using strong covalent bonds in the template-monomer assembly process.

20

Strong covalent bonds, notably ester bonds, were used to assemble the polymerizable template monomer species, followed by incorporation into divinylbenzene-based matrices using suitable reagent or by hydrolysis. Rebinding to the polymers involves esterification reaction of an acyl chloride with the alcohol or displacement of bromide by a salt of the carboxylic acid.

Also semi-covalent imprinting can be affected by making use of a condensation of amides and esters together. In the simplest form a (meth) acrylate ester of the template is copolymerized with the matrix-forming monomer mixture. The template is subsequently removed by hydrolysis, with rebinding of the unesterified template to the polymer. Rebinding process could be attributed to interaction of the template hydroxyl(s) with (meth)acrylic acid residues introduced into the imprinted site [11].

2. Non-covalent imprinting method is based on weak forces of attraction between molecules such as hydrogen bonds, dipole-dipole interactions, and Van der Waals weak forces to generate adducts between template molecules and functional monomers used in the polymerization process. In non-covalent imprinting mechanism, a pre-polymerization adduct (Complex) is formed between the molecular assembly being imprinted and the monomer molecule before polymerization process in the solution mixture. The functional monomers used in the imprinting process arrange themselves around the template molecules depending on the electrostatic interaction which can build between the complementary functional groups. For example hydroxyl groups orients toward oxygen or nitrogen in such a way that a hydrogen bonds arise using the lone pair of electrons of these atoms. After that polymerization process binds the building blocks together around the template molecules creating a network. Upon leaching of the template molecules out of the polymeric network, molecular cavities of the same size and geometry as the template molecules, are left behind. With these molecular cavities the template can rebind through the same forces adopted during the imprinting process.

To the best of our knowledge it was referred for the first time to non-covalent imprinting in 1980 by Mosbach's group [27, 28]. Non covalent imprinting can be affected by a single monomer or a combination of monomers .The first type is the simplest and the most widespread approach. During the last two decades many functional monomers have been used in non-covalent imprinting for different applications. According to their nature these monomers were classified as acidic, basic or neutral. Carboxylic acid based monomers have been widely utilized. Their widespread comes from the fact that they have few bonds with few rotational degrees of freedom, so they have different ways to interact with the template as H-bond donor, H-bond acceptor and through formal ion pair formation as well as weak dipole-dipole interactions. For instance, the group of methacrylic acid derivatives of enhanced properties including, triflouromethyl acrylic acid, methacrylic acid, methyl methacrylate, etc are examples of monomers of acidic properties. Basically vinyl-pyridines is an electron rich compounds, interact strongly with electron deficient molecules having aromatic rings [29]. A variety of neutral monomers include acrylamide which showed superior H-bonding ability compared to methacrylic acid under conditions of low polarity [30].

Imprinting with a combination of monomers have been very attractive to chemists as it introduced a probability of interaction with a wide varieties of molecular assemblies having wide range of chemical structures which in turn gives the chemist much more options in tailoring good recognition matrices able to rebind the template efficiently and selectively. In order that the imprinting process to be a successful one, the adducts formed between the template and the functional monomers need to be stronger than any interaction between the functional monomers themselves. Amino acid derivatives were imprinted in a mixture of methacrylic acid derivatives [31]. A variety of molecules were imprinted in a mixture of acrylamide and 2-vinyl pyridine [32].

1.3. Monomers Selection and Type of Template Molecule

At this point a key question refers to the selection of the building blocks used in synthesis of imprinted materials and different factors affecting the process. Most important in the molecular imprinting process is the selection of the appropriate monomer for a specific template molecule. Monomer selection plays a very critical role for the molecular recognition characteristics of polymerization complex with template molecules without which the starting step for imprinting process and without it, template molecules will not be settled within the host polymer. The polymerization complex stability determines the regioselectivity of the functional groups of both the template and functional monomers. This will consequently affect the ability of the imprinted molecular cavities created within the polymeric matrix structure, to rebind again the template molecule. A very informative example in this aspect is the imprinting of quinine molecule in a series of monomers. Methacrylic acid (MAA), acrylic acid (AA), and 2-vinylpyridine (2-Vpy) were synthesized in the presence of quinine. All the polymers, after leaching out of the template were analyzed for rebinding ability of the template molecule with respect to the molecular cavities created during the polymerization process [33]. The combination of MAA, AA and 2-Vpy, used in synthesis process produced imprinted material with the highest binding affinity for the imprinted template. On the other hand if MAA is dropped from the mixture very low binding affinity was observed. The difference in binding ability of both imprinted mixtures comes from the functionality of Methacrylic acid. It was found that MAA is not only a proton donor, but also it is a proton acceptor which augments its interaction ability. Another example is the imprinting of dopamine and tyramine molecules with a mixture of inorganic orthosilicate monomers. Atta et al used a combination of hybrid monomer solution composed of TEOS and PTEOS which was hydrolyzed. The phenyl group inserted in the sol–gel material is of great importance. This group allows for non-polar interaction with the imprinted molecules which all carry benzene rings so it could act as a π–π interaction center with the template molecules through further steps. Also, it was reported that a combination of functional monomers enhances and promotes the rebinding ability of template molecules to the imprinted films [26].

In this aspect we should refer to the molar ratio of monomer units and template molecules. The template molar ratio plays very determining role in the characteristics of tailored materials [33]. Determination of this ratio depends largely on the type of binding forces building between the functional monomers and template molecules during imprinting process. [34]. In cases where chemical bonds are formed between monomer and template it is not necessary to determine this ratio, since the template dictates the number of functional monomer that can be covalently attached in a stoichiometric manner forming a complex which then polymerizes producing the imprinted polymeric material. For non-covalent imprinting the optimized T/M ratio is achieved empirically by evaluating several polymers made with different formulations with an increasing template concentration [35]. A very informative example about the role of the molar ratio between the template molecules is the imprinting of doxazosin mesylate into methacrylic acid as the functional monomer and triallylisocyanurate as the cross-linker in methanol solution using 2, 20-azobis-isobutyronitrile as the initiator for polymerization process. As indicated in Table 1.1, the rebinding capacity Y for the imprinted polymer toward its template increases with the increase in the template concentration in the polymerization mixture reaching an optimum value at mixture P5 and levels off with higher concentrations. When the amount of template is low there is no chance for complexation with the functional monomer and as a result of that the binding sites formed in the polymer matrix would not be of appreciable importance. On the other hand using higher concentration of the template more than the optimum value has no benefit to the imprinting process, because any excessive amount would not have enough monomers to assemble with it.

Table 1.1. Preparation of MIPs for doxazosin mesylate
and binding capacity of MIPs [36].

No.	Doxazosin mesylate (mmol)	MAA (mmol)	TAIC (mmol)	Doxazosin mesylate:MAA:TAIC	Y (lmol/g)
1.	0	50	200	0:1:4	9.8
2.	20	0	200	1:0:10	6.4
3.	20	10	200	1:0.5:10	25.7
4.	20	40	200	1:2:10	48.3
5.	20	120	200	1:6:10	76.2
6.	20	140	200	1:7:10	77.4

Y is the rebinding capacity of the ratio of doxazosin mesylate (μmol/L) to the amount of imprinted polymer(g) undergoing the rebinding process.

Also temperature during synthesis process is very critical and should be controlled in a way that the interaction between the monomer units is neither too strong nor too weak beside the necessity to keep reasonable rate of polymerization to secure the growth of a three dimensional net work of the polymer around template molecules. Low degrees of temperature are favorable from the point of the stability of the template-functional monomer complex, however higher temperatures are necessary to keep suitable polymerization rates which in turn improve the quality of MIPs recognition sites created by the molecular assemblies during the polymerization process [37]. Lu et al [38] reported that the amount of quinine imprinted in thermally polymerized methacrylic acid, able to rebind to the imprinted film, is greatly affected by the variation of temperature during synthesis process. Polymerization at 15 °C and 55 °C respectively produced specific binding amounts of quinine equal to 65.8 µmol/g and 59.89 µmol/g. This observation might be explained by the fact that at low temperatures, the template-monomer complex is formed better than that at higher temperatures [34]. Moreover several studies have shown that polymerization at lower temperatures forms polymers with higher selectivity versus polymers made at elevated temperatures [38].

Specific functional monomers suitable for imprinting of various templates have been tried and tested in a lot of studies. The factor governing the selection of a specific functional monomer depends on its ability to assemble with the template forming the pre-polymerization complex around which a three-dimensional network grows creating molecular cavity of template molecules. Table 1.2 summarizes different pairs of template-monomers which produced successful imprinting characteristics. The group on molecules include both organic and in organic polymerizable monomers with a wide range of template molecules.

1.4. Control Factors of Recognition Process on Imprinted Materials

1.4.1. Recognition Media and Progen

The solvent used in the process of molecular imprinting is always called as progen. The importance of progen in molecular imprinting mechanism comes from the fact that it plays the central role in formation of the pre-polymerization complexes between the template molecules and functional monomer units by dissolving them together. A good progen should i) dissolve both the template and the functional monomer to the same extent, ii) interaction between the progen and either the template and /or ought not to be stronger than the interaction between the template and the monomer; iii) does not undergo any chemical reactions with either the template and/or the monomer during the imprinting process; iv) be easily evaporated. One important observation to be mentioned in this section is that the molecularly imprinted polymers exhibited higher rebinding effect in the solvent from which it was polymerized (progen). As an example of that, Metsulfuron-methyl (MSM) imprinted in a copolymer of 2-(triflouromethyl) acrylic acid (TF-MAA) and divinylbenzene (DVB) showed the highest binding capacity for MSM [39]. The rebinding experiments characterizing the rebinding capacity of MSM depicted that the MIP binds only MSM in dichloromethane, which was used as a progen during polymerization process.

Table 1.2. Functional monomers used for the imprinting
of wide range of templates.

Template Molecule	Functional Monomers /Cross-linkers used
4-L-phenylalanylaminopyridine (4-LpheNHPy)	Mixture of ethylene glycol dimethacrylate (EGDMA) and methacrylic acid
Tetracycline	Methacrylic acid (MAA) and ethylene glycol dimethacrylate as a cross-linker
D- and L-tyrosine	Polypyrrole
L-tryptophan	Acrylamide (AM) and trimethylacrylate
Ibuprofen	Methylmethacrylate (MMA) or 2-vinylpyridine in the presence of ethylene glycol dimethacrylate (EGDMA) as cross-linker
Uric acid	Acrylonitrile–acrylic acid
Propanolol	Polymer micro-spheres [p-(divinylbenzene)-co-methacrylic acid]
Morphine	Methacrylic acid and trimethylpropane trimethacrylate
Tobacco mosaic virus	Polyethylene glycol hydro-gels
Monocrotophos	Nylon-6 polymer
Doxazosin mesylate	Methacrylic acid as a functional monomer and triallylisocyanate as a cross-linker
2,4-dichlorophenoxyacetic acid	4-vinylpyridine (4-VP) and ethylene glycol dimethacrylate
HIV related protein	Dopamine
D- and L- 3, 4-dihydroxyphenylalanine (D- and L-Dopa) and R-and S-N,N'-dimethylferocenylethylamine (R-FC)	Tetramethylorthosilicate (TMOS) and phenyltrimethylorthosilicate (PMOS)
hemoglobin (Hb)	3-aminopropyl- trimethylorthosilicateda (APTMS), and trimethylpropylorthosiilcate (TMPS)
Caffeine	3-aminopropyltrimethoxysilane (APTMS), was used with tetraethylorthosilicate (TEOS) monomer
Cholesterol assembled with β-cyclodextrine)	Tetraethylorthosilicate (TEOS)
Silica scaffolds	Protein (lysozyme or RNas-A)
L- and D-phenylalanine anilide	Methacrylic acid as a monomer and glycol dimethacrylate (EGDMA)

One more example is the effect of solvent on the interaction between theophylline and methacrylic acid as a functional monomer studied by Zheng Liu and his coworkers [40]. The interaction energy between theophylline and chloroform, tetrahydrofuran (THF) and dimethyl sulfoxide (DMSO) were calculated using B3LYP/6-31+G** level of calculations as indicated in Table 1.3. In parallel to these computations the imprinted polymer was synthesized from the three solvent. 1H NMR spectroscopic investigations of the polymerization mixture showed that theophylline interacts most strongly with methacrylic acid in chloroform which showed the lowest interaction energy with theophylline while DMSO which had the highest interaction energy, showed the minimum interaction in the NMR. This result could be justified by the fact that whenever the solvent interaction with the template is smaller, that would assist to increasing the interaction between the template and the functional monomer.

Table 1.3. ΔE (a.u.) MAA in different solvents [40].

Environment	Energy (a.u.)	ΔE (a.u.)	ΔE (kJ/mol)
Vacuum	−306.49774	-	-
Chloroform	−306.50841	0.01067	28.00253
THF	−306.51003	0.01228	32.25112
DMSO	−306.51305	0.01530	40.17724

However, the role of the progen be visualized much more clearly as a controller of the morphological properties of the imprinted polymer like surface area and porosity of the imprinted material. Porosity controls the diffusion of the template molecules and rebinding efficiency of the imprinted materials. Almost in a lot of cases porosity arises from phase separation of the solvent (progen) and the growing polymer. Progens with low solubility separate early and tend to form larger pores with lower surface area. Conversely, progens with higher solubility phase were found to separate later in the polymerization providing materials with smaller pore radii distribution and greater surface area of the imprinted material [37].

1.4.2. pH of the Rebinding Mixture

The pH of solution from which rebinding process is conducted has great influence on the rebinding process of the template molecule to their imprinted polymers. The solution pH used for the rebinding process affects the rebinding process in two ways. On one hand the electrostatic charge the template carries in solution is greatly controlled by the medium pH and varies with its change especially for molecules carrying functional groups susceptible for the change in pH like carboxylic groups, phenolic groups, amino groups, etc. Consequently, the interaction between the template and the imprinted site would be affected. On the other hand, the change in the electrostatic charge of the molecular species might result in a change in the spatial configuration of the functional groups and their orientation in space with respect to the oriented functional groups in the polymer matrix. As a result of that these molecules might not be able to interact with the imprinted sites in an appropriate way and so on the efficiency of rebinding of the molecules to the molecular cavities will decline [42].

Most important to be discussed here is the effect of solution pH on rebinding process of neurotransmitter and drug like molecules which have been extensively studied during the last few years. Virtually all drug-like molecules are weak acids or bases. This means that they contain at least one site that can reversibly disassociate or associate a proton (a hydrogen ion) to form a negatively charged anion or positively charged cations. Molecules that disassociate protons are acids, and those that associate protons are bases. The reversibility means that a sample is always in equilibrium with some fraction protonated and the rest deprotonated [HA $H^+ + A^-$ or HB $H^+ + B^-$].

By varying the availability of protons, i.e. the acidity of the media, the balance of the equilibrium can be shifted. Alternatively, the pK_a values of a site can be thought of the pH at which the protonated and deprotonated fractions are equal. Experiments were carried out from different pH solutions on dopamine imprinted surface. All pH measurements were carried out from 100_mol L^{-1} with respect to dopamine. As illustrated in Fig. 1.2, the current response increases with the increase in pH value.

At pH-values lower than pK_{a1} (8.57), the protonated form of dopamine predominates than the unprotonated zwitter ion form. On the other hand, at pH-values higher than pK_{a2} (10.08) the deprotonated (phenoxide ion) form is higher than unprotonated zwitter ion. Between pK_{a1} and pK_{a2}, the zwitter ion form predominates over the charged form [29]. This argument clarifies the fact that the protonated dopamine molecule is less interacting with the oriented functional groups in the molecular cavities while the neutral form is best fitted to the imprinted sites so by increasing the pH values current response enhances. On the other hand, there was an increase in the oxidation potential value of the adsorbed species on the imprinted surface with the increase of pH value. This result could be explained from the fact that, positively charged molecules due to protonation of dopamine at lower pH values needs higher polarization potential (overvoltage) for oxidation. This is because the average electronic charge on the molecules is lowered, therefore it is more difficult to withdraw electrons at such a lower potential, while at higher pH values the molecule is either neutral or negatively charged which increases its electro-oxidation. This study was not extended to pH values higher than 9 because the glassy silicon film could decompose under the effect of strong alkaline medium.

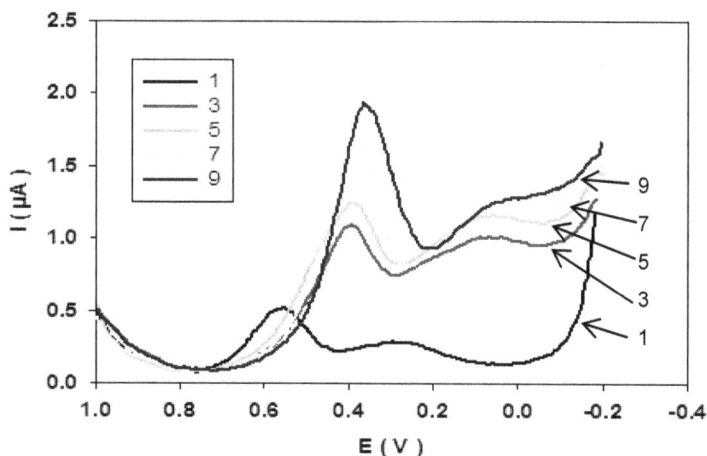

Fig. 1.2. Square Wave Voltammetry (SWVs) of dopamine imprinted film after dipping in 100 µM of dopamine solutions at different values of pH.

1.4.3. Molecular Template Recovery

One of the most important characteristics have to be considered seriously when discussing the performance of molecularly imprinted materials is the extent to which it is able to rebind the template once again into their imprinted sites. The sensitivity of the imprinted material to recover the template to the imprinted sites depends largely on some of the factors discussed in the foregoing sections including in particular the monomer selection, progen, template concentration in particular the post-treatment of the imprinted materials during the extraction step but most important to be discussed here, in this section is the impact of the rebinding efficiency on the molecular recognition characteristics of the imprinted materials. The sensitivity of the imprinted materials is related directly to the number of imprinted sites created during the polymerization process of the molecularly imprinted material. As mentioned above the number of these active sites in the polymer matrix is greatly affected by the extraction of the template and influence of this step is governed by two important factors. On one hand, the solvent shall interact with the polymeric material in such a way that it would change the conformational structure of the molecular cavities rendering them into incompatible structures unable to interact with template in the appropriate way. On the other hand, and most important to be considered is the affinity of the polymeric materials for cross-linking after the extraction process. After the extraction of template, most polymeric materials are aged for certain period of time. The aging process is adopted to cure the polymeric material from any deformation which might occur during the post-imprinting treatments. This period of time would sufficient for extensive polymerization of the imprinted material which assists in blocking of the imprinted sites.

Moreover, the method adopted for the polymer preparation would affect the ability of the polymer to rebind its template. The use of different methods for the polymerization would absolutely vary the morphology of the polymer which results in a different reaction of these surfaces toward the template molecules. One example is the molecular imprinting of morphine as a template molecule in 3,4-ethylenedioxythiophene (EDOT) as a functional monomer using two different polymerization methods. The polymer was polymerized using traditional polymerization conditions and precipitation polymerization conditions. EDOT monomers were used to immobilize the molecularly imprinted particles in a second step. Rebinding experiments revealed that the polymer prepared using precipitation technique could recover higher concentration of morphine. Additionally the imprinted material showed higher selectivity for morphine in the presence of codeine molecules [43].

As a matter of fact, the decline in the number of imprinted sites would take place whatever the procedures being used during the extraction or aging processes but we can decrease its effect by some choices including the following:

- Selecting a solvent for the extraction process, able to leach the template out from the matrix with a minimum interaction with the polymer.

- Keeping the imprinted material in an isolated and dry place under low temperature to decrease the rate of polymerization process.

- Avoiding the use of rebinding solution containing any species which would accelerate the polymerization process as much as possible.

1.4.4. Selectivity Imprinted Materials

One of the challenging difficulties facing the measurement protocols involving an extraction, separation or even detection step using molecularly imprinted materials is the ability to discriminate between different molecular species, especially when they have very similar molecular structures. The selectivity of molecularly imprinted is suppressed by two factors. On one hand, the non-specific adsorption on the non-imprinted sites in the polymer matrix contributes an appreciable interference during the recognition process. Non-specific adsorption can still be considered to be acceptable within the limits which would not affect the molecular selectivity of the imprinted materials especially both the interfering moieties and template molecules which interfere almost to the same extent. Moreover, the effect of non-specific adsorption can be avoided completely by conducting a control experiment. Non-imprinted polymer, prepared using the same conditions of polymerization in the absence of the template, is tested in a solution containing the template molecule in the same concentration used in the routine rebinding experiments then the response of the non-imprinted polymer is subtracted from the basic rebinding experiments. On the other hand, the interference coming from the anchoring of molecules of similar structures partially or completely into the imprinted sties presents the most challenging problem for molecular recognition characteristics of the imprinted materials. The molecular size in conjunction with spatial configuration of the functional groups of the interfering molecules with respect to the imprinted cavities of the original template plays the most influential role on enhancing or deteriorating the molecular selectivity of these materials. In principle, the imprinted materials are tailored so that the imprinted sites would be able to recognize only their templates having the appropriate size and structure. But some molecules may have similar structure which might help them to be partially adsorbed into the molecular cavities in the polymeric materials. Moreover, competitive molecules could be adsorbed completely into these sites when their molecular sizes are slightly smaller than the molecular size of the template. The difference in size facilitates the diffusion of these molecules more easily into the polymer matrix which decreases the selectivity toward the template molecule itself.

As an example the stereoselectivity of sol-gel imprinted with L-histidine. As template, L-histidine was imprinted into a hybrid sol-gel material synthesized from a mixture of functionalized organosilicon precursors (phenyltrimethoxysilane and methyltrimethoxysolane). The L-histidine imprinted and the control sol-gel films were exposed to a series of interfering substances. The sol-gel film has an excellent selectivity for L-histidine. Moreover, the imprinted sol-gel film showed a pronounced selectivity for L-histidine against its stereoisomer D-histidine [44].

On the light of these two examples, it can be concluded that the selectivity of molecularly imprinted polymers originates from the rigid three-dimensional structure created with the polymer matrix during the imprinting process and can be preserved as long as these formation were kept without any deformation of both size and configuration of their internal functional groups spatially oriented with the template molecules.

1.4.5. Repeatability of Rebinding Process

Here in this section, we will discuss the repeatability of the rebinding process on the imprinted films. Repeatability of rebinding measurements means how the repeated measurements could be close to each other. In other words, rebinding process should be repeated more than one time with respect to its imprinted material with an acceptable value of standard deviation of these measurements from the mean value. Repeatability of the rebinding process is very crucial from the point of view of long term use of the imprinted materials and how trustable are the results we get. Repeatability of results is affected by all the foregoing factors described in the previous sections especially those related to the synthesis process and post-treatment of the imprinted material. It can be discussed in terms of stability of molecular cavity created within the imprinted materials. Preserving the number and structure of these active sites without any deformations can be accessed various factors involved in the imprinting and rebinding process. The stability of results obtained depends also to a large extent on the type of imprinted material and its resistance for the external factors like temperature changes, long term exposure to solvents and post-treatment of the imprinted materials. Silica based sol-gel synthesized materials is an example of an excellent material showing good extent of results repeatability. One example is the imprinting of dopamine and tyramine molecules into a hybrid sol-gel material synthesized by acid hydrolysis of a mixture of tetraethylorthosilicate (TEOS) and phenyltriethylorthosilicate (PTEOS) [42]. Dopamine molecularly imprinted film cast on glassy carbon electrodes were tested for repeatability of their measurements. The rebinding experiments were carried out from a solution containing 50 μM dopamine prepared in PBS (pH 7.2 and 10 mM). The batch solutions from which adsorption experiments were prepared freshly for each experiment being done. The rebinding experiment was carried out five times for the same electrode with a period between each two experiments not more than 15 minutes. Results (current responses) obtained from dopamine oxidation on dopamine molecularly imprinted film indicated that the measurements are highly stable and repeatable. The standard deviation from the mean of measurements values calculated for the repeated experiments is very reasonable (7.4 %) with an accepted value 1.440×10^{-8} A, so we can write the mean value of these results as $(1.958\pm0.144)\times10^{-7}$ A. The stability of measurements confirms the fact that adsorption occurs only into well defined molecular sites (molecular cavities created during the imprinting process). Measurements were carried out on three different electrodes each time.

1.5. Modern Applications of Imprinted Materials

1.5.1. Molecular Recognition and Sensing Applications

The growing interest in these classes of materials as sensory elements is associated with their possible potential applications as artificial antibodies able to selectively and with high sensitivity to recognize the respective templated molecular species. Due to their promising characteristics recognition elements tailored using molecularly imprinted polymers find nowadays numerous applications in a lot of fields including medicine, science, engineering, and all fields of science and technology. MIP have been applied as trapping elements for separating and extraction of molecules, antibody analogues and receptors, molecular catalysts, as recognition elements for sensory applications, and as membranes for separation of molecules and different processes which are diffusion dependent.

Moreover, MIPs have been recently applied for a few important sensory applications. This technique has been ameliorated for the chiral recognition and separation of stereo-isomers of the same compound. Also sensing of various biologically active molecules ranging in size from small molecular species like neurotransmitters and amino acids to big size molecules like some specific proteins, enzymes, and hormones is now feasible [45]. This category of huge molecules is very difficult to be recognized by ordinary technique. We will discuss here a little bit in more details the foregoing applications. In particular we will focus on three basic applications of molecularly imprinted materials in biosensors including proteins, chiral compounds and widely used therapeutic compounds.

1.5.1.1. Molecular Recognition of Proteins

Molecularly imprinted polymers (MIPs) have considerable potential for applications in the areas of clinical analysis, medical diagnostics; environmental monitoring and drug delivery [46]. Sensing of big molecules like proteins has been always a challenge for chemists. Recently molecular imprinting introduced a smart solution for this problem through synthesis of different materials with antibody characteristics of desired proteins. Glad et al was the first group to introduce the idea of molecular imprinting of protein molecules [47]. A combination of polysiloxane copolymers imprinted with the dyes rhodanile blue or safranine 0 showed preferential binding of the respective compound. The observed recognition is believed to occur because cavities containing specific binding groups for the dyes at defined positions are developed during the polymerization procedure. Nevertheless, the progress in protein mimicking with imprinting technology faced some difficulties due to: i) Proteins are water-soluble compounds that are not always compatible with mainstream MIP technology, which relies on using organic solvents for the polymer preparation; ii) Proteins have a flexible structure and conformation, which can be easily affected by changes in temperature or in the environment; iii) Proteins have large number of functional groups available for the interaction with functional monomers and similar molecules which make it difficult to tailor specific selective matrix; iv) A selectivity challenge arises from the close similarity with other protein molecules [46]. In principle, the imprinting of proteins can be affected either by complete recognition of the whole molecular species by adsorption into a molecular cavity complementary and shape and size to this molecule or partial recognition of a small sequence of the molecule through surface sorption or partial filling of molecular cavities near to the surface of imprinted material. The first approach is like most of recognition methodologies of different molecules which is based on monomer-template direct interaction during the imprinting process. But most interesting is the second approach involving the templating of the polymeric matrix with a small molecular fragment of the whole molecule which is extracted after that (epitope imprinting). Rachkov et al introduced this principle by imprinting a compound, whose structure represents a small exposed fragment of a larger molecule (as an epitope represents an antigen) [48]. As indicated in Fig. 1.3 the template binds to the imprinted material only with the arm similar to the imprinted fragment.

The progress in this area included recently imprinting of a wide range of protein molecules ranging from small enzymes to viruses and viruses related proteins. A few recent examples protein molecules include imprinting of glucose oxidase in poly(dimethylsiloxane) (PDMS)-modified silica sol–gel (TEOS/PDMS Ormosil) glass [43], Hemoglobin immobilized on electrogenerated silica gel on glassy carbon electrodes [47].

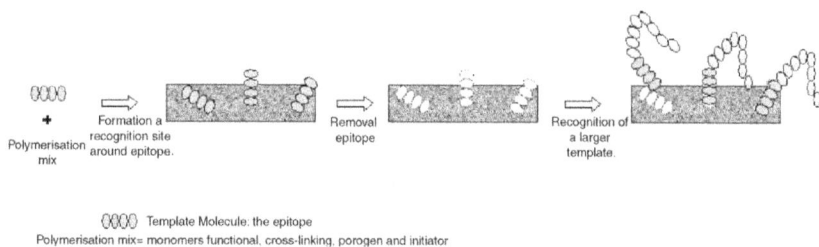

Fig. 1.3. Schematic representation of the epitope approaches [46].

Sensors for viruses and their related proteins have also been recently developed with imprinting technique. HIV related proteins imprinted based on epitope imprinting technique [46]. Hayden et al improved some synthetic polymer receptors for the online monitoring of bioanalytes [45]. The imprinted materials are capable of enriching whole cells, viruses and enzymes on the sensor layer surface. Also Wang et al applied surface molecular imprinting using self-assembled monolayers to design sensing elements for the detection of cancer biomarkers and other proteins. The imprinted elements consist of a gold-coated silicon chip onto which hydroxyl-terminated alkanethiol molecules and template biomolecule are co-adsorbed, where the thiol molecules are chemically bound to the metal substrate and self-assembled into highly ordered monolayers, the biomolecules can be removed, creating the foot-print cavities in the monolayer matrix for this kind of template molecules [50]. The human rhinovirus (HRV) and the foot-and-mouth disease virus (FMDV), which are two representatives of picornaviruses were imprinted in polyurethane layers coated on Quartz crystal microbalance (QCM) [51].

Enzymes are also interesting category of proteins which have been imprinted for various applications ranging from biocatalysis to chemical sensing. Rich et al developed strategies that overcome limitation of water solubility of various monomer molecules and that thus expand its applicability beyond water-soluble ligands. The solubility problem can be addressed either by converting the ligands into a water-soluble form or by adding relatively high concentrations of organic cosolvents, such as tert-butyl alcohol and 1,4-dioxane, to increase their solubility in the lyophilization medium [52].

1.5.1.2. Molecular Recognition of Chiral Compounds

Chiral compounds with one or more asymmetrical carbons are very crucial from the biological point of view since one of the two isomers of a certain compound can be used as a drug for certain disease while the other enantiomer can act as poison. The difficulty in sensing of chiral compounds is just equivalent to difficulty to differentiate between a twines dressed in the same way. The progress in molecular imprinting could also meet the aspirations to analyze chiral compounds with the smart sensing elements, imprinted with one of those asymmetric compounds, with extreme selectivity in presence of other compound. In an earlier work about the use of imprinted materials in recognition of racemates, Hart et al could create synthetic receptor sites for a series of chiral benzodiazepines. A detailed HPLC analysis of binding properties using molecularly imprinted polymers (MIPs) as the stationary phase showed that binding, as measured by chromatographic retention, shows significant dependence on the chiral match or mismatch [53]. A few years later Lu et al [54] introduced a valuable study about the chiral recognition mechanism with molecularly imprinted polymers (MIPs) in aqueous environment. The system used included a group of monomers (ethylene glycol dimethacrylate (EGDMA), methacrylic acid (MAA), and 4-l-phenylalanylamino-pyridine (4-l-PheNHPy)) as the cross-linking monomer, functional monomer and template respectively to assemble the imprinted polymer. In self assembly mechanism a group of functional monomers are organized around the template molecule before polymerization. Chiral compounds of biological activity have been intensively imprinted in the last few days.An acoustic wave sensor detector was developed by Stanly et al for the enantioselective detection between D- and L-serine. The imprinted film showed a detection limit of 2 ppb with a linear detection range up to 0.4 ppm. D- and L-Phenylalanine was imprinted in polyanaline electropolymerized on pt disc electrodes. Discrimination between l- and d-phenylalanine was seen with the application of differential pulse voltammetry [55]. Zhang et al developed a piezoelectric sensor coated with a thin molecularly imprinted sol-gel film for the determination of L-histidine in the liquid phase. Another example is the imprinting of L-tryptophan. A new quartz crystal microbalance (QCM) sensor that provides enantioselectivity to tryptophan enantiomers, with a high selectivity and sensitivity, was fabricated by the use of the molecularly imprinted polymers (MIPs) as the artificial biomimetic recognition material [37]. Molecularly imprinted conducting polymer of pyrrole monomer was developed to enantioselectively differentiate between L- and D- tyrosine amino acid. The imprinted polypyrrole film coated on nickel electrodes imprinted was evaluated by coulometry using an applied positive potential to induce adsorption of the target molecules. All these examples points out to a high potential in selective sensing of racimic mixture with the appropriate selection of imprinting functional monomers able to create selective molecular cavities for one of the enantiomers.

1.5.1.3. Imprinting of Drug Compounds

A lot of drugs of very high medical importance have been imprinted for the purpose of chemical sensing. The antibiotic tetracycline was imprinted in methacrylic acid and ethylene glycol dimethacrylate (EGDMA) crosslinker [56]. The results clearly indicate that synthesized MIPs have higher tetracycline binding ability as compared with corresponding non-imprinted polymers. Non-steroidal anti-inflammatory ibuprofen molecule was also imprinted in methacrylic acid and ethylene glycol dimethacrylate (EGDMA) crosslinker [57]. The imprinted polymer showed recoveries of template molecule of about 80 % with very high selectivity against structurally similar compounds in the rebinding mixture.

1.5.2. Imprinted Membranes

In separation technology imprinted materials provided also a chance to create selective membranes for specific molecules. Polymeric materials are imprinted and applied as membranes which are permeable only for the target molecules being imprinted. A series of enantioselective imprinted polymer membranes for amino acids and peptide derivatives were prepared using oligopeptide as functional monomers [58-61]. Surface grafting of porous membranes with molecularly imprinted polymers showed great enhancement in the efficiency of these porous materials than before. This was clear from the increased diffusion ability of the imprinted molecule through membranes. An interesting example of molecularly imprinted porous membranes was introduced by Piletska et al. In this study commercial porous polypropylene membranes were photografted, using benzophenone as photoinitiator, with the functional monomer 2-acrylamido-2- methylpropanesulfonic acid and the cross-linker *N,N¢*-methylenebis(acrylamide) in water [62]. It was found that molecularly imprinted polymer (MIP) membranes can be obtained which possess group affinity for the template and other triazine herbicides. Photografting methodology of the membrane with template molecule is indicated in Fig. 1.4. The molecularly assembled template-functional monomer complex diffuses into the pores of the original membrane. This step is followed by irradiation of the membrane with light in order to initiate the polymerization process of functional monomers around template molecule and with the walls of membrane at the same time, allowing for the building of stable and robust structure. This work presents the first successful example for molecular imprinting by in situ polymerization in water and on the surface of a commercially available synthetic polymer.

The experimental results for filteration experiments carried out for both the photo-grafted membrane and the non-modified membrane using several different triazines, the *s*-triazines desmetryn, terburtryn, terbumeton, and terbuthylazine, and the triazinone metribuzin; indicated higher filteration capacity for the modified membrane (see Fig. 1.5). In all tests a solute concentration of 10^{-5} M in water was applied. For unmodified PP membranes, 48 % uptake for desmetryn and 100 % for terbuthylazine and terbutryn were observed.

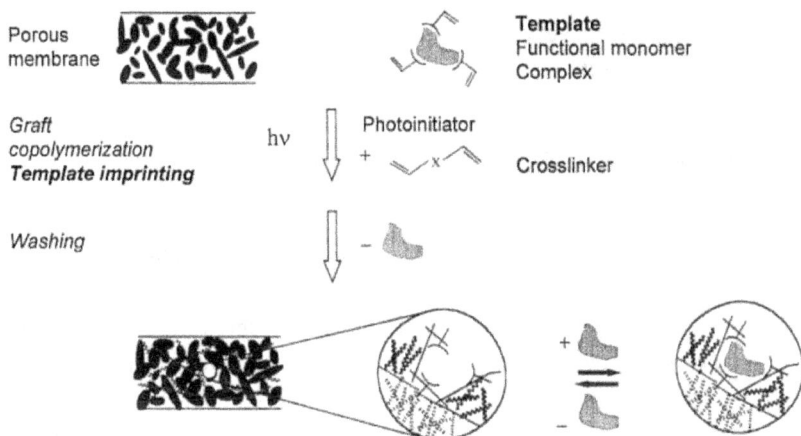

Fig. 1.4. MIP Synthesis via Surface Photografting onto Porous Polymer Membranes [62].

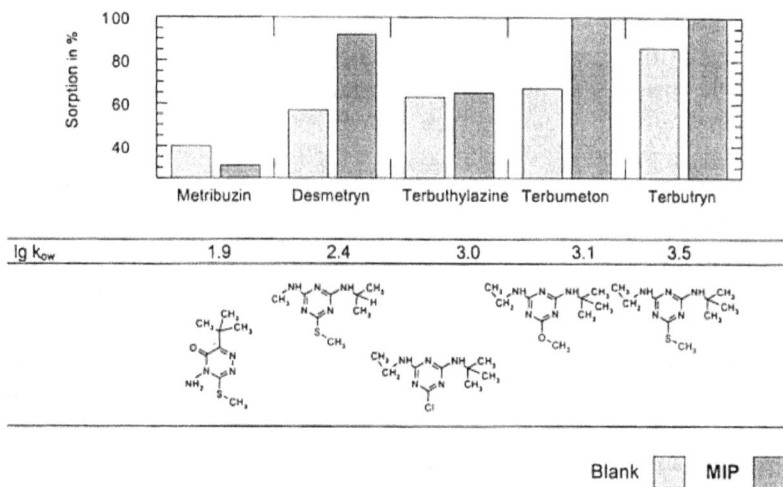

Fig. 1.5. Herbicide sorption for MIP and Blank membranes in filtration experiments with 10 mL of 10^{-5} M herbicide solution in water at a flow rate of 10 mL/min [62].

The novel MIP membranes can be used in a fast preconcentration step, solid-phase extraction, by a simple microfiltration for the determination of herbicides in water. The possibility to introduce specific binding sites into porous membranes by surface imprinting polymerization opens the door for wide range of applications in membrane technologies with very restricted diffusion ability for specific compounds in presence of others.

1.5.3. Preparation of Transition State Antibody Analogues

One of the very smart and promising applications of molecularly imprinted materials is the use as artificial antibody analogues. Recent studies have shown that the efficiency of antibodies and artificial receptors prepared by the molecular imprinting technique is relatively promising. Also these studies demonstrated the possibility of their application in therapeutic trials (clinical trials) [63, 64] which have always been one of the challenging aspects in drug delivery systems. Robinson et al reported that Molecular imprinting of p-nitrophenyl methylphosphonate, a transition state analogue, in poly[4(5)-vinylimidazole] leads to a polymer which hydrolyses p-nitrophenol acetate at an increased rate and which can then be inhibited by addition of the p-nitrophenol methylphosphonate. The print molecule, p-nitrophenol methylphosphonate is structurally similar to the substrate molecule p-nitrophenyl acetate, but contains a tetrahedral phosphoryl group in place of the carboxy group. Therefore, as with other phosphonate esters 4 p-nitrophenol methylphosphonate, is expected to serve as a transition state analogue for the hydrolysis of p-nitrophenyl acetate. The transition state analogue imparted largely similar structural and spatial conformation characteristics for the transition state involved in the hydrolysis of p-Nitrophenol methylphosphonate which could consequently stabilize the transition state species (with respect to the ester; forward rate increased) and increase the total reaction rate. From these examples we might draw a conclusion that antibody analogues (clinical trials) synthesized using molecular imprinting technique can be designed for various molecules which are of biological interest with a very selective rebinding capacity.

Also a synthesized receptor for morphine and for leu-enkephaline was found to be highly efficient not only in organic media but also in aqueous solutions [65]. Diazepam, a tranquilizer drug, synthesized receptors using molecular imprinting technology, showed also high selectivity to those of monoclonal antibodies when compared to similar drugs (benzodiazepines) [65]. From these examples we might draw a conclusion that antibody analogues (clinical trials) synthesized using molecular imprinting technique can be designed for various molecules which are of biological interest with a very selective rebinding capacity.

1.5.4. Chromatographic Applications

Imprinted materials could be used as stationary phases in chromatographic separation equipments especially for enantiomeric separations. The MIP based stationary phases exhibited better characteristic compared to the ordinary separation elements. Imprinted stationary phase materials does not depend only on polarity difference between the stationary and mobile phases or the size of the molecules relative to embedded particles but largely depends on the shape size complementarities with respect to the imprinted sites. This can increase the sensitivity and decreases the separation concentration. Consequently the separation processes will more sharp than in the ordinary cases. These chromatographic materials have allowed the separation of numerous compounds such as naproxen, anti-inflammatory drug [66], timolol, a beta-blocker-A [67], and nicotine [68]. Separation performance can be expressed by a factor α, which is only dependent on the retention times of the two enantiomers, or the resolution factor R_s which takes into account the breadth of the chromatographic peaks. The higher those factors are, the better is the separation. Noticeable separations have been obtained. For example, Rs = 4.3 for a racemic mixture of phenyl-α-mannopyranoside [69]. Krlz et al developed a Thin-layer chromatography plates were made based on the molecular imprinting technique of L-and D-phenylalanine anilide. Strategy adopted by Krlz et al depends on utilizing the higher affinity of imprinted materials toward their templates while the rebinding of other molecules remains limited. So that the retention time of templated molecules on the surface of thin layer supported imprinted materials is longer than other molecules which facilitates their separation [70].

Moreover, it has been reported on many occasions that the extraction on the surface of imprinted polymers can give better results than standard techniques such as liquid-liquid extraction or extraction on C18 phase. An example of these is the extraction of some analgesic drug, from human serum [71]. The extraction with the imprinted polymers showed high selectivity toward the extracted drug.

References

[1]. A. F. H. Dickey, The Preparation of Specific Adsorbents, *Proc. Natl. Acad. Sci.,* Vol. 35, Issue 5, 1949, pp. 227-229.

[2]. F. H. Dickey, Specific Adsorption, *J. Phys. Chem,* Vol. 59, Issue 8, 1959, pp. 695-707.

[3]. K. Raman, M. T. Anderson, C. J. Brinker, Template-Based Approaches to the Preparation of Amorphous, Nanoporous Silicas, *Chem Mater,* Vol. 8, 1996, pp. 1682-1701.

[4]. K. Mosbach, O. Ramstorm, The Emerging Technique of Molecular Imprinting and Its Future Impact on Biotechnology, *Nat. Biotechnol.,* Vol. 14, 1996, pp. 163-170.

[5]. S. Wei, M. Jakusch, B. Mizaikoff, Capturing molecules with templated materials Analysis and rational design of molecularly imprinted polymers, *Anal. Chim. Acta,* Vol. 578, 2006, pp. 50-58.

[6]. S. Mann, S. L. Burkett, S. A. Davis, C. E. Fowler, N. H. Medelson, S. D. Sims, D. Walsh, N. T. Whilton, Sol-gel synthesis of organized matter, *Chem. Mater.,* Vol. 9, 1997, pp. 2300-2310.

[7]. D. Kriz, O. Ramstrom, K. Mosbach, Peer Reviewed: Molecular Imprinting: New Possibilities for Sensor Technology, *Anal. Chem,* Vol. 69, 1997, pp. 345A-349A.

[8]. Q. F. H. Sambe, C. Kagawa, K. K. Kunimoto, J. Haginaka, Uniformly Sized Molecularly Imprinted Polymer for (S)-Nilvadipine. Comparison of Chiral Recognition Ability with HPLC Chiral Stationary Phases Based on a Protein, *Anal. Chem,* Vol. 75, 2003, pp. 191-198.

[9]. O. Ramstrom, K. Skuder, J. Haines, P. Patel, O. Bruggeman, Food analyses using molecularly imprinted polymers, *J. Agric. Food Chem.,* Vol. 49, 2001, pp. 2105-2114.

[10]. P. Spegel, Schweitz, S. Nilsson, Selectivity toward multiple predetermined targets in nanoparticle capillary electrochromatography, *Anal. Chem.,* Vol. 75, 2003, pp. 6608-6613.

[11]. G. Mayes, M. J. Whitcombe, Synthetic strategies for the generation of molecularly imprinted organic polymers, *Adv. Drug Delivery Rev.,* Vol. 57, 2005, pp. 1742-1778.

[12]. G. Wulff, A. Sarhan, Use of polymers with enzyme-analogous structures for the resolution of racemates, *Angew. Chem. Int. Ed. Engl,* Vol. 11, 1972, pp. 341-344.

[13]. G. Wulff, R. Vesper, R. Grobe Einsler, A. Sarhan, On the synthesis of polymers containing chiral cavities and their use for the resolution of racemates, *Makromol. Chem,* Vol. 78, 1977, pp. 2799-2816.

[14]. G. Wulff, S. Schauhoff, Enzyme-analog-built polymers. 27. Racemic resolution of free sugars with macroporous polymers prepared by molecular imprinting. Selectivity dependence on the arrangement of functional groups versus spatial requirements, *J. Org. Chem,* Vol. 56, 1991, pp. 395-400.

[15]. Kugimiya, J. Matsui, T. Takeuchi, K. Yano, H. Muguruma, A. V. Elgersma, J. Karube, Recognition of Sialic Acid Using Molecularly Imprinted Polymer, *Anal. Lett,* Vol. 28, Issue 13, 1995, pp. 2317-2323.

[16]. Kugimiya, J. Matsui, H. Abe, M. Aburatani, T. Takeuchi, Synthesis of castasterone selective polymers prepared by molecular imprinting, *Anal. Chim. Acta,* Vol. 365, 1998, pp. 75-79.

[17]. G. Wulff, J. Vietmeier, Enzyme-analogue built polymers, 26. Enantioselective synthesis of amino acids using polymers possessing chiral cavities obtained by an imprinting procedure with template molecules, *Makromol. Chem*, Vol. 190, 1989, pp. 1727-1735.

[18]. N. Sallacan, M. Zayats, T. Bourenko, A. B. Kharitonov, I. Willner, Imprinting of Nucleotide and Monosaccharide Recognition Sites in Acrylamidephenylboronic Acid–Acrylamide Copolymer Membranes Associated with Electronic Transducers, *Anal. Chem*, Vol. 74, 2002, pp. 702-712.

[19]. G. Wulff, W. Best, A. Akelah, Ion Exchangers, Sorbents, *React. Polymer*, Vol. 2, 1984, pp. 167-174.

[20]. K. J. Shea, T. K. Dougherty, Molecular recognition on synthetic amorphous surfaces. The influence of functional group positioning on the effectiveness of molecular recognition, *J. Am. Chem. Soc*, Vol. 108, 1986, pp. 1091-1093.

[21]. K. J. Shea, D. Y. Sasaki, On the control of microenvironment shape of functionalized network polymers prepared by template polymerization, *J. Am. Chem. Soc*, Vol. 111, 1989, pp. 3442-3444.

[22]. K. J. Shea, D. Y. Sasaki, An analysis of small-molecule binding to functionalized synthetic polymers by 13CP/MAS NMR and FT-IR spectroscopy, *J. Am. Chem. Soc*, Vol. 113, 1991, pp. 4109-4120.

[23]. J. Damenm, D. C. Neckers, On the memory of synthesized vinyl polymers for their origins, *Tetrahedron. Lett*, Vol. 21, 1980, pp. 1913-1916.

[24]. J. Damenm, D. C. Neckers, Memory of synthesized vinyl polymers for their origins, *J. Org. Chem*, Vol. 45, 1980, pp. 1382-1387.

[25]. J. Damenm, D. C. Neckers, Stereoselective syntheses via a photochemical template effect, *J. Am. Chem. Soc*, Vol. 102, 1980, pp. 3265-3267.

[26]. N. F. Atta, M. M. Hamed, A. M. Abdel-Mageed, Computational investigation and synthesis of a sol-gel imprinted material for sensing application of some biologically active molecules, *Anal Chim. Acta*, Vol. 667, 2010, pp. 63-70.

[27]. R. Arshady, L. Glad, K. Mosbach, Synthesis of Substrate-Selective Polymers by Host-Guest Polymerization, *J. Chromotogr*, Vol. 299, 1981, pp. 687-692.

[28]. O. Norrlöw, M. Glad, K. Mosbach, Imprinting of Amino Acid Derivatives in Macroporous Polymers, *J. Chromatogr*, Vol. 299, 1984, pp. 29-41.

[29]. Pietrzyk, R. Wiley, D. McDaniel, Base Strength of Monovinylpyridines, *J. Org. Chem*, Vol. 22, Issue 1, 1957, pp. 83-84.

[30]. Yu, K. Mosbach, Insights into the Origins of Binding and the Recognition Properties of Molecularly Imprinted Polymers Prepared Using an Amide as the Hydrogen-Bonding Functional Group, *J. Mol. Recognit*, Vol. 11, 1998, pp. 69-74.

[31]. O. Ramström, L. I. Andersson, K. Mosbach, Recognition Sites Incorporating Both Pyridinyl and Carboxy Functionalities Prepared by Molecular Imprinting, *J. Org. Chem,* Vol. 58, Issue 26, 1993, pp. 7562-7564.

[32]. Z. H. Meng, J. F. Wang, L. M. Zhou, Q. H. Wang, High Performance Cocktail Functional Monomer for Making Molecule Imprinting Polymer, *Anal. Sci,* Vol. 15, Issue 2, 1999, pp. 141-144.

[33]. D. Yang, N. Takahara, S. W. Lee, T. Kunitake, Preparation of functionalized copper nanoparticles and fabrication of a glucose sensor, *Sens. Actuators, B,* Vol. 130, 2008, pp. 379-386.

[34]. Wensheng, B. Ram, Gupta, Molecularly-imprinted polymers selective for tetracycline binding, *Sep. Purif. Technol.,* Vol. 35, Issue 3, 2004, pp. 215-222.

[35]. F. Liu, X. Liu, S. Ng, N. S. Chan, Enantioselective molecular imprinting polymer coated QCM for the recognition of L-tryptophan, *Sens. Actuators, B,* Vol. 113, Issue 1, 2006, pp. 234-240.

[36]. H. Y. Wang, J. G. Jiang, L. Y. Ma, Y. L. Pang, Syntheses of molecularly imprinted polymers and their molecular recognition study for doxazosin mesylate, *React. Funct. Polym.,* Vol. 64, 2005, pp. 119-126.

[37]. D. Silvestri, N. Barbani, C. Cristallini. P. Giusti, G. Ciardelli, Molecularly imprinted membranes for an improved recognition of biomolecules in aqueous medium, *J. Membr. Sci.,* Vol. 282, 2006, pp. 284-295.

[38]. Y. Lu, C. X. Li, X. D. Wang, P. C. Sun, X. H. Xing, Influence of polymerization temperature on the molecular recognition of imprinted polymers, *J. Chromatogr., B,* Vol. 804, 2004, pp. 53-59.

[39]. D. A. Spivak, Optimization, evaluation, and characterization of molecularly imprinted polymers, *Adv. Drug Delivery Rev.,* Vol. 57, 2005, pp. 1779-1794.

[40]. W. Donga, M. Yan, Z. Liu, G. Wu, Y. Li, Removal of trivalent chromium by electrocoagulation, *Sep. Purif. Technol.,* Vol. 53, 2007, pp. 183-188.

[41]. H. Liang, T. Ling, J. F. Rick, J. Choub, Molecularly imprinted electrochemical sensor able to enantroselectivly recognize D and L-tyrosine, *Anal. Chim. Acta,* Vol. 542, 2005, pp. 83-89.

[42]. N. F. Atta, A. M. Abdel-Mageed, Smart electrochemical sensor for some neurotransmitters using imprinted sol–gel films, *Talanta,* Vol. 80, 2009, pp. 511-518.

[43]. K. C. Hoa, W. M. Yeh, T. S. Tung, J. Y. Liao, Amperometric detection of morphine based on poly(3, 4-ethylenedioxythiophene) immobilized molecularly imprinted polymer particles prepared by precipitation polymerization, *Anal. Chim. Acta,* Vol. 542, 2005, pp. 90-96.

[44]. Z. Zhang, H. Liao, H. Li, L. Nie, S. Yao, Stereoselective histidine sensor based on molecularly imprinted sol-gel films, *Anal. Biochem.,* Vol. 336, 2005, pp. 108-116.

[45]. O. Hayden, R. Bindeus, C. Haderspöck, K. J. Mann, B. Wirl, F. L. Dickert, Mass-sensitive detection of cells, viruses and enzymes with artificial receptors, *Sens. Actuators, B*, Vol. 91, 2003, pp. 316-319.

[46]. A. Bossi, F. Bonini, A. P. F. Turner, S. A. Piletsky, Molecularly imprinted polymers for the recognition of proteins: the state of the art, *Biosens. Bioelectron.*, Vol. 22, 2007, pp. 1131-1137.

[47]. O. Nadzhafova, M. Etienne, A. Walcarius, Direct electrochemistry of hemoglobin and glucose oxidase in electrodeposited sol–gel silica thin films on glassy carbon, *Electrochem. Commun.*, Vol. 9, 2007, pp. 1189-1195.

[48]. M. Glad, O. Norrlow, B. Sellergren, N. Siegbahn, K. Mosbach, Use of Silane Monomers for Molecular Imprinting and Enzyme Entrapment in Polysiloxane-Coated Porous Silica, *J. Chromatogr.*, Vol. 347, 1985, pp. 11-23.

[49]. A. Rachkov, N. Minoura, Recognition of oxytocin and oxytocin-related peptides in aqueous media using a molecularly imprinted polymer synthesized by the epitope approach, *J. Chromatogr.*, A, Vol. 889, 2000, pp. 111-118.

[50]. Y. Wang, Z. Zhang, V. Jain, J. Yi, S. Mueller, J. Sokolov, Z. Liu, K. Levon, B. Rigas, M. H. Rafailovich, Potentiometric sensors based on surface molecular imprinting: Detection of cancer biomarkers and viruses, *Sens. Actuators, B*, Vol. 146, 2010, pp. 381-387.

[51]. M. Jenik, R. Schirhagl, C. Schirk, O. Hayden, P. Lieberzeit, D. Blaas, G. Paul, F. L. Dickert, Sensing Picornaviruses Using Molecular Imprinting Techniques on a Quartz Crystal Microbalance, *Anal. Chem.*, Vol. 81, 2009, pp. 5320-5326.

[52]. J. O. Rich, V. V. Mozhaev, J. S. Dordick, D. S. Clark, Y. L. Khmelnitsky, Molecular Imprinting of Enzymes with Water-Insoluble Ligands for Nonaqueous Biocatalysis, *J. Am. Chem. Soc.*, Vol. 124, 2002, pp. 5254-5255.

[53]. B. R. Hart, D. J. Rush, K. J. Shea, Discrimination between Enantiomers of Structurally Related Molecules: Separation of Benzodiazepines by Molecularly Imprinted Polymers, *J. Am. Chem. Soc.*, Vol. 122, 2000, pp. 460-465.

[54]. Y. Lu, C. Li, Study on the mechanism of chiral recognition with molecularly imprinted polymers, *Anal. Chim. Acta*, Vol. 489, 2003, pp. 33-43.

[55]. E. M. Sheridan, C. B. Breslin, Enantioselective Detection of D- and L-Phenylalanine Using Optically Active Polyaniline, *Electroanalysis*, Vol. 17, 2005, pp. 532-537.

[56]. W. Cai, R. B. Gupta, Molecularly-imprinted polymers selective for tetracycline binding, *Sep. Purif. Technol.*, Vol. 35, 2004, pp. 215-220.

[57]. K. Farrington, F. Regan, Investigation of the nature of MIP recognition: The development and characterisation of a MIP for Ibuprofen, *Biosens. Bioelectron.*, Vol. 22, 2007, pp. 1138-1146.

[58]. J. M. Krotz, K. J. Shea, Imprinted Polymer Membranes for the Selective Transport of Targeted Neutral Molecules, *J. Am. Chem. Soc.,* Vol. 118, 1996, pp. 8154-8155.

[59]. M. Yoshikawa, J. Izumi, T. Kitaao, Enantioselective Electrodialysis of N-α-Acetyltryptophans through Molecularly Imprinted Ploymeric Membranes, *Chem. Lett.,* 1996, pp. 611-612.

[60]. M. Yoshikawa, J. Izumi, T. Kitao, Sakamotos, Alternative molecularly imprinted polymeric membranes from a tetrapeptide residue consisting of D- or L-amino acids, *Makromol. Rapid. Commun.,* Vol. 18, 1997, pp. 761-767.

[61]. M. Ykoshikawa, T. Fujisawa, J. Izumi, T. Kitao, S. Sakamoto, Molecularly imprinted polymeric membranes involving tetrapeptide EQKL derivatives as chiral-recognition sites toward amino acids, *Anal. Chim. Acta,* Vol. 365, 1998, pp. 59-67.

[62]. S. A. Piletsky, H. Matuschewski, U. Schedler, A. Wilpert, E. V. Pilteska, E. T. A. Thiele, M. Ulbricht, Surface Functionalization of Porous Polypropylene Membranes with Molecularly Imprinted Polymers by Photograft Copolymerization in Water, *Macromolecules,* Vol. 33, 2000, pp. 3092-3098.

[63]. G. Vlatakis, Li. Andersson, R. Müller, K. Mosbach, Drug Assay Using Antibody Mimics Made by Molecular Imprinting, *Nature,* Vol. 361, 1993, pp. 645-647.

[64]. O. Ramstörm, L. Ye, K. Mosbach, Artificial antibodies to corticosteroids prepared by molecular imprinting, *Chem. Biol.,* Vol. 3, 1996, pp. 471-477.

[65]. M. Kempe, K. Mosbach, Receptor Binding Mimetics: A Novel Molecularly Imprinted Polymer, *Tetrahedron Lett.,* Vol. 36, 1995, pp. 3563-3566.

[66]. Katz, D. ME, Molecular imprinting of bulk, microporous silica, *Nature,* Vol. 403, 2000, pp. 286-289.

[67]. L. Fischer, R. Müller, B. Ekberg, K. Mosbach, Direct enantioseparation of . beta. -adrenergic blockers using a chiral stationary phase prepared by molecular imprinting, *J. Am. Chem. Soc.,* Vol. 113, 1991, pp. 9358-9360.

[68]. A. Zander, P. Findlay, T. Renner, B. Sellenger, Analysis of Nicotine and Its Oxidation Products in Nicotine Chewing Gum by a Molecularly Imprinted Solid-Phase Extraction, *Anal. Chem.,* Vol. 70, 1998, pp. 3304-3314.

[69]. G. Wulff, M. Minarik, Template Imprinted Polymers for HPLC Separation of Racemates, *Liq. Chromatogr.,* Vol. 13, 1990, pp. 2987-3000.

[70] D. Krlz, C. Berggren, Li. Anderson, K. Mosbach, Thin-Layer Chromatography Based on the Molecular Imprinting Technique, *Anal. Chem.,* Vol. 66, 1994, pp. 2636-2639.

[71] Li. Anderson, A. Paprica, T. Arvidsson, A highly selective solid phase extraction sorbent for pre-concentration of sameridine made by molecular imprinting, *Chromatographia,* Vol. 46, 1997, pp. 57-62.

Chapter 2

Graphene as Electrochemical Sensor and Biosensor: Synthesis, Characterization and Applications

Ahmed Galal, Nada F. Atta and Hagar K. Hassan

2.1. Introduction

2.1.1. Graphene, the Mother of all Carbon Allotropes

Graphene was considered as the missing allotrope of pure carbon materials, after the discovery of graphite, diamond, fullerenes and carbon nanotube [1]. Graphene is a member of 2-dimentional materials discovered by Andre Geim's research group at University of Manchester in 2004, and prepared by the so-called "scotch-tape" technique [2]. In this approach, single- or few-layered graphene sheets can be obtained by repeatedly peeling highly oriented pyrolytic graphite (HOPG) using scotch tapes [3]. As a result of their work, Geim and his colleague Novoselov were awarded the 2010 Nobel Prize in physics. Graphene is a one atom thick structure that consists of a hexagonal array of Sp^2- bonded carbon [4-6] atoms and looks like a honey comb [4, 7, 8]. It can be considered as the mother of all carbon materials where, single walled carbon nanotubes (SWCNTs) can be viewed as the result of rolling up a sheet of graphene [9, 10] in the same time graphite, one of the oldest known carbon allotropes, can be considered as a stacked graphene sheets while graphene can be coalesced to form fullerene as clearly presented in Fig. 2.1 [11].

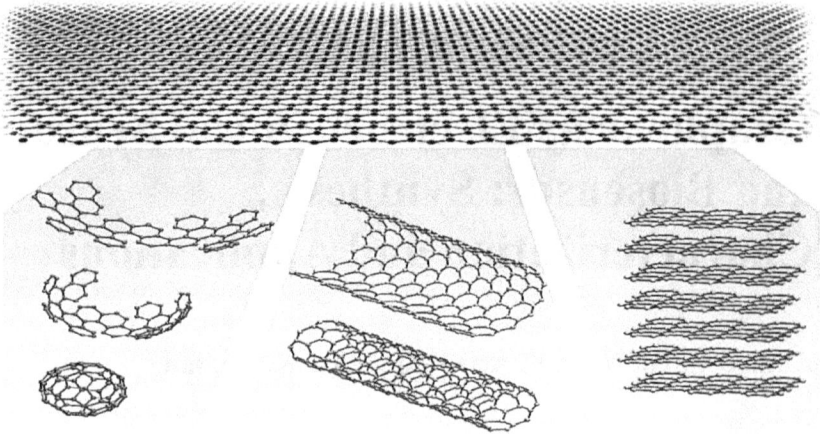

Fig. 2.1. Schematic diagram shows how graphene can be the mother of all carbon materials [11].

2.1.2. Properties of Graphene

Nowadays,graphene has some unique properties which make it one of the most interesting materials even compared to carbon nanotubes (CNTs). The optical and electrical properties of reduced graphene oxide depend on the spatial distribution of the functional groups and structural defects [12]. Some of its properties are: has large theoretical surface area about 2620 m^2/g [7, 10, 13], chemically stable and almost impermeable to gases, withstand large current densities , high thermal [1] and chemical conductivity [14,15] including extremely high charge (electrons and holes) mobility (230,000 cm^2/Vs) with 2.3 % absorption of visible light [16, 17] and thermal conductivity (3000 W/mK), outstanding mechanical properties [15, 18], with highest strength (130 GPa), large amount of edge planes/defects[14], in addition to its cheap production cost compared to CNTs [15]. One of the most interesting aspects of the graphene is its highly unusual nature of charge carriers,which behave as massless relativistic particles (Dirac fermions) [19, 20]. At these dirac points the valence and conductionbands are degenerated, making graphene a zero band gap semiconductor and useful as an electronic material. These properties make graphene promising for potential application in electrochemical field [14] which represents a promising catalyst carrier in the next generation of carbon-based supports [15].

2.1.3. Graphene Preparation

To understand the electrochemical properties of graphene, one must first understand the way that graphene is synthesized. Several typical methods have been developed and reviewed for graphene preparation, such as chemical vapor deposition (CVD) [21], micromechanical exfoliation of graphite [2, 9], "peeling-off" highly oriented pyrolytic graphite (HOPG) [22]. Other methods are: epitaxial growth on an electrically insulating surface [23] which can be used in addition to the peeling off HOPG, to study the electronic properties of graphene, solvothermal synthesis [23], and the reduction of GO are also known.

The reduction of GO can be carried out by chemical methods, using different reductants such as hydrazine, dimethylhydrazine, hydroquinone or $NaBH_4$ or by thermal methods which is believed to be a green method in which no hazardous reductants are used but this process requires rapid heating up to 1050 °C in an oven under argon atmosphere or up to 800 °C using hydrogen gas [24]. The first successful attempt to prepare a single sheet of graphene is that performed by Professor Andre Geim and his colleague using "scotch-tape" technique. Although the graphene sheets thus produced are poor in the production yield and controllability on the geometry, this mechanical exfoliation technique features high reliability and offers graphene sheets very high quality. Even nowadays present, the simple but efficient "scotch-tape" technique is still a common one to fabricate high-quality samples [3] applications, produce ~99 % multi-layer graphene which is called graphene nanoplatelets (GNPs) and only 1 % of monolayer graphene sheets [25]. Here, it is worth to mention that the electrochemistry of GNPs is as interesting and important as graphene (monolayer graphene). On the other hand, some of other preparation methods for graphene sheets will be mentioned briefly in the next subsections.

2.1.3.1. Unzipping of CNTs

One of the methods for graphene preparation is "unzipping" of CNTs. It is possible to use hypermanganate chemical oxidation of CNTs, which open up after the treatment [26], or plasma etching [27] of multiwalled CNTs (MWCNTs) to produce graphene nanoribbons starting from CNTs. However, this method has many disadvantages for example:the graphene sheets synthesized from MWCNTs brings with it the danger of introducing metallic impurities intercalated in MWCNTs with all sorts of negative consequences leading to the unpredictable electrochemical behavior of such materials as well as toxicity. Therefore, these methods should be avoided when using graphene for electrochemistry unless the MWCNTs are free of impurities that can also represent another challenge due to the high cost of purification of CNTs.

2.1.3.2. Arc-discharge Method

The arc-discharge method does not require a metallic catalyst during the preparation of graphene as in case of unzipping of CNTs. However, otherwise it is timesaving, facile method and a high quality graphene can be synthesized in a large scale. In this method, two graphite rods are put close to each other (about 1-2 mm) in a steel chamber which is cooled by water. A constant current about 100-150 A is applied in H_2, NH_3, He or air atmosphere which result in formation of plasma. The arc-discharge method is useful to prepare pure, B- and N-doped graphene. If the discharge occurred in presence of H_2 and B_2H_6 the resulting graphene will be B- doped graphene [28]. While using a mixture of He and NH_3 [29] or H_2 and pyridine [30] atmosphere leads to formation of N-doped graphene. The role of H_2gas here is terminating the dangling carbon bonds and prevents formation of closed structure at the same time it prevents rolling of graphene sheets into CNTs and graphitic polyhedral particles. On the other hand, NH_3 also plays another important role for the formation of N-doped graphene besides the as-described function. It decomposes to nitrogen and hydrogen under the high temperature during discharge process. The highly reactive hydrogen terminates the dangling carbon bonds at the edge of graphene sheets thereby preventing the graphene sheets from closing [28]. Some of applications of doped-graphene will be discussed later in this chapter.

2.1.3.3. Chemical Vapor Deposition (CVD) Method

Chemical vapor deposition is a simple, scalable and cost-efficient method to prepare single and few-layer graphene films on various substrates; it opens a new route to large-area production of high-quality graphene films for practical applications. Evaporated Ni film on SiO_2/Si wafers or copper foils are ideal substrates for graphene synthesis. It is assumed that the carbon atoms dissolve into the Ni crystalline surface, and at certain temperatures, they arrange epitaxial on the Ni (111) surface to form graphene. In this method the source of carbon is usually methane and it was found that using of diluted methane leads to formation of few layers of graphene (about 5 layers) while using of concentrated methane leads to formation of multilayers of graphene. Other carbon precursors such as ethylene, acetylene, ethanol, and isopropanol can be used also [31]. On the other hand, CVD technique has some disadvantages where, the control of regular parameters such as temperature, pressure and time compared to other deposition techniques also inevitably encounter is not easy. In a successful CVD process the transport kinetics of gas species tends to be complicated with convection and diffusion being dominating in different regions of a reactor [3]. Hence CVD is a more complex process than other processes [3].

2.1.3.4. Thermal Reduction of Graphene Oxide into Graphene

This method based on oxidation of graphite to prepare graphene oxide (GO) as starting material for graphene synthesis i.e. GO was reduced thermally to produce graphene. The thermal method is believed to be a green method in which no hazardous reductants are used but this process requires a rapid heating (>2000 °C min^{-1}) up to 1050 °C [32] in an oven under argon gas or up to 800 °C under hydrogen gas [33]. The thermal reduction of GO is accompanied by elimination of O_2, CO, CO_2 and H_2O as predicted from TGA data. Some other literatures reveal thatGO can be thermally reduced into graphene at lower temperature using octadecylamin (150-200 °C) [34], reducing solvents such as butanol (120-200 °C) [35] or N, N-dimethylacetamide (DMAc) by which graphene was prepared at atmospheric pressure and at temperature below 150 °C [25].

As pointed out by Kaner and co-workers, the thermal reduction techniques have some obvious drawbacks, namely, the incompatibility of the thermal reduction process under some conditions and the long processing time is required [36].

2.1.3.5. Chemical Reduction of GO into Graphene

The chemical reduction methods of exfoliated GO are usually performed in presence of reducing agents such as hydrazine hydrate, ammonium hydroxide, ethylenediamine or sodium boronhydride. These methods provide a promising approach for the efficient large scale production of chemically converted graphene (CCG) sheets [37-40]. Several researches proved that hydrazine hydrate is the best reducing agent that can be used for chemical reduction of GO and it produces very thin graphene layers [41, 42].

Microwave irradiation (MWI) has been used for the synthesis of a variety of nanomaterials including metals, metal oxides, bimetallic alloys and semiconductors with controlled size and shape without the need for high temperature or high pressure [39-45]. It has also been used for the synthesis of soluble single wall carbon nanotube derivatives [46] and for the exfoliation of graphite intercalation compounds [47]. The main advantage of MWI over other conventional heating methods is heating the reaction mixture uniformly and rapidly. Due to the difference in the solvent and reactant dielectric constants, selective dielectric heating can provide significant enhancement in the transfer of energy directly to the reactants which causes an instantaneous internal temperature rise. It also allows the simultaneous reduction of GO and a variety of metal salts thus resulting in the synthesis of metallic and bimetallic nanoparticles supported on the graphene sheets [41].

2.1.3.6. Electrochemical Preparation of Graphene

Electrochemical reduction of GO to graphene has drawn great attention due to its fast and green nature [48-50] and no hazardous materials are required. The electrochemical method was adopted as an effective and controllable alternative technique for the modification of electronic states. This is done by adjusting the external power source to change the Fermi energy level of the electrode surface [48], which reduces GO in the presence of direct current (DC) bias [51]. It is predicted that electrochemical reduction can be used to fabricate highly ordered and controllable graphene sheets on electrode materials and that it may be feasible to establish a "green" and fast method [15].

The electrochemical reduction of graphene can be carried out via several approaches; the first approach is by immobilization of GO on the surface of electrode then a negative potential is applied or by using cyclic voltammetry leading to the electroreduction of GO into graphene [52]. Another methods are either by electrophoretic deposition of GO followed by its electroreduction [15] or via one step preparation in which a simultaneous electrodeposition of GO and its electroreduction can be carried out in one step where the electrolyte is GO itself [53]

Fig. 2.2 displayed the TEM image of GO suspension, CV obtained
during the simultaneous electrodeposition and electroreduction, CV of
the as-prepared graphene in PBS solution in addition to SEM image of
the obtained graphene.

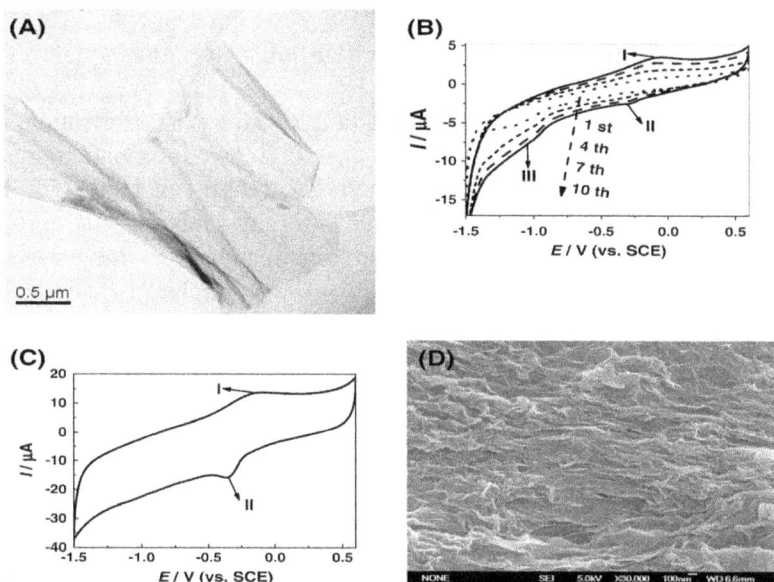

Fig. 2.2. (A) TEM image of GO dispersed in 0.067 M, pH 9.18 PBS;
(B) CVs depicting electrochemical reduction of 1.0 mg mL^{-1} GO in PBS
(0.067 M, pH 9.18) on a GCE at 10 mV s^{-1}; (C) CV of an EG/GCE in PBS
(0.067 M, pH 9.18) at 10 mV s^{-1}, and (D) SEM image of EG film
(10 electrodeposition cycles) modified GCE. (with permission from [53]).

2.1.3.7. Reduction of GO by Laser Radiation

Recently, new methods for graphene preparation based on irradiation of
GO have been reported such as laser reductions of graphite oxide [54]
and photothermal deoxygenation of graphene oxide [55] by camera.
El-Kady et al. [56] used a standard LightScribe DVD optical drive to
do the direct laser reduction of graphite oxide films to graphene. They
reported that, the resulted graphene reveals high mechanical property,
electrical conductivity (1738 siemens per meter) and specific surface
area (1520 square meters per gram), which helped them to use it as
powerful electrochemical capacitor.

53

During this chapter we will try to make a comparison between the electrochemistry of graphene prepared by different approaches because as we mentioned before the method of graphene preparation not only affects the number of layers obtained but also alters its morphology and purity that play an important role in the electrochemical processes.

2.1.4. Surface and Spectral Characterization of Graphene

Surface techniques are very important class for characterization of graphene and the modified graphene since they give information about the nature of graphene. Transmission electron microscope (TEM), atomic force microscope (AFM) and scanning electron microscope (SEM) represent the widely used techniques for the surface characterization of graphene and functionalized graphene. Besides the surface techniques, the spectral methods of analysis such as Raman spectroscopy and x-ray diffraction (XRD) also play an important role for both elucidation of the successful preparation of graphene and for its characterization.

2.1.4.1. Transmission Electron Microscope (TEM)

Transmission electron microscopy is a two-dimensional microscopy technique. It is difficult to probe the third dimension by this technique. The only way to measure the thickness of graphene layers is by performing cross-sectional TEM which can be accomplished by mounting the sample such that the electrons from the gun run parallel to the surface of the graphite layer. Naturally, graphene edges tend to fold and often show wrinkles in the sheet. Close analysis of the folded edge allows for the ability to count the number of layers [11].

2.1.4.2. Atomic Force Microscope (AFM)

By atomic force microscopy (AFM) technique, one is able to view the graphene layer in all the three dimensions. Several studies using AFM (Fig. 2.3) showed that, graphene thickness will range from 0.5 nm to 1.5 nm depending on the chemical contrast between the graphene layer and the substrate. For this reason, determination of the thickness of monolayer of graphene by AFM has to be complemented by other available techniques such as TEM as mentioned earlier [11]. The AFM might also use in quantitative measurements of surface forces, relating these to surface charge and other interactions [57].

2.1.4.3. Scanning Electron Microscope

It is one of the most versatile instruments available for the examination and analysis of the microstructure morphology and chemical composition characterizations [58]. A scanning electron microscope with a conventional secondary electron detector (SEM) and an energy dispersive x-ray analyser (EDX) were used to examine the surface microstructure and atomic composition of the surface region of graphene [59].

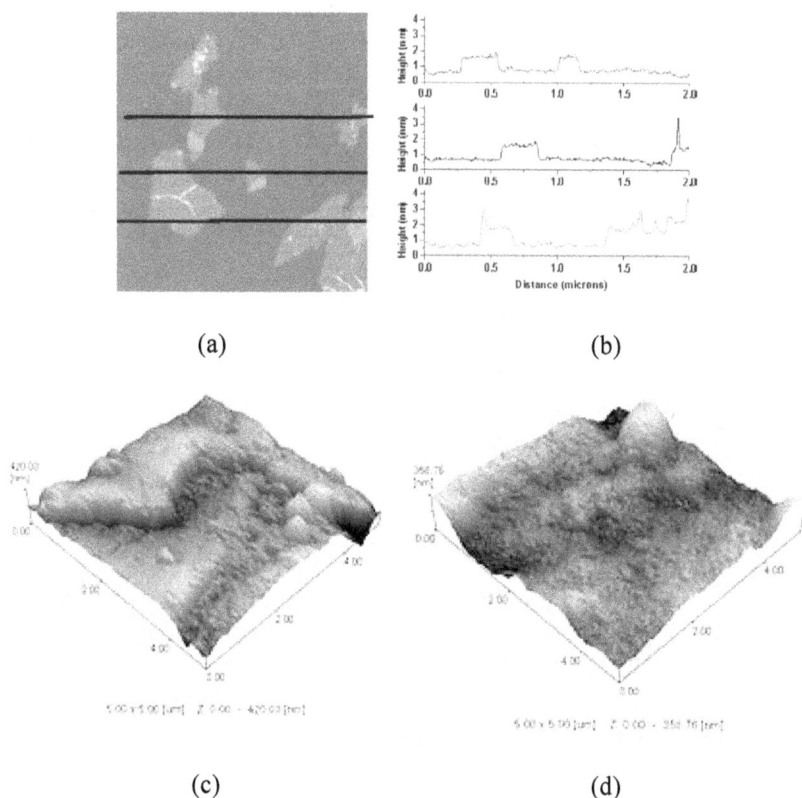

(a)   (b)

(c)   (d)

Fig. 2.3. A non-contact mode AFM image of exfoliated GO sheets (A)
with three height profiles acquired in different locations (B) (with permission
from [38]) and 3D-AFM images for CCG (C); ECG (D)
(with permission from [60]).

2.1.4.4. Raman Spectroscopy

Raman spectroscopy is one of the most widely used techniques to characterize the structural and electronic properties of graphene including disorder and defect structures, defect density and doping levels [61, 62]. Raman spectroscopy of carbon materials offers a very important versatility in distinguishing the structures of carbon material such as amorphous carbon (a-C), tetrahedrally amorphous carbon (ta-C), graphite, highly oriented pyrolytic graphite (HOPG), carbon nanotubes, carbon fullerens, diamond or graphene. Since graphene is a single layer of graphite, its Raman signature should contain most of the features that are contained in the graphite's Raman signature (Fig. 2.4) [11].

The Raman spectrum of graphene is characterized by three main features, the G mode arising from emission of zone-center optical phonons (usually observed at 1575 cm^{-1}), the D mode arising from the doubly resonant disorder-induced mode (1350 cm^{-1}) and the symmetry-allowed 2D overtone mode (2700 cm^{-1}) [63, 64]. The shift and line shape associated with these modes have been used to distinguish single, free-standing graphene sheets from bilayer and few-layer graphene (FLG) [61, 62].

There are two distinct features that stand out in a monolayer graphene that one cannot see in even two-layered graphene: (1) the slight upshift of the G peak by about 5 cm^{-1} and (2) the enhancement of graphite's shoulder peak to second order D peak (named traditionally as G`) to the expense of the main G` peak [11].

2.1.4.5. X-Ray diffraction (XRD)

XRD is a very powerful complementary technique to confirm the successful preparation of graphene. When graphene is prepared by the reduction of GO, XRD can distinguish between graphite, GO and finally graphene where, the XRD patterns of these materials are completely different (Fig. 2.5). In addition to the confirmation of the successful reduction of graphene; XRD used for the modified graphene especially with nanoparticles to identify the phases, orientation and the crystalline structure of the supported nanoparticles.

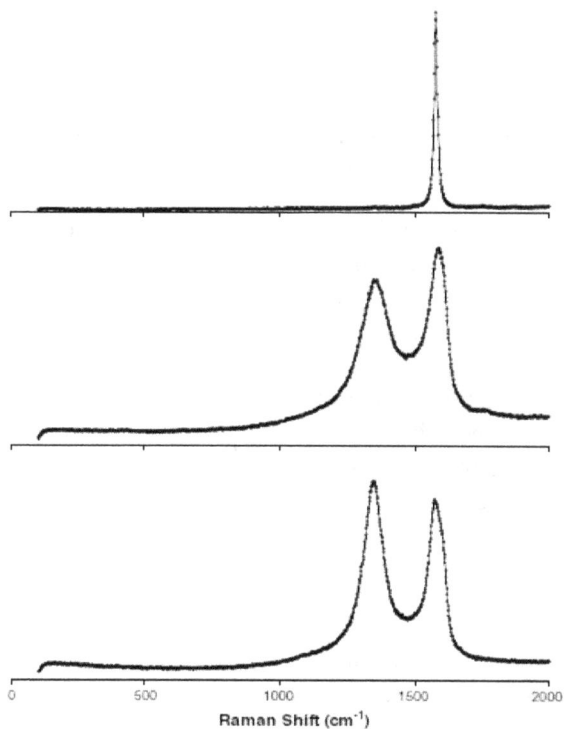

Fig. 2.4. Raman shift of graphite(top), GO (middle), and reduced graphene oxide (bottom). (With permission from [38]).

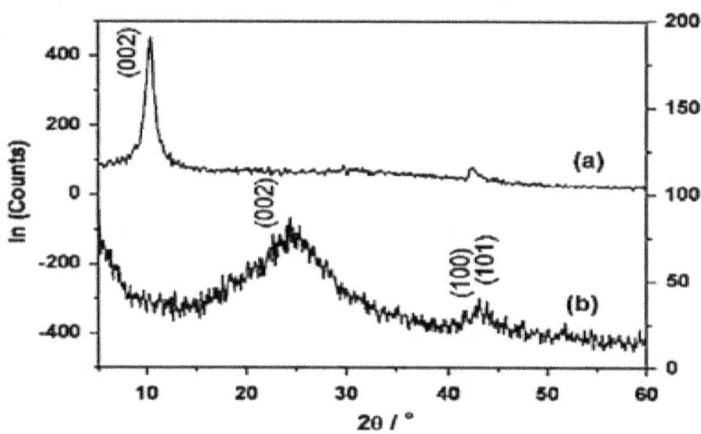

Fig. 2.5. XRD patterns of graphite oxide (a) and graphene (b). (With permission of [65]).

2.1.5. Applications of Graphene

The unique properties of graphene make it one of the most interesting and attractive material nowadays. Due to its high conductivity and large surface area, graphene is the best candidate to be a catalyst or a catalyst support for many processes. In addition to its applications as catalyst support for electrochemical oxidation of small organic molecules such as methanol [7, 10] and ethanol [10] graphene, was used also for oxidation of larger organic molecules such as glucose where, Yaojuan Hu et al. used the electrodeposited gold nanoparticles on graphene sheets for glucose oxidation [66]. Another example of using graphene as a catalyst or catalyst support is that carried by Ki Rak Lee et al. where they obtained a high catalytic activity toward oxygen reduction by using nitrogen doped graphene [67]. Besides its using in the field of electrocatalysis, graphene is widely used in sensing applications including gas sensors [68, 69], chemical sensors [70, 71], biosensors [72-79] and metal sensors [80-82].

Other applications of graphene based on its high conductivity and high capacitive property have been achieved in the field of electronics [83] energy storage and conversion (supercapacitors [84], batteries [85], fuel cells [86-91], solar cells [83, 92], and bioscience/biotechnologies.The next generation of optoelectronic devices requires transparent conductive electrodes to be lightweight, flexible, cheap, environmentally attractive, and compatible with large-scale manufacturing methods. Graphene is becoming a very promising candidate due to its unique electrical and optical properties. recently, Junbo Wu *et al.*, researchers at Stanford University, successfully demonstrated the application of graphene in organic light-emitting diodes (OLEDs) for the first time. OLEDs consist of active organic luminescent structure sandwiched between two electrodes, one of which must be transparent [93].

The extraordinary properties of graphene may lead to revolution in many technology areas. In this chapter we are focusing on the applications of graphene and modified graphene in sensing and biosensing areaand through the next lines we will follow up the use of graphene in these interesting areas upto date.

2.1.6. Electrochemical Sensors

The electrochemical sensors (ECS) have been proven as an inexpensive and simple analytical method with remarkable detection sensitivity, reproducibility, and ease of miniaturization rather than other instrumental analysis methods. A chemical sensor is a small device that can be used for direct measurement of the analyte in the sample matrix. Ideally, such a device is capable of responding continuously and reversibly and does not perturb the sample. By combining the sample handling and measurement steps, sensors eliminate the need for sample collection and preparation.

Electrochemical sensors represent an important subclass of chemical sensors in which an electrode is used as the transduction element. Such devices hold a leading position among sensors presently available, have reached the commercial stage, and have found a vast range of important applications in the fields of clinical, industrial, environmental, and agricultural analyses.

There are different methods for classification of electrochemical sensors. One of them is based on output signal from electrochemical sensor. In this manner electrochemical sensors classified into two major classes; potentiometric and voltammetric sensors [94].

2.1.6.1. Potentiometric Sensors

It is a type of chemical sensor that may be used to determine the analytical concentration of some components of the analyte gas or solution. These sensors measure the electrical potential of an electrode when no current is flowing. In potentiometry, information on the composition of a sample is obtained through the potential appearing between two electrodes. Selective potentiometric electrodes are currently widely used in many fields, including clinical diagnostics, industrial process control, environmental monitoring, and physiology. For example, such devices are used in nearly all hospitals around the globe for assessing several physiologically important blood electrolytes (K^+, Na^+, Ca^{2+}, Mg^{2+}, H^+, Cl^-) relevant to various health problems. The equipment required for direct potentiometric measurements includes an ion-selective electrode, a reference electrode, and a potential-measuring device (a pH/millivolt meter that can read 0.2 mV or better) [95].

2.1.6.2. Voltammetric Sensors

In voltammetry a time-dependent potential is applied to an electrochemical cell, and the current flowing through the cell is measured as a function of that potential. A plot of current as a function of applied potential is called a voltammogram, providing quantitative and qualitative information about the species involved in the oxidation or reduction reaction. Direct voltammetry of more substance at bare electrodes takes place in high overpotential and at nearly potential close to each other [96] In addition, the direct oxidation results in electrode surface contamination (fouling) due to the adsorption of oxidized products. These cases result in poor sensitivity, poor selectivity and unstable analytical signals. To avoid the above obstacles in electroanalytical methods, modified electrodes have been developed [97].

There is undesirable trend in electroanalytical articles of calling any electrochemical sensor detecting a biologically important compound (e.g., neurotransmitter, cofactor of enzyme, DNA base or free DNA strand) an "electrochemical biosensor". According to the IUPAC name, the electrochemical biosensor is the device that used for quantitative or semi-quantitative analysis in which a biological recognition element used in contact with an electrochemical transduction element [98]. Other than this condition we just call it as a "sensor".

2.2. Graphene as an Electrochemical Sensor

Papakonstantinou and co-workers [99] were the first researchers who used graphene-based nanomaterials as electrochemical sensor. Where, they prepared graphene on the surface of silica substrate in shape of nanoplatelets that known nowadays with graphene nanoplatelets (GNPs). Graphene nanoplatelets are several hundred stacked graphene sheets which have a thickness of several tens of nanometer. They reviled that, GNPs have fast electron-transfer kinetics for ferro/ferricyanide redox system and provide a good performance for simultaneous determination of a mixture of dopamine, ascorbic acid and uric acid compared to the bare GCE and edge-plane pyrolytic graphite (EPPG) electrode.

2.2.1. Graphene as Neurotransmitters Sensor

Dopamine (DA), an important neurotransmitter, plays a significant role in the function of the central nervous, renal and hormonal systems [100]. Dopamine detection attracts great of interest reanalysis due to its electrochemical activity [101-103]. Deficiency of DA may cause several diseases and neurological disorders such as Schizophrenia, Huntington's disease, and Parkinson's disease [104].

The most common compounds that are always coexisted with DA in biological samples are Ascorbic acid (AA) and uric acid (UA). Ascorbic acid (AA) is well known for its antioxidant property [105] While, uric acid is a relatively water-insoluble end product of purine metabolism in humans and is excreted via urine. An abnormally high level of UA (DA and/or AA) is usually a symptom of illness[106].Both AA and UA have very close oxidation potential to this of DA which results in overlapping their electrochemical signals [107]. So, they represent the most common interfering materials during the electrochemical detection of DA that make the discrimination of DA in their presence is difficult at bare electrodes [105, 107]. Carbon materials including carbon nanotubes [108], boron-doped diamond [109], carbon nanofibers and graphite [111] and recently grapheneRepresent a way to solve the problem of interference due to their promising characteristics such as their resistance to acid/basic media, possibility to control up to certain limits, their porosity and surface chemistry and easy recovery of precious metals when they used as metal support which result a low environmental impact [72].

As we mentioned before the unique properties of graphene make it the most interesting material nowadays that can be used in electrochemical sensing field. Since it was discovered in 2004, till now a lot of researches that use graphene and modified graphene for many applications have been published. However, the first article used graphene in the field of electrochemical sensing was in 2008 by Papakonstantinou and co-workers [98] as we mentioned before. Starting from this time, 2008, till now many papers concern using graphene in this area have been published. For example: Ying Wang et al. [72] used CVD technique to prepare graphene nanoflakes and dispersed the prepared graphene in chitosan solution to form graphene-Cs homogeneous mixture. Graphene-Cs modified GCE was used as a selective dopamine sensor where ascorbic acid was completely eliminated by using this graphene modified electrode.

The selectivity of graphene to DA comes from the π- π interaction between phenyl structure of DA and planar hexagonal carbon structure of graphene which makes the electron transfer is much feasible.Unlike DA, AA shows different molecular structure which results in weak interaction with graphene sheets [72]. Yang-Rae Kimet al. [65] used graphene modified glassy carbon electrode for the determination of DA in presence of ascorbic acid and obtained a lower detection limit 2.64 µM. Pt nanoparticles/polyelectrolyte-functionalized ionic liquid/graphene sheets (Pt/PFIL/GS) nanocomposite was synthesized in one pot by Fenghua Li, et al. Then the resulting Pt/PFIL/GS was used for the simultaneous determination of DA and AA [112].

Recently, graphene modified with Polyvinylpyrrolidone (PVP) was also used for determination of dopamine in presence of large concentration of AA amperometrically by Qin Liu et al. [113] a wide linear range (5×10^{-10} to 1.13×10^{-3} mol/L) with lower detection limit (0.2 nM) was obtained. It is worth to mention that graphene in this work was prepared electrochemically by CV from 0.0 to -1.5 V where, GO that is immobilized on GCE was the working electrode and electroreduced in 0.1 M PBS (pH 5.0). On the other hand, the simultaneous determination of ternary mixture of DA, AA and UA using graphene and graphene modified electrodes was also investigated in many published papers for example: Han et al. [114] used a chitosan-Gra modified electrode to resolve the oxidation potentials of AA, DA and UA. Similarly, Liu et al. [115] demonstrated that Gra–Au nanocomposite is a promising material for simultaneous electroanalysis of AA, DA and UA. Moreover, Xianqing Tian et al. [116] used gold nanoparticles-β-cyclodextrin- graphene modified electrode for the simultaneous determination of this ternary mixture using square wave voltammetry with lower detection limit of 10, 0.15 and 0.21 µM, for AA, DA and UA, respectively and the electro-oxidation of AA, DA and UA at AuNPs-CD-Gra/GCE is displayed in Fig. 2.6.

Doping of graphene sheets with foreign atoms changes the intrinsic properties of them. One of the doping atoms used is nitrogen due to its atomic size and strong valence bonds which is appropriate for carbon materials [117]. As we mentioned previously, nitrogen-doped graphene sheets (NGS) can be obtained using the chemical vapor deposition (CVD) method [29], the arc discharge method [118], and the nitrogen plasma process [116]. Zhen-Huan Sheng group used nitrogen doped graphene for the simultaneous determination of DA, AA and UA that showed a wide linear response for AA, DA and UA in the concentration range of 5.0×10^{-6} to 1.3×10^{-3} M, 5.0×10^{-7} to 1.7×10^{-4} M and 1.0×10^{-7} to 2.0×10^{-5} M with detection limit of 2.2×10^{-6} M, 2.5×10^{-7} M and 4.5×10^{-8} M (S/N = 3), respectively.

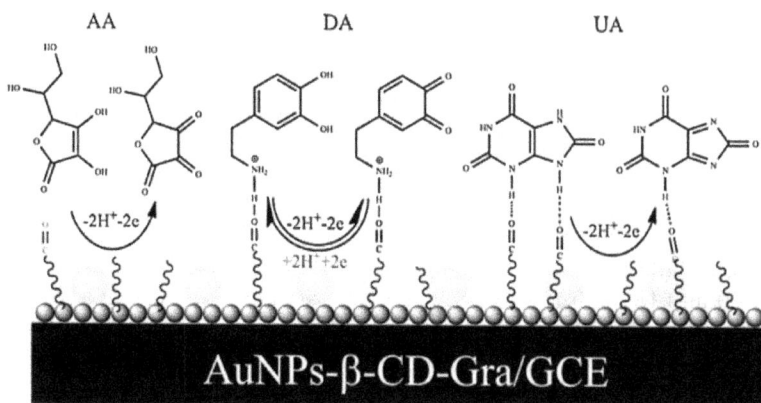

Fig. 2.6. The electro-oxidation of AA, DA and UA at AuNPs–β-CD-Gra/GCE (with permission from [116]).

This represents one of the most recent approaches for the sensitive determination of DA with low limit of detection (LOD). The highly electrocatalytic activity of NGS could be due to the fact that nitrogen atoms in NGS may interact with these molecules via hydrogen bond, which can activate the hydroxy and amine groups and accelerate the charge transfer kinetics of these molecules at NGS surface [119]. They revealed the enhancement of the reversibility of DA signals compared to bare GCE not only due to the π- π interaction between graphene sheets and DA but also to the hydrogen bond that is formed between the hydroxyl or amine group of DA and nitrogen atoms doped in graphene sheets. The second reason is also responsible for the enhancement of the electron transfer of AA at GNS. Additionally they concluded that GNS accelerates the electron transfer and decrease the overpotential of UA. All of these results leading to the simultaneous determination of DA, AA and UA with sufficient peak separation as shown in Figs. 2.7 and 2.8.

Away from DA, some other researchers studied the determination of epinephrine (EP) as an importance neurotransmitter in mammalian central nervous systems. EP is an electroactive molecule, whose electrochemical behaviors have been studied [120-126] and some methods have also been reported for its determination [127–129].

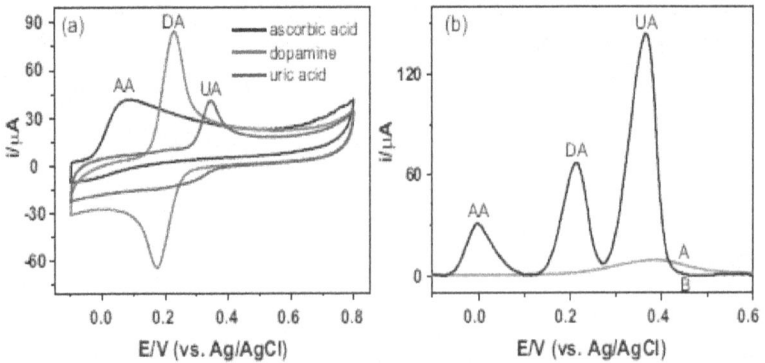

Fig. 2.7. (a) Cyclic voltammograms (CVs) of 1.0 mM AA, 1.0 mM DA and 1.0 mM UA in 0.10 M PBS (pH 6.0) at NG modified GCE at a scan rate of 100 mV/s; (b) Differential pulse voltammograms (DPVs) for 1.0 mM AA, 0.05 mM DA and 0.10 mM UA in a 0.1 M PBS (pH 6.0) at bare GCE (A) and NG/GCE (B), respectively(with permission from [117]).

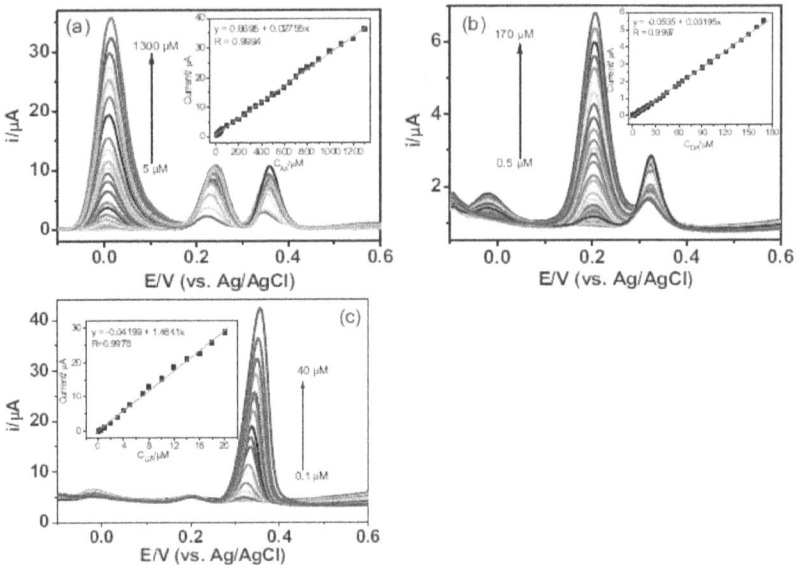

Fig. 2.8. DPV profiles at NG/GCE in 0.1 M PBS (pH 6.0) (a) containing 4 μM DA, 2 μM UA and different concentrations of AA from 5 μM to 1300 μM, (b) containing 100 μM AA, 5 μM UA and different concentrations of DA from 0.5 μ M to 170 μM, and (c) containing 100 μM AA, 5 μM DA and different concentrations of UA from 0.1 μM to 40 μM. Inset: plots of the anodic peak current as a function of AA, DA and UA concentrations. (with permission from [117]).

Graphene/Au modified GCE was used as a chemical sensor for determination of EP by F. Cui and X. Zhang [130]. Theydetermined EP concentration in a linear range 5.0×10^{-8} - 8.0×10^{-6} mol L^{-1} with lower detection limit as low as 7.0 nM. They mentioned that GR/Au nanocomposites could be synthesized in one pot by chemical co-reduction. In this case nano-Au layers separate the GR sheets and prevent graphene sheets from aggregation upon π-π stacking interaction. Table 2.1 shows a comparison between different surfaces including graphene and graphene modified electrodes for detection of some neurotransmitters.

2.2.2. Graphene as Electrochemical Sensor for Drug Ingredients

Paracetamol (acetaminophen, N-acetyl p-aminophenol) is widely used as an antipyretic and analgesic drug. It is used to reduce fever, relieve coughing, colds, and pain including muscular aches, chronic pain, migraine headache, backache, and toothache [136, 137]. Using of paracetamol with limited amount does not exhibit any harmful side effects. However, overdosing and the chronic use of paracetamol results in toxic metabolite accumulation that leads to kidney and liver damage [138]. Xinhuang Kang et al. [139] used graphene modified glassy carbon electrode for the direct electrochemical determination of paracetamol in $NH_3.H_2O$-NH_3Cl buffer/pH 9.3 with detection limit as low as 3.2×10^{-8} M. they found that paracetamol shows an irreversible behavior with relatively weak redox currents on the surface of unmodified GC. While it shows two well-defined redox peaks on using graphene modified GCE with peak separation about 42 mV. They found that in addition to the presence of graphene which enhances the reversibility of paracetamol, it also accelerates the electrochemical reaction leading to more enhancing of the current response compared to bare GCE. They attributed the lower overpotential of paracetamol on graphene-modified GCE to the π-π interaction between graphene and hence, paracetamol becomes easier to be oxidized.

Yang Fan, et al. [77] modified the graphene sheets with TiO_2 nanoparticles and used nafion as an antifouling coating to form Nafion/TiO_2–GR modified GC electrode. The results indicated that, inclusion of TiO_2 nanoparticles greatly enhanced the electrochemical performance of graphene toward paracetamol where, the presence of TiO_2 increases the current response to 1.9 fold compared to the case of Nafion/GR/GC electrode. The measured detection limit was found to be 2.1 × 10^{-7} M which is still higher than the unmodified graphene surface that is reported by Xinhuang Kang et al. Another sensor based on TiO$_2$-graphene nanocomposite but in this time in presence of poly (methyl red) (PMR/TiO$_2$–GR/GCE) was fabricated by Chun-Xuan Xu et al [140].

Table 2.1. A comparison between different surfaces including graphene and graphene modified electrodes for detection of neurotransmitters.

Electrode	Analyte	Detection limit	Linear range (µM)	Reference
OT-HDT/Au[a]	DA	90 µM	200-1200	(131)
	AA	90 µM	300-1400	
PPy-TDS/Au[b]	DA	0.4 µM	1-500	(132)
	AA	0.4 µM	1-500	
	UA	0.6 µM	1-500	
MWCNT/CCE [c]	DA	0.3 µM	0.5-100	(133)
	AA	7.71 µM	15-800	
	UA	0.42 µM	0.55-90	
OMC/Nafion[d]	DA	0.5 µM	1-90	(134)
	AA	20 µM	40-800	
	UA	4.0 µM	5-80	
Chitosan-Gra/GCE[e]	DA	1.0 µM	1.0-24	(135)
	AA	2.0 µM	2.0-45	
	UA	50 µM	50-1200	
AuNPs– β-CD–Gra/GCE[f]	DA	0.15 µM	0.5-150	(116)
	AA	0.21 µM	0.5-60	
	UA	10	30-2000	
NGS	DA	0.25 µM	0.5-170	(117)
	AA	2.2 µM	5-1300	
	UA	0.045 µM	0.1-20	
Graphene modified GCE	DA	2.64 µM	4-100	(72)
PVP/GR/GCE[g]	DA	0.2 nM	0.0005-1130	(113)
Au/GR/GCE	EP	7.0 nM	0.05-8	(130)

a) Au nanoparticles/ self-assembled monolayer of 1,6-hexanedithiol and 1-octanethiol /gold electrode.
b) Polypyrrole–tetradecyl sulfate film modified gold electrode.
c) Multiwalled carbon nanotube modified glassy carbon electrode.
d) Ordered Mesoporous Carbon/Nafion Composite Film.
e) Chitosan-graphene modified GCE.
f) Gold nanoparticles–β-cyclodextrin–graphene-modified electrode.
g) Polyvinylpyrrolidone graphene modified electrode.

Methyl-red (2-[4-(dimethylamino)phenylazo]benzoic acid) with molecular formula NC_6H_4COOH is an organic dye and typical aromatic azo compound. It has a dimethyl amine group, an azo group, and a carboxyl group, which is promising to adsorb paracetamol. They found that the electrochemical oxidation of paracetamol was improved by TiO_2–GR/PMR composite film. Where, a remarkable enhancement of the peak current beside a negative shift in the value of the peak potential was observed at TiO_2–GR/PMR. By comparing between different modified electrodes, they found that the oxidation peak shifted negatively by 41 mV, 123 mV and 139 mV at the TiO_2–GR/GCE, PMR/GCE and PMR/TiO_2–GR/GCE, respectively compared to bare GCE. The results revealed that the enhancement in the electrochemical response of PMR/TiO_2–GR/GCE compared to the other surfaces is due to the adsorption of paracetamol from the solution to the modified electrode surface through hydrogen bond and physical adsorption. It is worth to mention here that the detection limit is still slightly higher than that is reported by Xinhuang Kang et al [140] but it is better than that of Nafion/TiO_2/GR reported by Yang Fane et al. [77].

Morphine (MO), a group of narcotic pain relieve drugs, is a highly effective and preferred drug for moderate treatment of severe pains [141]. So, it is frequently used to relieve severe pain for patients, especially for those who undergo a surgical procedure or carcinomatosis. Usually morphine used in biological samples to monitor therapeutic levels in patients and drug concentrations in human and animal pharmacokinetics studies [142]. Also MO can be used to investigate of opiate abuse for epidemiological purposes of drug abuse control as well as for forensic cases as an indicator of heroin usage [143]. However, when overdosed or abused, MO is toxic and can cause disruption in the central nervous system [141, 144-147].

Bare electrodes always suffer from lack of sensitivity, reproducibility and stability during the electrochemical determination of MO [148] so; modification of the bare electrode with active substance, chemical mediators or using appropriate catalyst and catalyst support can solve this problem. For example, Jin and co-workers prepared a cobalt hexacyanoferrate modified carbon paste electrode combined with HPLC and successfully detected morphine in vivo [149]. A Prussian blue-modified indium tin oxide (ITO) electrode was used for morphine determination by Ho and his group [149] molecularly imprinted electrodes for morphine determination was also investigated [150]. Moreover, multiwalled carbon nanotubes (MWCNTs) [151] and

MWCNTs modified preheated glassy carbon electrode have also been used for the morphine detection [152]. Recently, graphene modified glassy carbon electrode was also used for determination of MO [153-155]. Aso Navaee et al. used graphene modified GCE for the first time for simultaneous detection of three major alkaloids in illicit heroin samples, heroine, morphine and noscapine [155]. In addition, an electrochemical sensor for codeine; highly interferent with morphine detection, was fabricated using composite graphene and Nafion film modified GCE. A detection limit of 1.5×10^{-8} mol L^{-1} was obtained and the proposed method was successfully applied to the determination of codeine in both urine samples and cough syrup [156].

Noscapine is the second most abundant constituents in opium after MO which is usually used as antitussive drug [155] while it has no analgesic activity or abuse potential just as codeine but unlike MO [157]. Heroin (3,6-diacetylmorphine, diamorphine) is a synthetic opiate and obtained from MO. It also has analgesic activity and introduced to medicine in 1898 as a cough suppressant [149]. Aso Navaee and his colleagues reported that the simultaneous determination of MO, noscapine and heroin is difficult due to the overlapping between their voltammograms and to use the voltammetric methods for their simultaneous determination a previous separation methods should be carried out [155].

On the other hand, Mohammad Hadi Parvin [158] succeeded in fabrication of graphene past electrode (GPE) then he used it for the voltammetric determination of chlorpromazine (CPZ). Chlorpromazine is a drug that can be used in management of psychotic conditions. He found that the electrochemical activity of graphene paste electrode toward both potassium ferricyanide and CPZ is greatly better than carbon paste as shown in Fig. 2.11. He attributed this higher electrochemical activity of graphene to increasing the electron transfer rate as results from its high conductivity. Moreover, the recorded lower detection limit of CPZ at GPE is 6.0 nM.

2.2.3. Graphene as Electrochemical Sensor for Biologically Active Compounds

Hydrogen peroxide (H_2O_2) is the general enzymatic product of oxidase which is important in biological processes. Additionally, it can be used as an essential mediator in food, pharmaceutical, clinical, industrial and environmental analyses [78]. So, it is important to develop a sensitive, reproducible and stable sensor for detection of H_2O_2 effectively. Construction of sensor for H_2O_2 based on decreasing the oxidation/reduction overpotential was reported by Yuyan Shao et al. [159]. Sen Liu et al. [160] used Ag nanoparticles/graphene (AgNPs/GN) composite for electrochemical determination of H_2O_2. They revealed that Ag can attach with graphene sheet through two ways, firstly through formation a coordination bond between Ag particle and -COO⁻ groups that are present at the edge of graphene sheet, the second possibility is by formation a true bond between -COO⁻ and Ag⁺ ion. AgNPs/GN showed remarkable catalytic activity toward H_2O_2 compared to bare GCE and limit of detection 0.5 μMin a linear range (0.1-100 mM H_2O_2). Zeng et al. [161] fabricated a glucose and maltose biosensor based on layer by layer assembly of graphene, glucose oxidase (GOx) and glucoamylase (GA). The detection limit and sensitivity were 0.168 mM (S/N = 3) and 0.261 μA mM^{-1} cm^{-2}, respectively.Yuanyuan Jiang et al. [162] stated that "Prussian blue (PB) has perfect catalytic activity towards some low molecular weight molecules (such as O_2, H_2O_2, hydrazine) due to its unique zeolite structure". So they used graphene modified PB for Amperometric determination of H_2O_2. The obtained linear range was 10-1440 μM with LOD = 3 μM. Moreover, Selvakumar Palanisamy and his colleagues [69] developed a nonenzymatic hydrogen peroxide sensor based on reduced graphene oxide/ZnO. They prepared this composite through the simultaneous electrodepostion of ZnO and electrochemical reduction of GO. This electrode showed limit of detection as low as 0.02 μM for the amperometric determination of H_2O_2. On the other hand, Lin Cui et al. [163] used nano-cobalt phthalocyanine modified graphene electrode for Amperometric determination of organic peroxide as tetra-butyl hydroperoxidase (TBHP) with detection limit of 5 μM.

The metabolic disorder of diabetes mellitus results in the deficiency of insulin and hyperglycemia and is reflected by blood glucose concentration higher or lower than the normal range of 80–120 mg dL^{-1} (4.4–6.6 mM) [164]. Therefore, determination of glucose level in human blood is very important to the care of diabetes. So, the researchers have searched for the best sensor for glucose determination that has high selectivity, sensitivity, stability and ease of construction [165]. The initial attempt to construct a device that can monitor glucose level is biosensor where, glucose oxidase immobilized on the surface of electrode or modified electrode. However the glucose oxidase is strongly affected by the environmental changes such as temperature, pH, humidity, etc. so, the nonenzymatic sensors have attracted a great of interest nowadays [166]. Yu-Wei Hsu et al. [167] stated that CuO/Cu(OH) nanoparticles that is deposited on graphene showed high sensitivity for glucose determination. The amperometric curve of glucose determination shows the synergism between CuO nanoparticles and graphene as shown in Fig. 2.9.

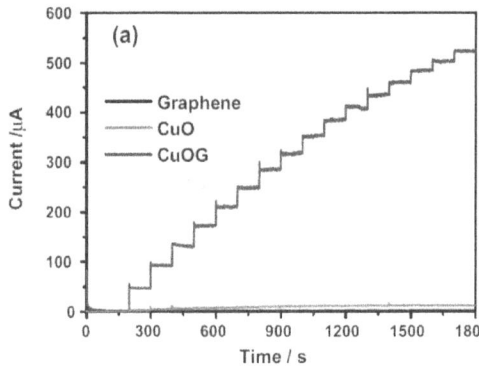

Fig. 2.9. Amperometric responses of the graphene-, CuO nanoparticle-, and CuO/graphene-modified GC electrodes after the subsequent addition of 0.5 mmol L−1 glucose in a 0.1 mol L−1 NaOH solution. (with permission from [176]).

Also, this sensor showed good stability where, the current response was 92 % of the original value after 14 days. The detection limit was found to be 1.0 µM in the linear range between 1.0 µM-8.0 mM. Liqiang Luo et al. used the same composite for determination of glucose but by using CuO nanocubes instead of nanoparticles [168]. They revealed that using CuO in shape of nanocubes leads to higher sensitivity and lower detection limit (1360 $\mu A.mM^{-1}cm^{-2}$ and 0.7 µM respectively.) compared to CuO nanoparticles that are used by Yu-Wei Hsu et al. Rutin (vitamin P) is a kind of flavonoid glycoside compounds, it has physiological functions as anti-flammatory, anti-tumor and anti-cancer [168]. Feng Gao et al. [169] used graphene-carbon ionic liquid composite for the voltammetric determination of rutin in a concentration range 0.070 to 100.0 µmol/L with the detection limit as 24.0 nmol/L. they revealed that the electrochemical activity of GR/CILE showed higher electrochemical activity toward rutin compared to bare carbon paste electrode (CPE) where, the redox current of GR/CILE was about 4.5 time that of CPE. Based on the influence of scan rate study they concluded that the mechanism of the electrochemical reaction of rutin on GR/CILE was two-proton/ two electron process and the calculated charge transfer number, electron transfer number and rate constant were as follow: 0.449, 2.10 and 2.39 s^{-1}, respectively. See Fig. 2.10 that shows the effect of scan rate on the electrochemical oxidation of rutin on GR/CILE. Table 2.2 contains some examples for graphene as electrochemical sensor for biologically active compounds; H_2O_2, glucose and rutin.

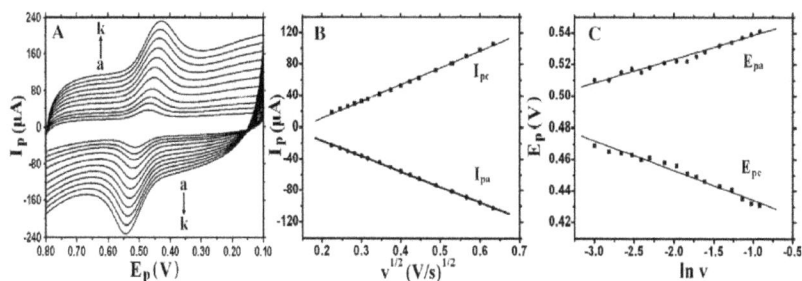

Fig. 2.10. (A) Cyclic voltammograms of 5.0×10^{-5} mol/L rutin with different scan rate (υ) for GR/CILE in pH 2.5 PBS (from a to k are 50, 70, 90, 120, 160, 200, 240, 280, 320, 360, and 400 mV/s, respectively); (B) linear relationship of cathodic and anodic peak current (I_p) versus $υ^{1/2}$; (C) linear relationship between peak potentials (Ep) and lnυ (with permission from [168]).

2.2.4. Graphene as Electrochemical Sensor for some Hazardous Organic Compounds

Hydroquinone (HQ), is an organic compound that has applications in many fields such as pharmaceutical, antioxidant, dye, photography and cosmetic industries [175], HQ has been considered as an environmental pollutant due to its toxicity and low degradability in the ecological environment [176]. Song Hu et al. [177] electrodeposited graphene and Au nanoparticles on the surface of carbon ionic liquid electrode (CILE) simultaneously and used the resulted electrode for the electrochemical determination of HQ. They revealed that Au-GR/CLILE displayed higher current response as well as smaller peak separation (0.077 V) compared to bare CILE, Au/CILE or even GR/CILE (0.298, 0.245 and 0.079 V, respectively). The calculated detection limit was 0.113 μM.

Table 2.2. Graphene as electrochemical sensor
for biologically active compound.

Electrode	Detected species	Medium	LOD (μM)	Reference
AgNPs/GN		0.1 M PBS of pH 6.5	0.5	[161]
Ag/MWCNT/GCE		0.2 M PBS of pH7.0	1.3	[170]
RGO/Fe3O4/GCE		PBS of pH 7.0	3.2	[171]
RGO/ZnO/GCE	H_2O_2	0.05 M PBS of pH 7.0	0.02	[163]
MnO2/GO		0.1 NaOH	0.8	[172]
PB-graphene/GCE		0.05 M PBS of pH 6.0	3.0	[69]
GN-Pt		0.1 M PBS of pH 7.4	0.5	[173]
CuO NP/graphene		0.1 M NaOH	1.0	[166]
CuONC/graphene	Glucose	0.1 M NaOH	0.7	[168]
CuNC/MWCNTs		0.1 M NaOH	1.0	[174]
GR/CILE[a]	Rutin	0.1 M PBS of pH 2.54	0.024	[169]

[a] graphene modified carbon ionic liquid electrode.

Haixia Fan et al. constructed an electrochemical sensor for detection of bisphenol A (BPA) based on N-doped graphene [178]. Bisphenol A is endocrine disruptor which could disrupt endocrine system and cause cancer. Bisphenol A is widely used in industry where it is used in plastic industry, dental fillings and lining of food cans [179]. However, BPA is very harmful to the human bens and can lead to breast cancer, prostate cancer, birth defect, infertility, precocious girls, diabetes and obesity [180].these harmful effects make the sensitive determination of BPA is very important. Haixia Fan et al [178] used chitosan modified N-doped graphene (CS-N-GR) for the amperometric determination of BPA with LOD of 5×10^{-9} M that was lower than the other reported methods such as [181] this sensor also showed high interfering resistance where, some ions such as 100-fold concentration of K^+, Na^+, Ca^{2+}, Mg^{2+}, Al^{3+}, Zn^{2+}, Cu^{2+} did not show any interference with the signal of PBA. High quality zirconia nanoparticles decorated graphene sheets (ZrO_2NPs-GNs) by environmentally friendly approach was successfully prepared by Jingming Gong et al. [182] they prepared ZrO_2NPs-GNs using simultaneous electrodeposition. It is worth to mention that, this method for preparation of the modified graphene, has attracted great of interest as we referred before that is because it does not involve any used hazardous chemicals during the chemical reduction of GO. Xiaet al. demonstrated that this approach has several clear advantages: no toxic solvents are used and therefore will not result in contamination of the product; the high negative potential can overcome the energy barriers for the reduction of oxygen functionalities, leading to efficiently reduced the exfoliated GO [183].The method of preparation of ZrO_2NPs-GNs is demonstrated in Fig. 2.11. They get a detection limit of 0.6 nM in the linear range 0.002 to 0.9 μ M for the voltammetric determination of methyl parathion. It is important to mention that this LOD is lower than that was obtained at CPE as in [184].

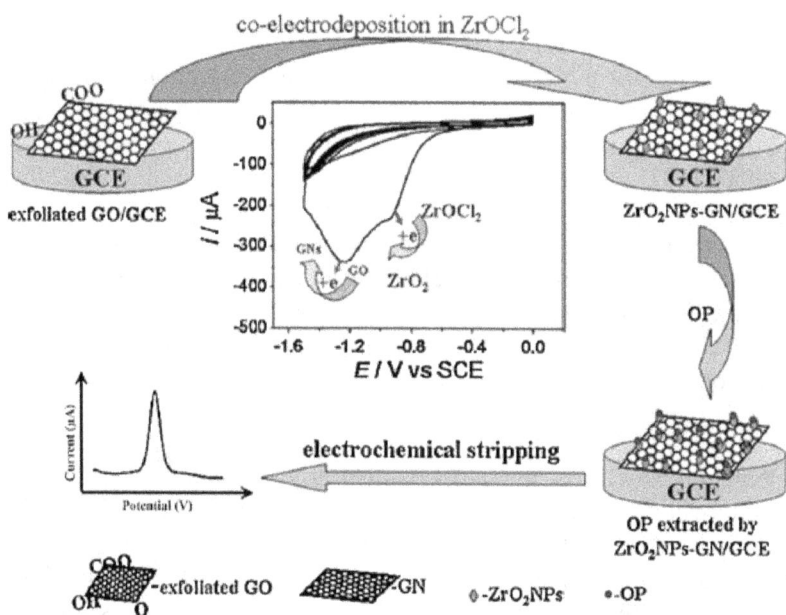

Fig. 2.11. Schematic illustration for green synthesis of ZrO_2NPs-GNs/GCE via a one-step electrochemical approach and electrochemical sensing nitroaromatic OP compound (with permission from [182]).

2.3. Graphene as Biosensors

As we mention before the term "biosensor" is used when a biological compound is used as part of the sensing material. The scope of this section is to record the different attempts of using graphene in such interesting area "biosensing" to detect some of the biologically important compounds.

As mentioned previously, direct electrochemistry of enzymes involves direct electron transfer (DET) between the electrode and the active center of the enzymes without the participation of mediators or other reagents [185]. Recent research has shown that graphene can enhance DET between enzymes and electrodes [165]. Cholesterol and its ester are essential constituents of all animal cells. They are precursors of bioanalytes such as bile acid and steroid hormones. However, increases of cholesterol levels can cause life-threatening coronary heart diseases, cerebral thromboses,and artherosclerosis [186], therefore accurate detection of cholesterol level is medically useful. Dey and Raj [186] developed a highly sensitive amperometricbiosensor based on a hybrid material derived from PtNP and graphene for the detection of $H_2O_2$and cholesterol. The cholesterol biosensor was developed by immobilizing cholesterol oxidase and cholesterol esterase on the surface of the GR/PtNP hybrid material. The sensitivity and detection limit of the electrode towards cholesterol ester were 2.07 ± 0.1 μA μM^{-1} cm^{-2} and 0.2 μM, respectively. Very recent, Revanasiddappa Manjunatha and co-workers [187] fabricated a cholesterol biosensor based on enzyme-functionalized graphene sheets. They found that graphene accelerate the electron transfer from the electrode surface to the immobilized enzyme also they found that cholesterol oxidase (ChOx)/ graphene shows a detection limit of 5 μM in a linear range of cholesterol from 50-350 μM while co-immobilization of ChEt (cholesterol esterase) and ChOx on graphene showed a detection limit of 15 μM in the linear range 50-300 μM.

Salbutamol (SAL) is the most widely used β2-adrenergic receptor agonist which induces bronchodilation, so it is effective for curing bronchial asthma, chronic obstructive pulmonary disease and other allergic diseases associated with respiratory pathway [188] The highly sensitive and selective method for determination of SAL is required because the residues of it can be toxic for the human [189, 190]. Jiadong Huang et al. [191] developed immunosensor based on nanogold particles (nano-Au), Prussian Blue (PB), polyaniline/poly (acrylic acid) (PANI/(PAA)) and Au-hybrid graphene nanocomposite (AuGN) for electrochemical determination of SAL. AuGN was prepared as core-shell composite where, chitosan-protected graphene was used as core and multi-nanogold particles as shell. In addition to these composites peroxidase–anti-SAL antibody (HRP–AAb) with AuGN (AuGN–HRP–AAb) was used as the label for the immunosensor. The method for preparation of this immunosensor is shown in Fig. 2.12. This sensor provided recovery from 94.8-103.6 and detection limit 0.04 ng mL^{-1} which is very low relative to the other reported methods for SAl determination [190, 192, 193].

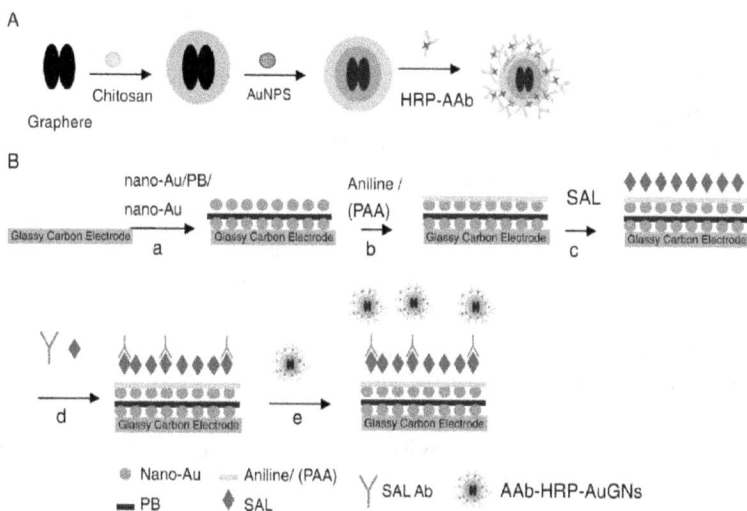

Fig. 2.12. (A) The synthesis process of the AAb-HRP-AuGN labels;
(B) Schematic illusteration of the stepwise immunosensor fabrication process.
(with permission from [191]).

2.3.1. Glucose Biosensor

Glucose biosensors based on glucose oxidase (GOD) are widely discussed along the last years. The general principle of the glucose biosensor is based on the amperometric detection of H$^+$ via the following chemical reactions [171]:

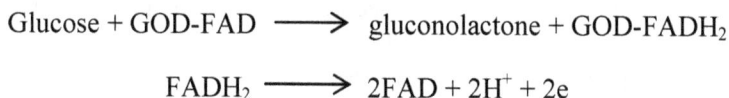

$$Glucose + GOD\text{-}FAD \longrightarrow gluconolactone + GOD\text{-}FADH_2$$

$$FADH_2 \longrightarrow 2FAD + 2H^+ + 2e$$

Xiaoping Chen et al. [194] used encapsulated glucose oxidase on graphene-nafion modified GCE for amperometric determination of glucose with limit of detection as low as 1.0 μM in a linear range 2-100 μM. Xinhuang Kang et al. [195] studied the electrochemical detection of glucose using glucose oxidase-chitosan- graphene modified GCE. This biosensor exhibits a linearity range from 0.08 mM to 12 mM glucose with a detection limit of 0.02 mM and a sensitivity of (37.93 μA mM^{-1} cm^{-2}). While Changsheng Shanet al. [118] investigated the influence of introducing gold nanoparticles to this composite to form Graphene/AuNPs/GOD/chitosan nanocomposites film for detection of glucose with a linear range from 2 to 10 mM at −0.2 V and from 2 to 14 mM at 0.5 V and a good reproducibility and detection limit of 180 μM was obtained. TiO$_2$ nanoparticles have attracted considerable interest due to their superior properties, such as their large specific surface area, high uniformity, and excellent biocompatibility. Ying Wang et al. [196] used GOD immobilized on nitrogen-doped graphene modified electrode in a linear range of 0.1-1.1 mM. Tessy Theres Baby et al. [197] fabricated Nafion/GOD/f-G-(Pt–Au) and Nafion/GOD/f-G-Au GC electrodes and used them as glucose biosensor. They revealed that the addition of nanocrystalline (Pt–Au) or (Au) metal nanoparticles act as spacers and eliminates the restacking of graphene sheets which has resulted in the increase in the electro active surface area of the electrode. They reported that f-G/Nafion/GOD/(Pt–Au) exhibits a linearity up to about 25 mM and that of f-G/Nafion/GOD/Au exhibits a linearity up to about 30 mM with a detection limit of 1 μM. Hee Dong Jang et al. [198] synthesized TiO$_2$-GR by evaporating a droplets composed of TiO$_2$ nanoparticles, GO and water by an aerosol-assisted self-assembly (AASA) method then they used the as-prepared composite was dispersed in GOD to form TiO$_2$–GR/GOD which was used as a biosensor for electrochemical determination of glucose. Another attempt to use graphene modified GOD for glucose determination is that performed by Yuanyuan Jiang et al. [199] they used GOD immobilized on Ionic liquid functionalized graphene (IL-graphene) composite for the amperometric determination of glucose. They explained that IL-graphene was successfully synthesized through an epoxide ring-opening reaction between graphene oxide (GO) and amine-terminated ionic liquid (NH2-IL). Albert Gutés et al. [200] used CVD for preparing a single layer graphene sheet on Cu substrate then the as-prepared graphene sheets were decorated with Au nanoparticles via electroless deposition. GOD-Nafion mixture was drop casted over graphene modified electrode to form GCE/graphene/AuNP/

GOD/Nafion. They revealed that this resulting composite exhibited limit of detection 4 μM in a linear range 10–366 μM glucose. Based on the interaction between GOD and PB forming pseudobienzyme system, Xia Zhong et al. [201] developed a biosensor for detection of glucosebased on chitosan-Prussian blue-graphene-Au nanoparticles and Concanavalin A (Con A). This biosensor provided a detection limit as low as 10 μM in a linear range 2.5×10^{-5} to 3.2×10^{-3} M. Xia Zhong said that "Con A can increase the immobilization quantity of enzyme through sugar-lectin biospecific interactions and improve the stability of the enzyme biosensor" where Con A is a lectin protein with a molecular weight: 104,000 and can be obtained from Canavalia ensiformis (jack bean). Con A contains four identical binding sites to the molecules containing α-D-glucosyl or α-D- mannosyl residues [202]. So, it can be effectively used in glucose biosensing field. The technique of preparing this biosensor is as follow: Preparing (CS-PB-GR) composite nanosheets then immobilize the CS-PB-GR composite nanosheets onto the surface of electrode, then gold nanoparticles (AuNps) were absorbed onto the amino-groups of CS-PB-GR composites. The first layer of GOD was immobilized on AuNPs for a capture of Con A. finally adding the second layer of GOD that is attached to the electrode through the interaction between Con A and GOD. Moreover, this biosensor provided good interfering resistance towardsascorbic acid(AA), uric acid (UA) and p-acetamidophenol (AP). Where, no interference in the current response was observed on adding 1.0 mM AA, 1.0 mM UA, and 1.0 mM AP to 2.0 mM glucose solution.

2.3.2. Graphene-based DNA Sensors

GR can provide a favorable microenvironment for deoxyribonucleic acid (DNA) and effectively accelerate the direct electron transfer rate from DNA to the electrode surface, which can be further applied to determine DNA with excellent sensitivity and long-term stability [203, 204].

The sensitive detection of DNA sequence has emerged as a hot subject of research due to its extensive applications in molecular diagnostics, food, environment, anti-terrorism and forensic science [205-207]. Zhou et al. [208] prepared CR-GO electrodes for label-free electrochemical detection of four DNA bases in ssDNA/dsDNA (single stand and double strand DNA). Su et al. [209] designed a new graphene-based immunosensing platform for the detection of carcinoembryoninc antigen (CEA) using ssDNA and gold nanoflowers. In this sensor Gold nanoflowers and single-stranded DNA (ssDNA) molecules were initially assembled onto the surface of graphene for the fabrication of the electrochemical immunosensor using layer-by-layer strategy. They reported that "this electrochemical immunosensor exhibited a wide dynamic range of 0.05 to 45 ng mL^{-1} with a relatively low detection limit of 0.01 ng mL^{-1} CEA at signal-to-noise ratio of 3. Xiaowei Qi et al. [210] detected *staphylococcus aureus nuc* gene sequence using ss-DNA immobilized on CTS-Co3O4-GR/CILE. In this work Co$_3$O$_4$ nanorods (nano-Co$_3$O$_4$), graphene (GR) and chitosan (CTS) were mixed together to form a novel nanocomposite material and casted on the surface of CILE then ss-DNA was immobilized on the previous composite. They reported that "the proposed electrochemical DNA sensor was applied to the detection of *nuc* gene sequence fragment from the *staphylococcus aureus* and further used to the PCR product of *staphylococcusaureus* endogenous gene". A detection limit of the *staphylococcus aureus nuc* gene sequence was estimated to be 4.3×10^{-13} M. Haifeng Donga and co-workers [211] constructed an electrochemical DNA biosensor by assembling probe labeled gold nanoparticles (ssDNA–AuNP) on electrochemically reduced graphene oxide (ERGO) modified electrode with thiol group tagged (GT) DNA strand (d(GT)29SH) and coupling with horseradish peroxidase (HRP) functionalized carbon sphere (CNS) as tracer as shown in Fig. 2.13. They revealed that this novel triplex signal amplification strategy can detect target DNA down to attomolar level with high selectivity to differentiate single-base mismatched and three-base mismatched sequences of DNA.

On the other hand, Wei Lv et al. [212] developed a non-enzymatic glucose biosensor based on single strand DNA-dispersed graphene/NiO modified GC electrode (GNS/NiO/DNA-GC). They revealed that this sensor provided good sensing performance toward glucose determination represented in its low detection limit of 2.5 μM, wide linear range1-200 μM and high sensitivity and stability. They concluded that ss-DNA form self-assembled monolayer on GNS when they subjected to sonication also DNA has hydrophilic part, sugar-phosphate backbone, which is exposed to the aqueous medium leading to a very stable suspension. They also tried to disperse GNS/Ni in an anionic surfactant as sodium dodecyl sulfate (SDS) and in a cationic surfactant as cetyltrimethylammonium bromide (CTAB) but GNS-NiO hybrid could not be dispersed in both of them as in ss-DNA. Fig. 2.14 shows the CV of GNS/NiO/DNA-GC at different glucose concentration as well as the effect of scan rate.

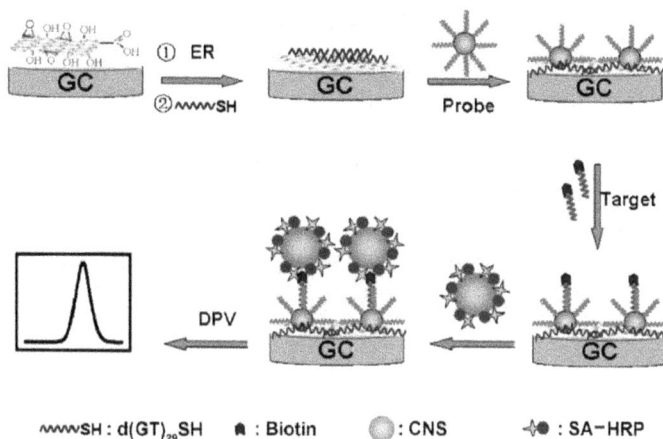

wwwSH : d(GT)₂₉SH   ♠ : Biotin   ⬤ : CNS   ⊸⊕ : SA−HRP

Fig. 2.13. Schematic presence of electrochemical DNA biosensing using probe labeled AuNPs–graphene modified electrode and enzyme functionalized carbon sphere for DPV tracing. (with permission from [211]).

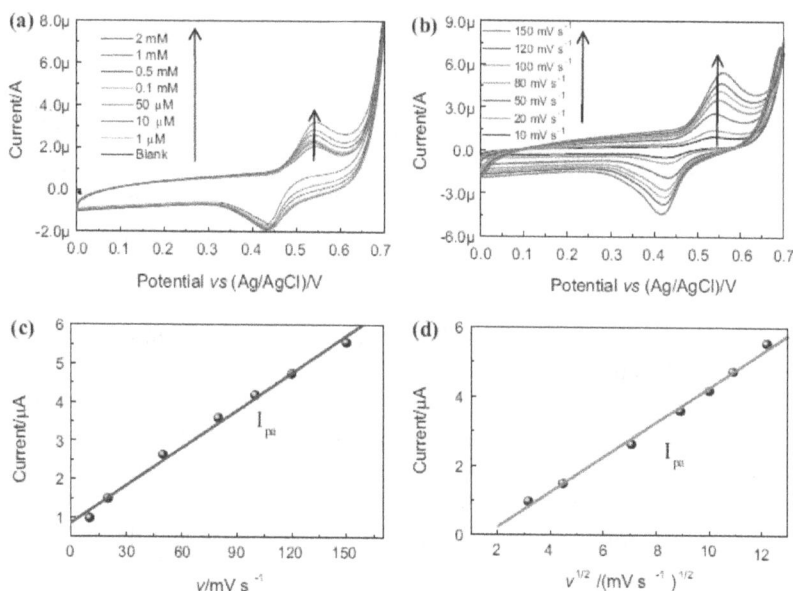

Fig. 2.14. Electrochemical performance of the activated GNS/NiO/DNA-GC electrode. (a) CV profiles of the activated non-enzymatic glucose sensors with different glucose concentrations at the scan rate of 50 mV s^{-1}; (b) CV profiles of the activated non-enzymatic glucose sensors at different scan rates with the presence of 0.5 mM glucose; (c) and (d) show the plots of peak current vs. v and v$^{1/2}$ (v denotes scan rate). (with permission from [212]).

Another DNA sensor based on thionine-graphene nanocomposite modified gold electrode was prepared by Limei Zhu [213]. This biosensor exhibited high sensitivity and low detection limit for detecting complementary oligonucleotide of 1.26×10^{-13} M (S/N = 3) in a linear range 1.0×10^{-12} to 1.0×10^{-7} M. Additionally, Meng Du et al. [214] constructed a novel sensor based on the ssDNA/ERGNO/PAN nanocomposites. They found that the biosensor discriminated mismatched DNA samples of similar lengths and used for the quantitative detection of CaMV35S.

2.3.3. Graphene-based Hemoglobin Biosensor

Hemoglobin (Hb), an important redox respiratory protein that contains the porphyrin complex of iron(II) or hemin(III) as a prosthetic group, is considered to be a protein with an enzyme-like activity [215, 216]. A lot of information about the electron transfer mechanisms between proteins in biological or physiological systems can be given by studying the direct electron transfer process of protein-electrode system. [217] several articles studied the electrochemistry of hemoglobin (Hb) on graphene modified electrodes as an example of important protein in the biological system. Kunping Liu et al. [79] studied the electrocatalysis of Hb on graphene poly(diallyldimethylammonium chloride) functionalized graphene sheets/room temperature ionic liquid (RMIL) composite film. Poly(diallyldimethylammonium chloride) (PDDA) is a linear positively charged polyelectrolyte that can be used to nanocovalently functionalize graphene sheets. Additionally PDDA has excellent binding capability with graphene sheets and can maintain their electronic structure. In addition to PDDA increase the solubility of graphene sheets for their applications in biosensing.As mentioned before RMIL can improve the conductivity [218] and the dispersion of graphene sheets [219] they studied the electrochemistry of the immobilized Hb on RTIL/PDDA-G/GCEby CV in 0.1 M PBS of pH 7.0. The results showed that both RTIL/PDDA-G and the PDDA-G modified electrodesare electroinactive while by combining with Hb, a pair of ill-defined and irreversible redox peaks was observed. These redox peaks were greatly enhanced with introducing RTIL and the resulting Hb/RTIL/PDDA-G/GCE showed a stable, well-defined and quasi-reversible redox peaks with a peak-to-peak separation of about 68 mV. Additionally,in order to test the electrocatalytic activity of Hb/RTIL/PDDA-G/GCE biosensor, they used it for determination of nitrite which exhibited a detection limit of 0.04 µM in a linear range 0.2-32.6 µM.Qingliang Feng et al. [220] replaced RTIL with gold nanoparticles and studied the electrochemistry of the resulting Hb/AuNPs/PDDA-G but at this time they used this biosensor for determination of hydrogen peroxide.additionally, they investigated the electrocatalytic activity of the different modified electrodes in 0.1 M PBS of pH 7.4 and they found that no obvious voltammetric peaks were observed at the bare GCE, PDDA-G modified electrode and AuNPs/PDDA-G film modified electrode. While the Hb/AuNPs/PDDA-G film-modified electrode showed a couple of well-definedand quasi-reversible redox peaks at −0.413 V and −0.330 V with a peak-to-peak separation of about 83 mV. Moreover, Hb/AuNPs/PDDA-G biosensor showed a limit of detection as low as 0.39 µM in a linear range of 6 µM to 1010 µM H_2O_2. Ferrocene (Fc) has attracted great of interest in the area of electroanalysis because it is one of the most important electroactive groups and has reversible Fc^+/Fc redox behavior [221, 222]. Ke-Jing Huang et al [223] developed hemoglobin based on chitosan–ionic liquid–ferrocene/graphene composite film and used it as H_2O_2 biosensor. They found that the reduction currents on the as-prepared Hb/Chit–IL–Fc/Gr/GCE increased linearly with the increasing H_2O_2 concentration with a limit of detection 3.8 µM in a linear range 50 to 1200 µM and a sensitivity of 14.6 µA mM^{-1}. This indicates that Hb immobilized in Chit–IL–Fc/Gr composite exhibited excellent bioelectrocatalytic activity to H_2O_2. Another graphene/hemoglobin-based biosensor for nitric oxide (NO) is that performed by Wei Wen et al. [224] where they prepared Hb–CS/GR–CTAB nanocomposite film and used it for determination of nitric oxide. Hb-Cs/GR-CTAB provided high electrocatalytic activity toward nitric oxide with a detection limit of 6.75 nM and a sensitivity of 615 µA mM^{-1}. The function of the surfactant CTAB is preventing the agglomeration of graphene sheets. X.H. Kang et al. [225] reported that, CS is a naturalbiopolymer with unique structural features possesses a primaryamine at the C-2 position of the glucosamine residues and is solublein aqueous acidic media. It has good hydrophilicity, excellent film forming ability, biocompatibility, nontoxicity andhigh mechanical strength. These features make it widely used to dispersenanomaterials and immobilize proteins.According to Wei Wen et al. [224], the presence of Hb increased the current response and reduced the overpotential of it.

2.4. Graphene as a Gas Sensor

Gas sensing by graphene usually based on the adsorption and desorption of gaseous molecules on the graphene surface, leading to change in conductance in the graphene.the high sensitivity of graphene sheets to the gaseous molecules make them suitable for sensing many gaseous molecules as CO, NH_3, H_2 and NO_2 [226]. Wei Wu et al. [227] fabricated graphene decorated palladium for hydrogen sensing application.The sensing device composed of graphene decorated with Pd as sensing material and $SiO2/Si$ as a substrate. They could measure the hydrogen sensing capability by measuring sensor resistance change when the sensor was exposed to different concentration of hydrogen [227]. Density functional theory was used to investigate the potential applications of B or N-doped graphene sheets as a gas sensor by Zhang and co-workers [228] they revealed that B-doped graphene gives the tightest binding with NH_3. Lu et al. [229] reported that hydrazine-reduced GO sensors are more effective than thermally reduced GO for the detection of NO_2 and NH_3.Lu et al. [230] used a macro graphene foam-like three-dimensional network to detect NO_2 and NH_3 which provided a high sensitivity toward gas molecules Fig. 2.19 shows SEM image and Raman shift of the as-prepared graphene foams (GF). W. Wu et al. deposited a thin Pd film on graphene prepared by CVD and used it as a hydrogen sensor and they found that The percentage of the sensor's resistance changed from 0.0025 to 1 for the change in hydrogen concentration (25–10,000 ppm) [227].

2.5. Graphene as a Heavy Metal Ions Sensor

Graphene-based electrochemical sensors have been developed for environmental analysis for the detection of heavy metal ions such as lead, cadmium, silver, mercury and arsenic that have severe environmental and medical effects [80]. Therefore development of a new sensitive sensor for detection of ultra- traces of metal ions attracted great of interest. Nafion/graphene modified electrode showed high sensitivity and selectivity toward Cd^{+2} and Pb^{+2} ions [81]. The detection limits (S/N = 3) are 0.02 µg L^{-1} for both Cd^{+2} and Pb^{+2} in a linear range of (0.5 mgL^{-1} – 50 mgL^{-1} and 1.5 mgL^{-1} – 30 mgL^{-1} for Pb^{+2} andCd^{+2}, respectively) compared to 0.09 µgL^{-1} for Cd(II) and 0.02 µgL^{-1} for Pb(II) on graphite nanofibers–Nafion composite modified bismuth filmelectrode [231]. Brownson and Banks [226] revealed that the commercially available graphene with addition of a surfactant during its fabrication inhibits the electro-analytical sensing of cadmium (II). Where, they demonstrated that the presence of surfactants used in the manufacturing process of commercially available graphene enhances nucleation of our target heavy metal but inhibits the stripping step. Moreover, Ramesha and sampath [82] prepared reduced graphene oxide–lead dioxide composite and used it for determination of arsenic with a detection limit of 10 nM. Table 2.3. contains some typical examples for graphene as biosensor.

Table 2.3. Graphene as biosensor.

Electrode	Detected substance	Linear range (μM)	Sensitivity (μA mM^{-1} cm^{-1})	LOD (μM)	Ref.
TiO$_2$–GR/GOD	Glucose	0-800	6.2	---	198
ChOx-ChEt/GR/PtNPs	Cholesterol		2.07 ± 0.1	0.2	192
ChOx-ChEt/ graphene	Cholesterol	50-300	---	15	193
Graphene/AuNPs/GOD/ chitosan	Glucose	2000- 10^4 and 2000- 14 × 10^3	----	180	118
GCE/graphene/AuNP/G OD/Nafion	Glucose	10–366	4.0	---	200
GNS/NiO/ss-DNA-GC	Glucose	1-200	2.5	---	212
Hb/RTIL/PDDA-G/GCE	Nitrite	0.2-32.6	15.7	0.04	79
Hb/AuNPs/PDDA-G	H$_2$O$_2$	6-1010	---	0.39	220
Hb/Chit–IL–Fc/Gr/GCE	H$_2$O$_2$	50-1200	14.6	3.8	223
Hb-Cs/GR-CTAB	Nitric	---	615	0.00675	224

2.6. Field Effect Transistor (FET)

Excellent electronic properties of graphene sheets with reported carrier mobilities between 3000 and 27000 cm²/Vs make it an extremely promising material for future nanoelectronic devices [233, 234]. FET consists of a semiconductors active layer in contact with two electrodes "source" and "drain" and there is a third electrode "gate" which is separated from the active layer by an insulating film. In this device the source drain voltage is applied and a source drain current is measured [235].

The FET relies on an electric field to control the shape and conductivity of a channel of one type of charge carrier in a semiconducting materialsee Fig. 2.15. Ohano et al. [235] carried out highly sensitive solution pH sensing and monitoring of the charge-type dependence of protein adsorptions onto the surface of graphene using single-layer G-FETs. As results, the lowest detection limit (signal/noise = 3) of the change in solution pH value was estimated to be 0.025 and the G-FETs could be electrically distinguished between positive and negative charged proteins in a buffer solution, indicating the high potential of G-FETs for applications in chemical and biological sensors. [236] studied the effect of AuNP on the in FET behavior of graphene. It was found that the AuNP doped FET exhibited much higher hole mobility. In addition, the AuNP provided readily functionalization sites for conjugating biomolecular probes on the graphene–AuNP nanocomposites.

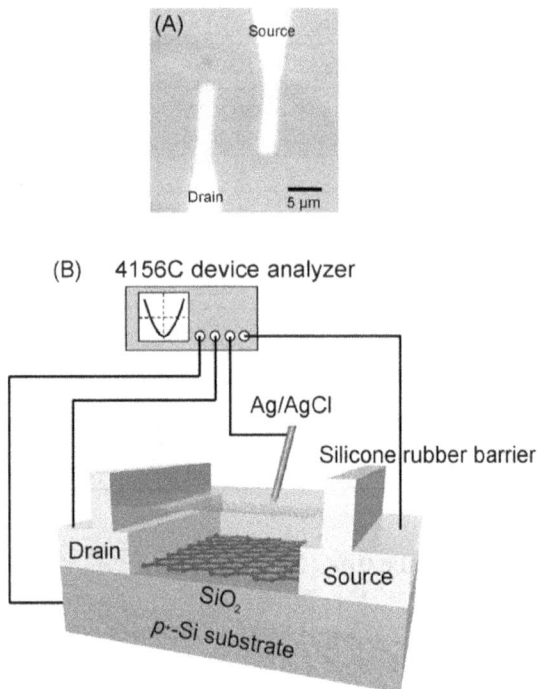

Fig. 2.15. (A) Optical micrograph of typical G-FET, and (B) schematic illustration of experimental setup with G-FETs (with permission from [235]).

According to Yeon Hwa Kwak et al. [237], the surface of the graphene-based FET on a flexible polyethylene terephthalate (PET) substrate was functionalized with glucose oxidase (GOD). They used this biosensor to measure hydrogen peroxide (H_2O_2) and glucose levels continuously as shown in Fig. 2.16.

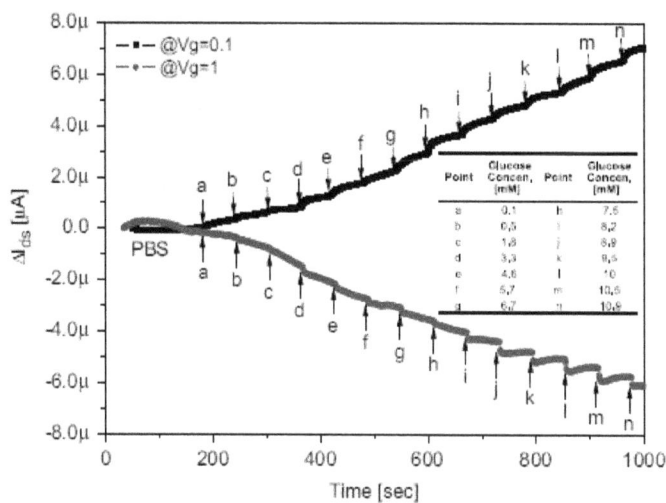

Fig. 2.16. The glucose response for the sequential concentration increase. Changes of drain-source current with Vg¼0.1 V and Vg¼1.0 V for the fixed drain-source voltage of Vd¼_0.2 V. Start in g with a pure PBS solution, the glucose was first injected at180 sand added every minute thereafter. The glucose response for the sequential concentration increaseshowedafairly enough resolution as well as continuity, which transcripts the efficacy as a clinically acceptable flexible glucose sensor. (with permission from [237]).

2.7. Smart Graphene-based Sensors

Using silk strands pulled from cocoons, gold wires thinner than a spider's web and graphene, researchers at Princeton University have created a removable tattoo that adheres to dental enamel and could eventually monitor a patient's health with unprecedented sensitivity. The sensor generated a response to the patient's breath and transmitted a signal to a nearby monitor. This device was built by first imprinted tiny graphene sensors onto an extremely thin film of water-soluble silk. Next, they used a stencil-mask assisted evaporation technique to pattern an antenna made of thin gold strand onto the silk film and connected it to the graphene sensors. When completed, the device resembles a common removable tattoo see Figs. 2.17 and 2.18.

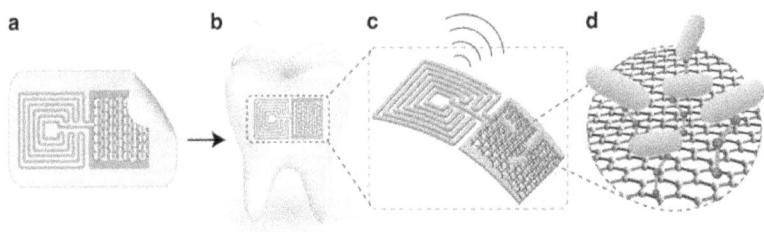

Fig. 2.17. Biotransferrable graphene wireless nanosensor: (a) Graphene is printed onto bioresorbable silk and contacts are formed containing a wireless coil; (b) Biotransfer of the nanosensing architecture onto the surface of a tooth; (c) Magnified schematic of the sensing element, illustrating wireless readout; (d) Binding of pathogenic bacteria by peptides self-assembled on the graphene nanotransducer. (with permission from [238]).

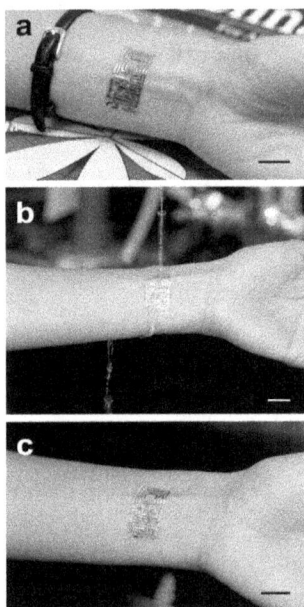

Fig. 2.18. Supplementary Fig. S4. Stability of sensor in running water. Optical images of (a) biotransferred sensor onto a human arm, (b) mild rinsing in running water, and (c) the sensor following exposure to running water. Scale bars are 1 cm (with permission from [238]).

To attach the sensor, the researchers place it against a tooth, or a person's skin, and wash it with water. The silk base dissolves in the water, but the graphene sensor and the antenna remain securely fastened to the spot. McAlpine, an assistant professor of mechanical and aerospace engineering at Princeton, said "one of the goals was to create a device that was small, flexible and passive, capable of providing an *in situ* detection". So the researchers designed the device without a power supply. "the system not only has the ability to supply fast results, but is able to detect very small amounts of bacteria – a feature that could prove critical in treating certain diseases" said Michael McAlpine [238].

References

[1]. O. Leenearts, B. Partoens, F. M. Peeters, Electric Field Effect in Atomically Thin Carbon Films, *J. Microelectr.*, Vol. 40, 2009, pp. 860.

[2]. K. S. Novoselov, A. K. Geim, S. V. Morozov, D. Jiang, Y. Zhang, S. V. Dubonos, I. V. Grigorieva, and A. A. Firsov, Electric Field Effect in Atomically Thin Carbon Films Science, Vol. 306, 2004, pp. 666-669.

[3]. S. Mikhailov (ed.), Physics and Applications of Graphene- Experiments, *InTech Janeza Trdine*, Rijeka, Croatia, 2011.

[4]. J. H. Chen, M. Ishigami, C. Jang, D. R. Hines, M. S. Fuhrer, E. D. Williams, Printed graphene circuits, *Adv. Mater.*, Vol. 19, 2007, pp. 3623-3627.

[5]. J. Wintterling, M.-L. Bocquet, Graphene on metal surfaces, *Surf. Sci.*, Vol. 603, 2009, pp. 1841-1452.

[6]. C. J. Pool, On the applicability of the two-band model to describe transport across n–p junctions in bilayer graphene, *Solid State Commun.*, Vol. 150, 2010, pp. 632-635.

[7]. Y. Li, L. Tang, J. Li, On the applicability of the two-band model to describe transport across n–p junctions in bilayer graphene, *Electrochem. Commun.*, Vol. 11, 2009, pp. 846-849.

[8]. Z. P. Inga, J. S. Murry, M. E. Grice, S. Boyd, C. J. O'Conner, P. Politzer, Computational characterization of surfaces of model graphene systems, *J. Mol. Struct. Theochem.*, Vol. 549, 2001, pp. 147-158.

[9]. L. Tang, Y. Wang, Y. Li, H. Feng, J. Lu, J. Li, Preparation, Structure, and Electrochemical Properties of Reduced Graphene Sheet Films, *Adv. Funct. Mater*, Vol. 19, 2009, pp. 2782-2789.

[10]. L. Dong, R. R. S. Gari, Z. Li, M. M. Craig, S. Hou, Graphene-supported platinum and platinum–ruthenium nanoparticles with high

electrocatalytic activity for methanol and ethanol oxidation, Carbon, Vol. 48, 2010, pp. 781-787.

[11]. http://arxiv.org/ftp/cond-mat/papers/0702/0702595.pdf

[12]. J. H. Chen, W. G. Cullen, C. Jang, M. S. Fuhrer, E. D. Williams, Defect scattering in graphene, *Phys Rev Lett*, Vol. 102, Issue 23, 2009, pp. 236805.

[13]. P. Lian, X. Zhu, S. Liang, Z. Li, W. Yang, H. Wang, Large reversible capacity of high quality graphene sheets as an anode material for lithium-ion batteries, *Electrochem. Acta*, Vol. 55, 2010, pp. 3909-3914.

[14]. J. Wu, Y. Wang, D. Zhang, B. Hou, Studies on the electrochemical reduction of oxygen catalyzed by reduced graphene sheets in neutral media, *J. Power Sources*, Vol. 196, 2011, pp. 1141-1144.

[15]. S. Liu, J. Wang, J. Zeng, J. Ou, Z. Li, X. Liu, S. Yang, "Green" electrochemical synthesis of Pt/graphene sheet nanocomposite film and its electrocatalytic property, *J. Power Sources*, Vol. 195, 2010, pp. 4628-4633.

[16]. K. I. Bolotin, K. J. Sikes, Z. Jiang, M. Klima, G. Fudenberg, J. Hone, et al., Ultrahigh electron mobility in suspended graphene, *Solid State Commun*, Vol. 146, 2008, pp. 351-355.

[17]. R. R. Nair, P. Blake, A. N. Grigorenko, K. S. Novoselov, T. J. Booth, T. Stauber, N. M. R. Peres, A. K. Geim, Fine structure constant defines visual transparency of graphene, *Science*, Vol. 320, 2008, pp. 1308-1320.

[18]. M. Zheng, K. Takei, B. Hsia, H. Fang, X. Zhang, N. Ferralis, H. Ko, Y.-L. Chueh, Y. Zhang, R. Mabudian, A. Javey, crystallization of amorphous carbon to graphene, *Appl. Phys. Lett.*, Vol. 96, 2010, pp. 063110-063113.

[19]. Y. B. Zhang, Y. W. Tan, H. L. Stormer, P. Kim, Experimental Observation of the Quantum Hall Effect and Berry's Phase in Graphene, *Nature*, Vol. 438, 2005, pp. 201-204.

[20]. K. S. Novoselov, D. Jiang, F. Schedin, T. J. Booth, V. V. Khotkevich, S. V. Morozov, A. K. Geim, Two-dimensional atomic crystals, *Proc Natl Acad Sci USA*, Vol. 102 2005, pp. 10451-10453.

[21]. F. S. Kim, Y. Zhao, H. Jang, S. Y. Lee, J. M. Kim, K. S. Kim, J. H. Ahn, P. Kim, J. Y. Choi, B. H. Hong, Large-scale pattern growth of graphene films for stretchable transparent electrodes, *Nature*, Vol. 457, 2009, pp. 706-710.

[22]. X. Lu, M. Yu, H. Huang and R. S. Ruoff, Tailoring graphite with the goal of achieving single sheets, *Nanotechnology*, Vol. 10, 1999, pp. 269-276.

[23]. C. Berger, Z. Song, X. Li, X. Wu, N. Brown, C. Naud, D. Mayou, T. Li, J. Hass, A. N. Marchenkov, E. H. Conrad, P. N. First and W. A. de Heer, Electronic Confinement and Coherence in Patterned Epitaxial Graphene, *Science*, Vol. 312, 2006, pp. 1191-1196.

[24]. W. Chen, L. Yan, Preparation of graphene by a low-temperature thermal reduction at atmosphere pressure, *Nanoscale*, Vol. 2, Issue 4, 2010, pp. 559-563.

[25]. P. K. Ang, S. Wang, Q. Bao, J. T. L. Thong, K. P. Loh, High-Throughput Synthesis of Graphene by Intercalation-Exfoliation of Graphite Oxide and Study of Ionic Screening in Graphene Transistor, *ACS Nano*, Vol. 3, 2009, pp. 3587-3594.

[26]. D. V. Kosynkin, A. L. Higginbotham, A. Sinitskii, J. R. Lomeda, A. Dimiev, K. Price, J. M. Tour, Longitudinal unzipping of carbon nanotubes to form graphene nanoribbons, *Nature (London)*, Vol. 458, 2009, pp. 872-876.

[27]. L. Jiao, L. Zhang, X. Wang, G. Diankov, H. Dai, Narrow graphene nanoribbons from carbon nanotubes, *Nature (London)*, Vol. 458, 2009, pp. 877-880.

[28]. A. Maiti, C. J. Brabec, C. Roland, J. Bernholc, Theory of carbon nanotube growth, *Phys. Rev. B*, Vol. 52, 1995, pp. 14850-14858.

[29]. N. Li, Z. Y. Wang, K. K. Zhao, Z. J. Shi, Z. N. Gu, S. N. Xu, Large scale synthesis of N-doped multi-layered graphene sheets by simple arc-discharge method, *Carbon*, Vol. 48, 2010, pp. 255-259.

[30]. K. S. Subrahmanyam, L. S. Panchakarla, A. Govindaraj, C. N. R. Rao, Simple Method of Preparing Graphene Flakes by an Arc-Discharge Method, *J. Phys. Chem. C*, Vol. 113, 2009, pp. 4257-4259.

[31]. J. R. Gong (ed.) Graphene synthesis, Characterization, Properties and Applications, *InTech Janeza Trdine*, Rijeka, Croatia, 2011.

[32]. M. J. Mcallister, J. L. Li, D. H. Adamson, H. C. Schnlepp, A. A. Abdalam, J. Liu, I. A. Aksay, Single Sheet Functionalized Graphene by Oxidation and ThermalExpansion of Graphite, *Chem. Mater.*, Vol. 19, 2007, pp. 4396-4404.

[33]. H. C. Schniepp, J. Li, M. J. Mcallister, H. Sai, D. H. Adamson, R. Car, D. A. Saville, I. A. Aksay, Functionalized Single Graphene Sheets Derived from Splitting Graphite Oxide, *J. Phys. Chem. B*, Vol. 110, 2006, pp. 8535-8539.

[34]. L. L. Chua, S. Wang, P. J. Chia, L. Chen, L. H. Zhao, W. Chen, A. T. S. Wee, P. K. H. Ho, Deoxidation of graphene oxide nanosheets to extended graphenites by "unzipping" elimination, *J. Chem. Phys.*, Vol. 129, 2008, pp. 114702-114708.

[35]. C. Nethravathi, M. Rajamathi, Chemically modified graphene sheets produced by the solvothermal reduction of colloidal dispersions of graphite oxide, *Carbon*, Vol. 46, 2008, pp. 1994-1998.

[36]. V. C. Tong, L. M. Chen, M. J. Allen, J. K. Wassail, K. R. Nelson, B. Kaner, Y. Yang, Low-Temperature Solution Processing of Graphene−Carbon Nanotube Hybrid Materials for High-Performance Transparent Conductors, *Nano Lett.*, Vol. 9, Issue 5, 2009, pp. 1949-1955.

[37]. H. Fukushima, L. T. Drzal, A carbon nanotube alternative: Graphite Nanoplatelets as reinforcements for polymers, *Annu. Tech. Conf. Soc. Plast. Eng.*, Vol. 61, 2003, pp. 2230-2235.

[38]. S. Stankovich, D. A. Dikin, R. D. Piner, K. A. Kohlhaas, A. Kleinhammes, Y. Jia, Y. Wu, S. T. Nguyen, R. S. Ruoff, Synthesis of

graphene-based nanosheets via chemical reduction of exfoliated graphite oxide, *Carbon*, Vol. 45, 2007, pp. 1558–1565.

[39]. S. Gilije, S. Han, M. Wang, K. L. Wang and R. B. Kaner, A Chemical Route to Graphene for Device Applications, *Nano Lett.*, Vol. 7, Issue 11, 2007, pp. 3394-3398.

[40]. B. X. Fan, W. Peng, Y. Li, X. Li, S. Wang, G. Zhang, F. Zhang, Deoxygenation of Exfoliated Graphite Oxide under Alkaline Conditions: A Green Route to Graphene Preparation, *Adv. Mater.*, Vol. 20, Issue 23, 2008, pp. 4490-4493.

[41]. Y. Shao, J. Wang, H. Wu, J. Liu, I. A. Aksay, Y. Lin, Graphene Based Electrochemical Sensors and Biosensors: A Review, *Electroanal.*, Vol. 22, issue 10, 2010, pp. 1027-1036.

[42]. H. M. A. Hassan, V. Abdelsayed, A. S. Khder, K. M. AbouZeid, J. Terner, M. Samy El-Shall, S. I. Al-Resayes, A. A. El-Azhary, Microwave synthesis of graphene sheets supporting metal nanocrystals in aqueous and organic media, *J. Mater. Chem.*, Vol. 19, 2009, pp. 3832-3837.

[43]. J. A. Gerbec, D. Magana, A. Washington and G. F. Strouse, Microwave-Enhanced Reaction Rates for Nanoparticle Synthesis, *J. Am. Chem. Soc.* Vol. 127, 2005, pp. 15791-15800.

[44]. A. B. Panda, G. P. Glaspell and M. S. El-Shall, Microwave Synthesis of Highly Aligned Ultra Narrow Semiconductor Rods and Wires, *J. Am. Chem. Soc.*, Vol. 128, 2006, pp. 2790-2791.

[45]. B. Panda, G. P. Glaspell and M. S. El-Shall, Microwave Synthesis and Optical Properties of Uniform Nanorods and Nanoplates of Rare Earth Oxides, *J. Phys. Chem. C.*, Vol. 111, 2007, pp. 1861-1864.

[46]. V. Abdelsayed, A. B. Panda, G. P. Glaspell, M. S. El-Shall, in Nanoparticles: Synthesis, Stabilization, Passivation, and Functionalization, R. Nagarajan and T. Alan Hatton (eds.), *ACS Symposium Series*, 996, 2008.

[47]. F. Della Negra, M. Meneghetti, E. Menna, Microwave-Assisted Synthesis of a Soluble Single Wall Carbon Nanotube Derivative, *Fuller. Nanotub. Car. N.*, Vol. 11, Issue 1, 2003, pp. 25-34.

[48]. E. H. L. Falcao, R. G. Blair, J. J. Mack, L. M. Viculis, C. Kwon, M. Bendikov, R. B. Kaner, B. S. Dunn, F. Wudl, Production of small single-wall carbon nanohorns by CO_2 laser ablation of graphite in Ne-gas atmosphere, *Carbon*, Vol. 45, 2007, pp. 1364-1367.

[49]. H. L. Guo, X. F. Wang, Q. Y. Qian, F. B. Wang, X. H. Xia, A Green Approach to the Synthesis of Graphene Nanosheets, *ACS Nano*, Vol. 3, 2009, pp. 2653-2659.

[50]. Y. Y. Shao, J. Wang, M. Engelhard, C. M. Wang, Y. M. Lin, Facile and controllable electrochemical reduction of graphene oxide and its applications, *J. Mater. Chem.*, Vol. 20, 2010, pp. 743-748.

[51]. M. Zhou, Y. L. Wang, Y. M. Zhai, J. F. Zhai, W. Ren, F. Wang, S. J. Dong, Controlled Synthesis of Large-Area and Patterned Electrochemically Reduced Graphene Oxide Films, *Chem. Eur. J.*, Vol. 15, 2009, pp. 6116-6120.

[52]. Y. Shao, J. Wang, M. Engelhard, C. Wang, Y. Lin, Facile and controllable electrochemical reduction of graphene oxide and its applications, *J. Mater. Chem.,* Vol. 20, 2010, pp. 743-748.

[53]. L. Chen, Y. Tang, K. Wang, C. Liu, S. Luo, Direct electrodeposition of reduced graphene oxide on glassy carbon electrode and its electrochemical application, *Electrochem. Commun.,* Vol. 13, 2011, pp. 133–137.

[54]. D. A. Sokolov, K. R. Shepperd, T. M. Orlando, Formation of Graphene Features from Direct Laser-Induced Reduction of Graphite Oxide, *J. Phys. Chem. Lett.,* Vol. 1, 2010, Issue 18, pp. 2633-2636.

[55]. S. Gilje, S. Dubin, A. Badakhshan, J. Farrar, S. A. Danczyk, R. B. Kaner, Photothermal Deoxygenation of Graphene Oxide for Patterning and Distributed Ignition Applications, *Adv. Mater.,* Vol. 22, 2010, pp. 419-423.

[56]. M. F. El-Kady, V. Strong, S. Dubin, R. B. Kaner, Laser scribing of high-Performance and flexible graphene-based electrochemical capacitors, *Science,* Vol. 335, 2012, pp. 1326-1330.

[57]. A. J. Bard, L. R. Faulkner, Electrochemical method, Fundamentals and applications, *John Wiley & Sons, Inc.,* New York, 2001.

[58]. A. Galal, N. F. Atta, H. K. Hassan, Graphene Supported-Pt-M (M = Ru or Pd) for Electrocatalytic Methanol Oxidation, *Int. J. Electrochem. Sci.,* 7, 2012, pp. 768 – 784.

[59]. J. Israelachvili, Intermolecular and Surface Forces, *Academic,* New York, 1992.

[60]. W. Zhou, Z. L. Wang, (eds.), Scanning Microscopy for Nanotechnology: Techniques and Applications, *Springer Science and Business Media,* 2006.

[61]. W. Barzyk, A. Kowal, A. Pomianowski, A. Rakowska, SEM/EDX and AFM study of gold cementation on copper (I) sulphide, *Physicochem. Probl. Mi.,* Vol. 36, 2002, pp. 9-20.

[62]. L. Baraton, Z. He, C. S. Lee, J.-L. Maurice, C. S. Cojocaru, A.-F. Gourgues-Lorenzon, Y. H. Lee, D. Pribat, Synthesis of few-layered graphene by ion implantation of carbon in nickel thin films, *Nanotechnology,* Vol. 22, 2011, pp. 085601.

[63]. Y. Lu, R. G. Reddy, Electrocatalytic properties of carbon supported cobalt phthalocyanine–platinum for methanol electro-oxidation, *Int. J. Hydrogen Energ.,* Vol. 33, Issue 14, 2008, pp. 3930-3937.

[64]. M. Wang, D.-J. Guo, H.-L. Li, High activity of novel Pd/TiO2 nanotube catalysts for methanol electro-oxidation, *J. Solid State Chem.,* Vol. 178, 2005, Issue 6, pp. 1996-2000.

[65]. Y. Wang, Y. Li, L. Tang, J. Lu, J. Li, Application of graphene-modified electrode for selective detection of dopamine, *Electrochem. Communi.,* Vol. 11, 2009, pp. 889–892.

[66]. Y. Hu, J. Jin, P. Wu, H. Zhang, C. Cai, Graphene–gold nanostructure composites fabricated by electrodeposition and their electrocatalytic activity toward the oxygen reduction and glucose oxidation, *Electrochim. Acta,* Vol. 56, Issue 1, 2010, pp. 491-500.

[67]. K. R. Lee, K. U. Lee, J. W. Lee, B. T. Ahn, S. I. Woo, Electrochemical oxygen reduction on nitrogen doped graphene sheets in acid media, *Electrochem. Commun.*, Vol. 12, 2010, Issue 8, pp. 1052-1055.

[68]. G. Ko, H. -Y. Kim, J. Ahn, Y. -M. Park, K. -Y. Lee, J. Kim, Graphene-based nitrogen dioxide gas sensors, *Curr. Appl. Phys.*, Vol. 10, 2010, pp. 1002-1004.

[69]. U. Lange, T. Hirsch, V. M. Mirsky, O. S. Wolfbeis, Hydrogen sensor based on a graphene – palladium nanocomposite, *Electrochem. Acta*, 56, 2011, pp. 3707-3712.

[70]. Y. Jiang, X. Zhang, C. Shan, S. Hua, Q. Zhang, X. Bai, L. Dan, L. Niu, Functionalization of graphene with electrodeposited Prussian blue towards amperometric sensing application, *Talanta*, Vol. 85, 2011, pp. 76–81.

[71]. S. Liu, J. Tian, L. Wang, X, Sun, Electrochemical detection of dopamine in the presence of ascorbic acid using graphene modified electrodes, *J. Nanopart. Res.*, Vol. 13, Issue 10, 2011, pp. 4539-4548.

[72]. F. Li, J. Chai, H. Yang, D. Han, L. Niu, Synthesis of Pt/ionic liquid/graphene nanocomposite and its simultaneous determination of ascorbic acid and dopamine, *Talanta*, Vol. 81, Issue 8, 2010, pp. 1063-1068.

[73]. Y.-R. Kim, S. Bonga, Y.-J. Kang, Y. Yang, R. K. Mahajan, J. Seung Kim, Hasuck Kim, Electrochemical detection of dopamine in the presence of ascorbic acid using graphene modified electrodes, *Biosens. Bioelectron.*, Vol. 25, 2010, pp. 2366–2369.

[74]. G. Shang, P. Papakonstantinou, M. Mcmullan, M. Chu, A. Stamboulis, A. Potenza, S. S. Dhesi, H. Marchetto, Catalyst-Free Efficient Growth, Orientation and Biosensing Properties of Multilayer Graphene Nanoflake Films with Sharp Edge Planes, *Adv. Funct. Mater.*, Vol. 18, 2008, pp. 3506-3514.

[75]. J.-F. Wu, M.-Q. Xu, G.-C. Zhao, Graphene-based modified electrode for the direct electron transfer of Cytochrome c and biosensing, *Electrochem. Commun.*, Vol. 12, 2010, pp. 175-177.

[76]. M. Pumera, A. Ambrosi, E. L. K. Chng, H. L. Poh, Graphene for electrochemical sensing and biosensing, *TRAC-Trend. Anal. Chem.*, Vol. 29, 2010, pp. 954-965.

[77]. X. Kang, J. Wang, H. Wu, J. Liu, I. A. Aksay, Y. Lin, A graphene-based electrochemical sensor for sensitive detection of paracetamol, *Talanta*, Vol. 81, 2010, pp. 754–759.

[78]. Y. Fan, J.-H. Liu, H.-T. Lu, Q. Z., Electrochemical behavior and voltammetric determination of paracetamol on Nafion/TiO2–graphene modified glassy carbon electrode, *Colloids and Surfaces B: Biointerfaces*, Vol. 85, 2011, pp. 289–292.

[79]. M. H. Parvin, Graphene paste electrode for detection of chlorpromazine, *Electrochem. Commun.*, Vol. 13, 2011, pp. 366–369.

[80]. K. Liu, J. Zhang, G. Yang, C. Wang, J.-J. Zhu, Direct electrochemistry and electrocatalysis of hemoglobin based on poly(diallyldimethylammonium chloride) functionalized graphene

sheets/room temperature ionic liquid composite film, *Electrochem. Commun.,* Vol. 12, 2010, pp. 402–405.

[81]. J. Li, S. Guo, Y. Zhai, E. Wang, High-sensitivity determination of lead and cadmium based on the Nafion-graphene composite film, *Anal. Chim. Acta,* Vol. 649, 2009, pp. 196–201.

[82]. G. K. Ramesha, S. Sampath, In-situ formation of graphene – lead oxide composite and its use in trace arsenic detection, *Sensor. Actuat. B-Chem.,* Vol. 160, Issue 1, 2011, pp. 306-311.

[83]. X. Wang, L. J. Zhi, N. Tsao, Z. Tomovic, J. L. Li, K. Mullen, Transparent Carbon Films as Electrodes in Organic Solar Cells, *Angew. Chem.-Int. Ed.,* Vol. 47, 2008, pp. 2990-2992.

[84]. J. Hass, W. A. de Heer, E. H. Conrad, The growth and morphology of epitaxial multilayer graphene, *J. Phys. Cond. Matter,* Vol. 20, 2008, pp. 323202.

[85]. M. D. Stoller, S. J. Park, Y. W. Zhu, J. H. An, R. S. Ruoff, Graphene-Based Ultracapacitors, *Nano Lett.,* Vol. 8, 2008, pp. 3498-3502.

[86]. E. Yoo, J. Kim, E. Hosono, H. Zhou, T. Kudo, I. Honma, arge Reversible Li Storage of Graphene Nanosheet Families for Use in Rechargeable Lithium Ion Batteries, *Nano Lett.,* Vol. 8, 2008, pp. 2277-2279.

[87]. D. H. Wang, D. W. Choi, J. Li, Z. G. Yang, Z. M. Nie, R. Kou, D. H. Hu, C. M. Wang, L. V. Saraf, J. G. Zhang, I. A. Aksay, J. Liu, Self-Assembled TiO2–Graphene Hybrid Nanostructures for Enhanced Li-Ion Insertion, *ACS Nano,* Vol. 3, 2009, pp. 907-914.

[88]. B. Seger, P. V. Kamat, Electrocatalytically Active Graphene-Platinum Nanocomposites. Role of 2-D Carbon Support in PEM Fuel Cells, *J. Phys. Chem. C,* Vol. 113, 2009, pp. 7990-7995.

[89]. R. Kou, Y. Shao, D. H. Wang, M. H. Engelhard, J. H. Kwak, J. Wang, V. V. Viswanathan, C. M. Wang, Y. H. Lin, Y. Wang, I. A. Aksay, J. Liu, Enhanced activity and stability of Pt catalysts on functionalized graphene sheets for electrocatalytic oxygen reduction, *Electrochem. Commun.,* Vol. 11, 2009, pp. 954-957.

[90]. Y. C. Si, E. T. Samulski, Exfoliated Graphene Separated by Platinum Nanoparticles, *Chem. Mater.,* Vol. 20, 2008, pp. 6792-6797.

[91]. Y. M. Li, L. H. Tang, J. H. Li, Preparation and electrochemical performance for methanol oxidation of pt/graphene nanocomposites, *Electrochem. Commun.,* Vol. 11, 2009, pp. 846-849.

[92]. Y. Wang, J. Lu, L. H. Tang, H. X. Chang, J. H. Li, Graphene Oxide Amplified Electrogenerated Chemiluminescence of Quantum Dots and Its Selective Sensing for Glutathione from Thiol-Containing Compounds, *Anal. Chem.,* Vol. 81, 2009, pp. 9710-9715.

[93]. R. Alcala, J. W. Shabaker, G. W. Huber, M. A. S.-Castillo, J. A. Dumesic, Experimental and DFT Studies of the Conversion of Ethanol and Acetic Acid on PtSn-Based Catalysts, *J. Phys. Chem. B,* Vol. 109, 2005, pp. 2074-2085.

[94]. L. A. Currie, Nomenclature in evaluation of analytical methods including detection and quantification capabilities (IUPAC Recommendations 1995), *Pure Appl. Chem.,* Vol. 67, Issue 67, 1995, pp. 1699-1723.

[95]. J. Wang, *Analytical Electrochemistry*, 3rd Edition, *John Wiley & Sons, Inc.*, New Jersey, 2006.

[96]. M. M. Ardakani, H. Beitollahi, M. A. S. Mohseni, H. Naeimi, N. Taghavinia, Novel nanostructure electrochemical sensor for electrocatalytic determination of norepinephrine in the presence of high concentrations of acetaminophene and folic acid, *Appl. Catal. A: Gen.*, Vol. 378, 2010, pp. 195-201.

[97]. A. Ambrosi, M. Pumera, Stacked graphene nanofibers for electrochemical oxidation of DNA bases, *Phys. Chem. Chem. Phys.*, Vol. 12, 2010, pp. 8943-8947.

[98]. N. G. Shang, P. Papakonstantinou, M. McMullan, M. Chu, A. Stamboulis, A. Potenza, S. S. Dhesi, H. Marchetto, Catalyst-Free Efficient Growth, Orientation and Biosensing Properties of Multilayer Graphene Nanoflake Films with Sharp Edge Planes, *Adv. Funct. Mater.*, Vol. 18, Issue 21, 2008, pp. 3506–3514.

[99]. M. Heien, A. Khan, J. Ariansen, J. Cheer, P. Phillips, K. Wassum, M. Wightman, Real-time measurement of dopamine fluctuations after cocaine in the brain of behaving rats, *Proc. Natl. Acad. Sci. USA*, Vol. 102, 2005, pp. 10023-10028.

[100]. W. Yeh, Y. Kuo, S. Cheng, Voltammetry and flow-injection amperometry for indirect determination of dopamine, *Electrochem. Commun.*, Vol. 10, 2008, pp. 66-70.

[101]. Z. Jia, J. Liu, Y. Shen, Electrochemical properties of ordered mesoporous carbon and its electroanalytical application for selective determination of dopamine, *Electrochem. Commun.*, Vol. 9, 2007, pp. 2739-2743.

[102]. N. Jia, Z. Wang, G. Yang, H. Shen, L. Zhu, Electrochemical properties of ordered mesoporous carbon and its electroanalytical application for selective determination of dopamine, *Electrochem. Commun.*, Vol. 9, 2007, pp. 233-238.

[103]. R. M. Wightman, L. J. May, A. C. Michael, Detection of dopamine dynamics in the brain, *Anal. Chem.*, Vol. 60, 1988, pp. 769A–779A.

[104]. G. C. Yen, P. D. Duh, H. L. Tsai, Antioxidant and pro-oxidant properties of ascorbic acid and gallic acid, *Food Chem.*, Vol. 79, 2002, pp. 307–313.

[105]. G. Z. Hu, Y. G. Ma, Y. Guo, S. J. Shao, Electrocatalytic oxidation and simultaneous determination of uric acid and ascorbic acid on the gold nanoparticles-modified glassy carbon electrode, *Electrochim. Acta*, Vol. 53, 2008, pp. 6610–6615.

[106]. A. P. dos Reis, C. R. T. Tarley, N. Maniasso, L. T. Kubota, Exploiting micellar environment for simultaneous electrochemical determination of ascorbic acid and dopamine, *Talanta*, Vol. 67, 2005, pp. 829-835.

[107]. L. Jiang, C. Liu, L. Jiang, Z. Peng, G. Lu, A Chitosan-Multiwall Carbon Nanotube Modified Electrode for Simultaneous Detection of Dopamine and Ascorbic Acid, *Anal. Sci.*, Vol. 20, Issue 7, 2004, pp. 1055-1058.

[108]. A. Fujishima, T. N. Rao, E. Popa, B. V. Sarada, I. Yagi, D. A. Tryk, Electroanalysis of dopamine and NADH at conductive diamond electrodes, *J. Electroanal. Chem.*, Vol. 473, 1999, pp. 179-227.

[109]. M. L. A. V. Heien, P. E. M. Phillips, G. D. Stuber, A. T. Seipel, R. M. Wightman, Overoxidation of carbon-fiber microelectrodes enhances dopamine adsorption and increases sensitivity Electronic supplementary information (ESI) available, *Analyst,* Vol. 128, 2003, pp. 1413-1419.

[110]. P. Ramesh, G. S. Suresh, S. Sampath, Selective determination of dopamine using unmodified, exfoliated graphite electrodes, *J. Electroanal. Chem.*, Vol. 561, 2004, pp. 173-180.

[111]. P. Serp, M. Corrias, P. Kalck, Carbon nanotubes and nanofibers in catalysis, *Appl. Catal. A-Gen.*, Vol. 253, 2003, pp. 337-358.

[112]. Q. Liu, X. Zhu, Z. Huo, X. He, Y. Liang, M. Xu, Electrochemical detection of dopamine in the presence of ascorbic acid using PVP/graphene modified electrodes, *Talanta*, Vol. 97, 2012, pp. 557–562.

[113]. D. X. Han, T. T. Han, C. S. Shan, A. Ivaska, L. Niu, Simultaneous Determination of Ascorbic Acid, Dopamine and Uric Acid with Chitosan-Graphene Modified Electrode, *Electroanalysis*, Vol. 22, 2010, pp. 2001–2008.

[114]. C. B. Liu, K. Wang, S. L. Luo, Y. H. Tang, L. Y. Chen, Direct Electrodeposition of Graphene Enabling the One-Step Synthesis of Graphene–Metal Nanocomposite Films, *Small,* Vol. 7, Issue 9, 2011, pp. 1203–1206.

[115]. X. Tian, C. Cheng, H. Yuan, J. Du, D. Xiao, S. Xie, M. M. F. Choi, Simultaneous determination of l-ascorbic acid, dopamine and uric acid with gold nanoparticles-cyclodextrin-graphene-modified electrode by square wave voltammetry, *Talanta*, Vol. 93, 2012, pp. 79– 85.

[116]. Z.-H. Sheng, X.-Q. Zheng, J.-Y. Xu, W.-J. Bao, F.-B. Wang, X.-H. Xia, Electrochemical sensor based on nitrogen doped graphene: Simultaneous determination of ascorbic acid, dopamine and uric acid, *Biosens. Bioelectron.*, Vol. 34, 2012, pp. 125– 131.

[117]. D. Wei, Y. Liu, Y. Wang, H. Zhang, L. Huang, G. Yu, Synthesis of N-Doped Graphene by Chemical Vapor Deposition and Its Electrical Properties, *Nano Lett.*, Vol. 9, 2009, pp. 1752-1758.

[118]. Y. Wang, Y. Y. Shao, D. W. Matson, J. H. Li, Y. H. Lin, Nitrogen-doped graphene and its application in electrochemical biosensing, *ACS Nano,* Vol. 4, 2010, pp. 1790–1798.

[119]. A. Sucheta, J. Rusling, Effect of background charge on estimating diffusion coefficients by chronocoulometry at glassy carbon electrodes, *Electroanal.*, Vol. 3, 1991, pp. 735–739.

[120]. X. Z. Wu, L. J. Mu, W. Z. Zhang, Impedance of the electrochemical oxidation of epinephrine on a glassy carbon electrode, *J. Electroanal. Chem.*, Vol. 352, 1993, pp. 295–300.

[121]. E. L. Ciolkowski, K. M. Maness, P. S. Cahill, R. M. Wightman, D. H. Evans, B. Fosset, C. Amatore, Disproportionation during

---

electrooxidation of catecholamines at carbon-fiber microelectrodes, *Anal. Chem.,* Vol. 66, 1994, pp. 3611–3617.

[122].M. D. Hawley, S. V. Tatawawadi, S. Piekarski, R. N. Adams, Electrochemical studies of the oxidation pathways of catecholamines, *J. Am. Chem. Soc.,* Vol. 89, 1967, pp. 447–450.

[123].H. S. Wang, D. Q. Huang, R. M. Liu, Study on the electrochemical behavior of epinephrine at a poly(3-methylthiophene)-modified glassy carbon electrode, *J. Electroanal. Chem.,* Vol. 570, 2004, pp. 83–90.

[124].H. M. Zhang, X. L. Zhou, R. T. Hui, N. Q. Li, D. P. Liu, Studies of the electrochemical behavior of epinephrine at a homocysteine self-assembled electrode, *Talanta,* Vol. 56, 2002, pp. 1081–1088.

[125].M. Zhu, X. M. Huang, J. Li, H. X. Shen, Peroxidase-based spectrophotometric methods for the determination of ascorbic acid, norepinephrine, epinephrine, dopamine and levodopa, *Anal. Chim. Acta,* Vol. 357, 1997, pp. 261–267.

[126].J. O. Schenk, E. Miller, R. N. Adams, Electrochemical techniques for the study of brain chemistry, *J. Chem. Educ.,* Vol. 60, 1983, pp. 311–315.

[127].Y. Hasebe, T. Hirano, S. Uchiyama, Determination of catecholamines and uric acid in biological fluids without pretreatment, using chemically amplified biosensors, *Sens. Actuators B,* Vol. 24, 1995, pp. 94–97.

[128].T. Łuczak, Comparison of electrochemical oxidation of epinephrine in the presence of interfering ascorbic and uric acids on gold electrodes modified with S-functionalized compounds and gold nanoparticles, *Electrochim. Acta,* Vol. 54, 2009, pp. 5863–5870.

[129].F. Cui, X. Zhang, Electrochemical sensor for epinephrine based on a glassy carbon electrode modified with graphene/gold nanocomposites, *J. Electroanal. Chem.,* Vol. 669, 2012, pp. 35–41.

[130].J. B. Raoof, A. Kiani, R. Ojani, R. Valiollahi, S. Rashid-Nadimi, Simultaneous voltammetric determination of ascorbic acid and dopamine at the surface of electrodes modified with self-assembled gold nanoparticle films, *J. Solid State Electrochem.,* Vol. 14, 2010, pp. 1171–1176.

[131].Z. Gao, H. Huang, Simultaneous determination of dopamine, uric acid and ascorbic acid at an ultrathin film modified gold electrode, *Chem. Commun.,* Vol. 1998, Issue 19, 1998, pp. 2107-2108.

[132].B. Habibia, M. H. Pournaghi-Azar, Simultaneous determination of ascorbic acid, dopamine and uric acid by use of a MWCNT modified carbon-ceramic electrode and differential pulse voltammetry, *Electrochim. Acta,* Vol. 55, 2010, pp. 5492–5498.

[133].D. Zheng, J. Ye, L. Zhou, Y. Zhang, C. Yu, Simultaneous determination of dopamine, ascorbic acid and uric acid on ordered mesoporous carbon/Nafion composite film, *J. Electroanal. Chem.,* Vol. 625, 2009, pp. 82–87.

[134].D. X. Han, T. T. Han, C. S. Shan, A. Ivaska, L. Niu, Simultaneous Determination of Ascorbic Acid, Dopamine and Uric Acid with Chitosan-Graphene Modified Electrode, *Electroanalysis,* Vol. 22, 2010, pp. 2001–2008.

[135].R. M. D. Carvalho, R. S. Freire, S. Rath, L. T. Kubota, Effects of EDTA on signal stability during electrochemical detection of acetaminophen, *J. Pharm. Biomed. Anal.,* Vol. 34, 2004, pp. 871-878.

[136].R. T. Kachoosangi, G. G. Wildgoose, R. G. Compton, Sensitive adsorptive stripping voltammetric determination of paracetamol at multiwalled carbon nanotube modified basal plane pyrolytic graphite electrode, *Anal. Chim. Acta,* Vol. 618, 2008, pp. 54-60.

[137].R. N. Goyal, V. K. Gupta, M. Oyama, N. Bachheti, Differential pulse voltammetric determination of paracetamol at nanogold modified indium tin oxide electrode, *Electrochem. Commun.,* Vol. 7, 2005, pp. 803-807.

[138].M. Li, L. H. Jing, Electrochemical behavior of acetaminophen and its detection on the PANI–MWCNTs composite modified electrode, *Electrochim. Acta,* Vol. 52, 2007, pp. 3250-3257.

[139].X. Kang, J. Wang, H. Wu, J. Liu, I. A. Aksay, Y. Lin, A graphene-based electrochemical sensor for sensitive detection of paracetamol, *Talanta,* Vol. 81, 2010, pp. 754–759.

[140].C.-X. Xu, K.-J. Huang, Y. Fan, Z.-W. Wu, J. Li, Electrochemical determination of acetaminophen based on TiO2–graphene/poly(methyl red) composite film modified electrode, *J. Mol. Liq.,* Vol. 165, 2012, pp. 32–37.

[141].P. Kalimuthu, S. A. John, S. A., Modification of electrodes with nanostructured functionalized thiadiazole polymer film and its application to the determination of ascorbic acid, *Electrochim. Acta,* Vol. 55, 2009, pp. 183-189.

[142].M. G. Khansari, R. Zendehdel, M. P. Hamedani, M. Amini, Determination of morphine in the plasma of addicts in using Zeolite Y extraction following high-performance liquid chromatography, *Clin. Chim. Acta,* Vol. 364, 2006, pp. 235-238.

[143].N. F. Atta, A. Galal, R. A. Ahmed, Direct and Simple Electrochemical Determination of Morphine at PEDOT Modified Pt Electrode, *Electroanal.,* Vol. 23, 2011, pp. 737-746.

[144]. R. Verpoorte, A. B. Svendsen, Chromatography of Alkaloids, *Elsevier,* 1984.

[145].P. Norouzi, M. R. Ganjali, A. A. M. Movahedi, B. Larijani, Fast Fourier transformation with continuous cyclic voltammetry at an Au microelectrode for the determination of morphine in a flow injection system, *Talanta,* Vol. 73, 2007, pp. 54-61.

[146].K. Ary, K. Róna, LC determination of morphine and morphine glucuronides in human plasma by coulometric and UV detection, *J. Pharm. Biomed. Anal.,* Vol. 26, 2001, pp. 179-187.

[147].D. Projean, T. M. Tu, J. Ducharme, Rapid and simple method to determine morphine and its metabolites in rat plasma by liquid chromatography–mass spectrometry, *J. Chromatogr., B,* Vol. 787, 2003, pp. 243-253.

[148].A. Navaee, A. Salimi, H. Teymourian, Graphene nanosheets modified glassy carbon electrode for simultaneous detection of heroine, morphine and noscapine, *Biosens. Bioelectron.,* Vol. 31, 2012, pp. 205– 211.

[149]. F. Li, J. Song, C. Shan, D. Gao, X. Xu, L. Niu, Electrochemical determination of morphine at ordered mesoporous carbon modified glassy carbon electrode, *Biosens. Bioelectron.*, Vol. 25, 2010, pp. 1408-1413.

[150]. K. C. Ho, C. Y. Chen, H. C. Hsu, L. C. Chen, S. C. Shiesh, X. Z. Lin, Amperometric detection of morphine at a Prussian blue-modified indium tin oxide electrode, *Biosens. Bioelectron.*, Vol. 20, 2004, pp. 3-8.

[151]. W. M. Yeh, K. C. Ho, Amperometric morphine sensing using a molecularly imprinted polymer-modified electrode, *Anal. Chim. Acta*, Vol. 542, 2005, pp. 76-82.

[152]. G. Sakai, K. Ogata, T. Uda, N. Miura, N. Yamazoe, A surface plasmon resonance-based immunosensor for highly sensitive detection of morphine, *Sens. Actuator B.*, Vol. 49, 1998, pp. 5-12.

[153]. C.-H. Weng, W.-M. Yeh, K.-C. Ho, G.-B. Lee, A microfluidic system utilizing molecularly imprinted polymer films for amperometric detection of morphine, *Sens. Actuators* B, Vol. 121, 2007, pp. 576-582.

[154]. K.-C. Ho, W.-M. Yeh, T.-S. Tung, J.-Y. Liao, Amperometric detection of morphine based on poly(3, 4-ethylenedioxythiophene) immobilized molecularly imprinted polymer particles prepared by precipitation polymerization, *Anal. Chim. Acta*, Vol. 542, 2005, pp. 90-96.

[155]. A. Salimi, R. Hallaj, G. R. Khayatian, Amperometric Detection of Morphine at Preheated Glassy Carbon Electrode Modified with Multiwall Carbon Nanotubes, *Electroanal.*, Vol. 17, 2005, pp. 873-879.

[156]. Y. Li, K. Li, G. Song, J. Liu, K. Zhang, B. Ye, Electrochemical behavior of codeine and its sensitive determination on graphene based modified electrode, *Sens. Actuators* B, Vol. 182 , 2013, pp. 401-407.

[157]. H. X. Sun, H. L. Ling, L. Wang, H. H. Wang, Y. H., Zhao, R. Dong, Development of ELISA Method for Residue of Salbutamol, *Chinese Journal of Animal Health Inspection*, Vol. 26, Issue 12, 2009, pp. 44-47.

[158]. M. Zhou, Y. M. Zhai, S. J. Dong, Electrochemical Sensing and Biosensing Platform Based on Chemically Reduced Graphene Oxide, *Anal. Chem.*, Vol. 81, 2009, pp. 5603-5613.

[159]. Y. Shao, J. Wang, H. Wu, J. Liu, I. A. Aksay, Y. Lin, Graphene Based Electrochemical Sensors and Biosensors: A Review, *Electroanal.*, Vol. 22, 2010, Issue 10, pp. 1027 – 1036.

[160]. S. Liu, J. Tian, L. Wang, X. Sun, Microwave-assisted rapid synthesis of Pt/graphene nanosheet composites and their application for methanol oxidation, *J. Nanopar. Res.*, Vol. 13, 2011, pp. 4731-4737.

[161]. G. Zeng, Y. Xing, J. Gao, Z. Wang, X. Zhang, Unconventional Layer-by-Layer Assembly of Graphene Multilayer Films for Enzyme-Based Glucose and Maltose Biosensing, *Langmuir*, Vol. 26, 2010, pp. 15022-15026.

[162]. S. Palanisamy, S.-M. Chen, R. Sarawathi, A novel nonenzymatic hydrogen peroxide sensor based on reduced graphene oxide/ZnO composite modified electrode, *Sens. Actuat. B*, Vol. 166– 167, Issue 1, 2012, pp. 372– 377.

[163].L. Cui, L. Chen, M. Xu, H. Su, S. Ai, Nonenzymatic amperometric organic peroxide sensor based on nano-cobalt phthalocyanine loaded functionalized graphene film, *Anal. Chim. Acta,* Vol. 712, 2012, pp. 64– 71.

[164].T. Kuila, S. Bose, P. Khanra, A. K. Mishra, N. H. Kim, J. H. Lee, Recent advances in graphene-based biosensors, *Biosen. Bioelectron, Biosen. Bioelectron.,* Vol. 26, 2011, pp. 4637-4648.

[165].Y.-W. Hsu, T.-K. Hsu, C. L. Sun, Y.-T. Nien, N.-W. Pud, M.-D. Ger, Synthesis of CuO/graphene nanocomposites for nonenzymatic electrochemical glucose biosensor applications, *Electrochim. Acta,* Vol. 82, 2012, pp. 152-157.

[166].S. Park, H. Boo, T. D. Chung, Electrochemical non-enzymatic glucose sensors, *Anal. Chim. Acta,* Vol. 556, 2006, pp. 46-57.

[167].L. Luo, L. Zhu, Z. Wang, Nonenzymatic amperometric determination of glucose by CuO nanocubes–graphene nanocomposite modified electrode, *Bioelectrochem.,* Vol. 88, pp. 156-163.

[168].F. Gao, X. Qi, X. Cai, Q. Wang, F. Gao, W. Sun, Magnetic softness and interparticle exchange interactions of $(Fe_{65}Co_{35})_{1-x}(Al_2O_3)_x$ (x = 0–0. 50) nanogranular films, *Thin Solid Films,* Vol. 520, 2012, pp. 5064–5069.

[169].Y. Shi, Z. Liu, B. Zhao, Y. Sun, F. Xu, Y. Zhang, Z. Wen, H. Yang, Z. Li, Carbon nanotube decorated with silver nanoparticles via noncovalent interaction for a novel nonenzymatic sensor towards hydrogen peroxide reduction, *J. Electroanal. Chem.,* Vol. 656 2011, pp. 29–33.

[170].Y. Ye, T. Kong, X. Yu, Y. Wu, K. Zhang, X. Wang, Enhanced nonenzymatic hydrogen peroxide sensing with reduced graphene oxide/ferroferric oxide nanocomposites, *Talanta,* Vol. 89 2012, pp. 417–421.

[171].L. Li, Z. Du, S. Liu, Q. Hao, Y. Wang, Q. Li, T. Wang, A novel nonenzymatic hydrogen peroxide sensor based on MnO2/graphene oxide nanocomposite, *Talanta,* Vol. 82, 2010, pp. 1637–1641.

[172].F. Xu, Y. Sun, Y. Zhang, Y. Shi, Z, Wen, Z. Li, Graphene-Pt nanocomposite for nonenzymatic detection of hydrogen peroxide with enhanced sensitivity, *Electrochem. Commun.,* Vol. 13 2011, pp. 1131–1134.

[173].J. Yang, W. D. Zhang, S. Gunasekaran, An amperometric non-enzymatic glucose sensor by electrodepositing copper nanocubes onto vertically well-aligned multiwalled carbon nanotube arrays, *Biosens. Bioelectron.,* Vol. 26, 2010, pp. 279–284.

[174].R. L. Blakley, D. D. Henry, C. J. Smith, Lack of correlation between cigarette mainstream smoke particulate phase radicals and hydroquinone yield, *Food Chem. Toxicol.,* Vol. 39, 2001, pp. 401–406.

[175].T. Xie, Q. Liu, Y. Shi, Q. Liu, Simultaneous determination of positional isomers of benzenediols by capillary zone electrophoresis with square wave amperometric detection, *J. Chromatogr. A,* Vol. 1109, 2006, pp. 317–321.

[176]. S. Hu, Y. Wang, X. Wang, L. Xu, J. Xiang, W. Sun, Electrochemical detection of hydroquinone with a gold nanoparticle and graphene modified carbon ionic liquid electrode, *Sens. Actuat. B*, Vol. 168, 2012, pp. 27– 33.

[177]. H. Fan, Y. Li, D. Wu, H. Ma, K. Mao, D. Fan, B. Du, H. Li, Q. Wei, Electrochemical bisphenol A sensor based on N-doped graphene sheets, *Anal. Chim. Acta*, Vol. 711, Issue 2, 2012, pp. 24– 28.

[178]. H. S. Yin, L. Cui, S. Y. Ai, H. Fan, L. S. Zhu, Electrochemical determination of bisphenol A at Mg–Al–CO3 layered double hydroxide modified glassy carbon electrode, *Electrochim. Acta,* Vol. 55, 2010, pp. 603–610.

[179]. L. N. Vandenberg, R. Hauser, M. Marcus, N. Olea, W. V. Welshons, Human exposure to bisphenol A (BPA), *Reprod. Toxicol.*, Vol. 24, 2007, pp. 139–177.

[180]. Y. Q. Wang, Y. Y. Yang, L. Xu, J. Zhang, Bisphenol A sensing based on surface molecularly imprinted, ordered mesoporous silica, *Electrochim. Acta*, Vol. 56, 2011, pp. 2105–2109.

[181]. J. Gong, X. Miao, H. Wan, D. Song, Facile synthesis of zirconia nanoparticles-decorated graphene hybrid nanosheets for an enzymeless methyl parathion sensor, *Sens. Actuat. B*, Vol. 162, 2012, pp. 341– 347.

[182]. H. L. Guo, X. F. Wang, Q. Y. Qian, F. B. Wang, X. H. Xia, A green approach to the synthesis of graphene nanosheets, *ACS Nano,* Vol. 3, 2009, pp. 2653–2659.

[183]. M. B. Pomfret, C. Stoltz, B. Varughese, R. A. Walker, Structural and compositional characterization of yttria-stabilized zirconia: evidence of surface-stabilized, low-valence metal species, *Anal. Chem.,* Vol. 77, 2005, pp. 1791–1795.

[184]. C. Leger, P. Bertrand, Direct Electrochemistry of Redox Enzymes as a Tool for Mechanistic Studies, *Chem. Rev.,* Vol. 108, 2008, pp. 2379–2438.

[185]. S. R. Dey, C. R. Raj, Development of an Amperometric Cholesterol Biosensor Based on Graphene−Pt Nanoparticle Hybrid Material, *J. Phys. Chem. C,* Vol. 114, 2010, pp. 21427–21433.

[186]. R. Manjunatha, G. S. Suresh, J. S. Melo, S. F. D'Souza, T. V. Venkatesh, An amperometric bienzymatic cholesterol biosensor based on functionalized graphene modified electrode and its electrocatalytic activity towards total cholesterol determination, *Talanta,* Vol. 99, 2012, pp. 302-309.

[187]. R. Denooz, N. Dubois, C. Charlier, Deux ans d'analyse de saisies d'héroïne en région liégeoise, *Rev. Med. Liege*, Vol. 60, 2005, pp. 724–728.

[188]. J. W. Landen, R. Lang, S. J. McMahon, N. M. Rusan, A. M. Yvon, A. W. Adams, M. D. Sorcinelli, R. Campbell, P. Bonaccorsi, J. C. Ansel, D. R. Archer, P. Wadsworth, C. A. Armstrong, H. C. Joshi, Noscapine Alters Microtubule Dynamics in Living Cells and Inhibits the Progression of Melanoma, *Cancer Res.,* Vol. 62, 2002, pp. 4109–4114.

[189]. W. Sneader, The discovery of heroin, *Lancet*, Vol. 352, 1998, pp. 1697–1699.

[190]. M. H. Spyridaki, P. Kiousi, A. Vonaparti, P. Valavani, V. Zonaras, M. Zahariou, E. Sianos, G. Tsoupras, C. Georgakopoulos, Doping control analysis in human urine by liquid chromatography–electrospray ionization ion trap mass spectrometry for the Olympic Games Athens 2004: Determination of corticosteroids and quantification of ephedrines, salbutamol and morphine, *Anal. Chim. Acta*, Vol. 573, 2006, pp. 242–249.

[191]. M. R. Ganjali, P. Norouzi, M. Ghorbani, A. Sepehri, Fourier transform cyclic voltammetric technique for monitoring ultratrace amounts of salbutamol at gold ultra microelectrode in flowing solutions, *Talanta*, Vol. 66, 2005, pp. 1225–1233.

[192]. J. Ouyang, J. L. Duan, W. R. G. Baeyens, J. R. Delanghe, A simple method for the study of salbutamol pharmacokinetics by ion chromatography with direct conductivity detection, *Talanta*, Vol. 65, 2005, pp. 1–6.

[193]. X. Chen, H. Ye, W. Wang, B. Qiu, Z. Lin, G. Chen, Electrochemiluminescence Biosensor for Glucose Based on Graphene/Nafion/GOD Film Modified Glassy Carbon Electrode, *Electroanal.*, Vol. 22, Issue 20, 2010, pp. 2347–2352.

[194]. X. Kang, J. Wang, H. Wu, I. A. Aksay, J. Liu, Y. Lin, Glucose Oxidase–graphene–chitosan modified electrode for directelectrochemistry and glucose sensing, *Biosen. Bioelectron.*, Vol. 25, 2009, pp. 901–905.

[195]. C. Shan, H. Yang, D. Han, Q. Zhanga, A. Ivaska, L. Niu, Graphene/AuNPs/chitosan nanocomposites film for glucose biosensing, *Biosens. Bioelectron.*, Vol. 25, 2010, pp. 1070–1074.

[196]. T. T. Baby, S. S. J. Aravind, T. Arockiadoss, R. B. Rakhi, S. Ramaprabhu, Metal decorated graphene nanosheets as immobilization matrix for amperometric glucose biosensor, *Sensor Actuator B*, Vol. 145, 2010, pp. 71–77.

[197]. H. D. Jang, S. K. Kim, H. Chang, K.-M. Roh, J.-W. Choi, A glucose biosensor based on TiO2–Graphene composite, J. Huang, *Biosens. Bioelectron.*, Vol. 38, Issue 1, 2012, pp. 184-188.

[198]. Y. Jiang, Q. Zhanga, F. Li, L. Niu, Glucose oxidase and graphene bionanocomposite bridged by ionic liquid unit for glucose biosensing application, *Sensor. Actuator. B*, Vol. 161, 2012, pp. 728– 733.

[199]. A. Gutés, C. Carraro, R. Maboudian, Single-layer CVD-grown graphene decorated with metal nanoparticles as a promising biosensing platform, *Biosens. Bioelectron.*, Vol. 33, 2012, pp. 56– 59.

[200]. X. Zhong, R. Yuan, Y.-Q. Chai, Synthesis of chitosan-Prussian blue-graphene composite nanosheets for electrochemical detection of glucose based on pseudobienzyme channeling, *Sensor. Actuator. B*, Vol. 162, 2012, pp. 334–340.

[201]. Z. Z Wen, B. Niemeyer, Evaluation of two different Concanavalin A affinity adsorbents for the adsorption of glucose oxidase, *J. Chromatogr. B*, Vol. 857, 2007, pp. 149–157.

[202]. S. Stankovich, D. A. Dikin, G. H. B. Dommett, K. M. Kohlhaas, E. J. Zimney, E. A. Stach, R. D. Piner, S. T. Nguyen, R. S. Ruoff, Graphene-based composite materials, *Nature*, Vol. 442, 2006, pp. 282-286.

[203]. H. W. Ch. Postma, Rapid sequencing of individual DNA molecules in graphene nanogaps, *Nano. Lett.*, Vol. 10, 2010, pp. 420-425.

[204]. X. C. Dong, C. M. Lau, A. Lohani, S. G. Mhaisalkar, J. Kasim, Z. X. Shen, X. N. Ho, J. A. Rogers, L. J. Li, Electrical Detection of Femtomolar DNA via Gold-Nanoparticle Enhancement in Carbon-Nanotube-Network Field-Effect Transistors, *Adv. Mater.*, Vol. 20, 2008, pp. 2389–2393.

[205]. H. F. Dong, W. C. Gao, F. Yan, H. X. Ji, H. X. Ju, Fluorescence Resonance Energy Transfer between Quantum Dots and Graphene Oxide for Sensing Biomolecules, *Anal. Chem.*, Vol. 82, 2010, pp. 5511–5517.

[206]. S. Bi, J. L. Zhang, S. S. Zhang, Ultrasensitive and selective DNA detection based on nicking endonuclease assisted signal amplification and its application in cancer cell detection, *Chem. Commun.*, Vol. 46, 2010, pp. 5509–5511.

[207]. A. Sassolas, B. D Leca-Bouvier, L. J. Blum, DNA Biosensors and Microarray, *Chem. Rev.*, Vol. 108, 2008, pp. 109–139.

[208]. M. Zhou, Y. Zhai, S. Dong, Electrochemical Sensing and Biosensing Platform Based on Chemically Reduced Graphene Oxide, *Anal. Chem.*, Vol. 81, 2009, pp. 5603–5613.

[209]. B. Su, J. Tang, H. Yang, G. Chen, J. Huang, D. Tang, A Graphene Platform for Sensitive Electrochemical Immunoassay of Carcinoembryoninc Antigen Based on Gold-Nanoflower Biolabels, *Electroanal.*, Vol. 23, Issue 4, 2011, pp. 832–841.

[210]. X. Qi, H. Gao, Y. Zhang, X. Wang, Y. Chen, W. Sun, Electrochemical DNA biosensor with chitosan-Co3O4 nanorod-graphene composite for the sensitive detection of staphylococcus aureus nuc gene sequence, *Bioelectrochem.*, Vol. 88, 2012, pp. 42–47.

[211]. H. Dong, Z. Zhu, H. Jua, F. Yan, Triplex signal amplification for electrochemical DNA biosensing by coupling probe-gold nanoparticles–graphene modified electrode with enzyme functionalized carbon sphere as tracer, *Biosens. Bioelectron.*, 33, 2012, pp. 228– 232.

[212]. W. Lva, F.-M. Jin, Q. Guoc, Q.-H. Yang, F. Kang, DNA-dispersed graphene/NiO hybrid materials for highly sensitive non-enzymatic glucose sensor, *Electrochim. Acta*, Vol. 73, 2012, pp. 129– 135.

[213]. L. Zhu, L. Luo, Z. Wang, Fabrication of DNA/graphene/polyaniline nanocomplex for label-free voltammetric detection of DNA hybridization, *Biosensor. Bioelectron.*, Vol. 35, 2012, pp. 507–511.

[214]. M. Du, T. Yang, X. Li, K. Jiao, Fabrication of DNA/graphene/polyaniline nanocomplex for label-free voltammetric detection of DNA hybridization, *Talanta*, Vol. 88, 2012, pp. 439–444.

[215]. F. Wang, X. X. Chen, Y. X. Xu, S. S. Hu, Z. N. Gao, Enhanced electron transfer for hemoglobin entrapped in a cationic gemini surfactant films

on electrode and the fabrication of nitric oxide biosensor, *Biosens. Bioelectron.,* Vol. 23, 2007, pp. 176–182.

[216].L. Zhang, G. C. Zhao, X. W. Wei, Z. S. Yang, A nitric oxide biosensor based on myoglobin adsorbed on multi-walled carbon nanotubes, *Electroanal.,* Vol. 17, 2005, pp. 630–634.

[217].K-J. Huang, Y.-X. Miao, L. Wang, T. Gan, M. Yua, L.-L. Wang, Direct electrochemistry of hemoglobin based on chitosan–ionic liquid–ferrocene/graphene composite film, *Process Biochem.,* Vol. 47, 2012, pp. 1171–1177.

[218].G. C. Zhao, M. Q. Xu, J. Ma, X. W. Wei, Direct electrochemistry of hemoglobin on a room temperature ionic liquid modified electrode and its electrocatalytic activity for the reduction of oxygen, *Electrochem. Commun.,* Vol. 9, Issue 5, 2007, pp. 920-924.

[219].C. S. Shan, H. F. Yang, J. F. Song, D. X. Han, A. Ivaska, L. Niu, Direct Electrochemistry of Glucose Oxidase and Biosensing for Glucose Based on Graphene, *Anal. Chem.,* Vol. 81, 2009, pp. 2378-2382.

[220].Q. Feng, K. Liu, J. Fua, Y. Zhanga, Z. Zheng, C. Wang, Y. Du, W. Yea, Direct electrochemistry of hemoglobin based on nano-composite film of gold nanopaticles and poly (diallyldimethylammonium chloride) functionalized graphene, *Electrochim. Acta,* Vol. 60, 2012, pp. 304– 308.

[221].V. S. Elanchezhian, M. A. Kandaswarny, ferrocene-based multi-signaling sensor molecule functions as a molecular switch, *Inorg. Chem. Commun.,* Vol. 12, 2009, pp. 161–165.

[222].J. D. Qiu, W. M. Zhou, J. Guo J, R. Wang, R. P. Liang, Amperometric sensor based on ferrocene-modified multiwalled carbon nanotube nanocomposites as electron mediator for the determination of glucose, *Anal Biochem.,* Vol. 385, 2009, pp. 264–269.

[223].K.-J. Huang, Y.-X. Miao, L. Wang, T. Gan, M. Yu, L.-L. Wang, Direct electrochemistry of hemoglobin based on chitosan–ionic liquid–ferrocene/graphene composite film, *Process Biochem.,* Vol. 47, 2012, pp. 1171–1177.

[224].W. Wen, W. Chen, Q.-Q. Ren, X.-Y. Hu, H.-Y. Xiong, X.-H. Zhang, S.-F. Wang, Y.-D. Zhao, A highly sensitive nitric oxide biosensor based on hemoglobin chitosan/graphene–hexadecyltrimethylammonium bromide nanomatrix, *Sensor. Actuator. B,* Vol. 166–167, 2012, pp. 444– 450.

[225].X. H. Kang, J. Wang, H. Wu, I. A. Aksay, J. Liu, Y. H. Lin, Glucose oxidase-graphenechitosan modified electrode for direct electrochemistry and glucose sensing, *Biosens. Bioelectron.,* Vol. 25, 2009, pp. 901–905.

[226].F. Schedin, A. K. Geim, S. V. Morozov, E. W. Hill, P. Blake, M. L. Katselson, K. S. Novoselov, Detection of individual gas molecules adsorbed on graphene, *Nature,* Vol. 6, Issue 9, 2007, pp. 652–655.

[227].W. Wu, Z. Liu, L. A. Jaureguic, Q. Yua, R. Pillai, H. Caoc, J. Bao, Y. P. Chenc, S.-S. Pei, Wafer-scale synthesis of graphene by chemical vapor deposition and its application in hydrogen sensing, *Sensor. Actuator. B,* Vol. 150, 2010, pp. 296–300.

[228]. Y. H. Zhang, Y. B. Chen, K. G. Zhou, C. H. Liu, J. Zeng, H. L. Zhang, Y. Peng, Improving gas sensing properties of graphene by introducing dopants and defects: a first-principles study, *Nanaotechnology*, Vol. 20, Issue 18, 2009, pp. 185504 (8 pp.).

[229]. J. Lu, I. Do, L. T. Drzal, R. M. Worden, I. Lee, Nanometal-Decorated Exfoliated Graphite Nanoplatelet Based Glucose Biosensors with High Sensitivity and Fast Response, *ACS Nano*, Vol. 2, 2008, Issue, 9, pp. 1825–1832.

[230]. G. Lu, L. E. Ocola, J. Chen, Gas detection using low-temperature reduced graphene oxide sheets, *Appl. Phys. Lett.*, Vol. 94, 2009, pp. 083111-083113.

[231]. D. Li, J. Jia, J. Wang, Simultaneous determination of Cd(II) and Pb(II) by differential pulse anodic stripping voltammetry based on graphite nanofibers–Nafion composite modified bismuth film electrode, *Talanta*, Vol. 83, 2010, pp. 332–336.

[232]. D. A. C. Brownson, C. E. Banks, Graphene electrochemistry: Surfactants inherent to graphene inhibit metal analysis, *Electrochem. Commun.*, Vol. 13, 2011, pp. 111–113.

[233]. C. Berger, Z. Song, X. Li, X. Wu, N. Brown, C. Naud, D. Mayou, T. Li, J. Hass, A. N. Marchenkov, E. H. Conrad, P. N. First, and W. A. de Heer, Electronic Confinement and Coherence in Patterned Epitaxial Graphene, *Science*, Vol. 312, 2006, pp. 1191-1196.

[234]. Nada F. Atta, Ahmed Galal and Maher F. El-kady, Chemical Sensors and Biosensors Based on Conducting Polymers Thin Films, *Nova*, 2011.

[235]. Y. Ohno, K. Maehashi, K. Matsumoto, Chemical and biological sensing applications based on graphene field-effect transistors, *Biosens. Bioelectron.*, Vol. 26, 2010, pp. 1727–1730.

[236]. X. Dong, W. Huang, P. Chen, In Situ Synthesis of Reduced Graphene Oxide and Gold Nanocomposites for Nanoelectronics and Biosensing, *Nanoscale Res. Lett.*, Vol. 6, 2011, pp. 60-66.

[237]. Y. H. Kwak, D. S. Choi, Y. N. Kim, H. k. Kim, D. H. Yoon, S.-S. Ahn, J.-W. Yang, W. S. Yang, S. K. Seo, Flexible glucose sensor using CVD-grown graphene-based field effect transistor, *Biosens. Bioelectron.*, Vol. 37, 2012, pp. 82–87.

[238]. M. S. Mannoor, H. Tao, J. D. Clayton, A. Sengupta, D. L. Kaplan, R. R. Naik, N. Verma, F. G. Omenetto, M. C. McAlpine, Graphene-based wireless bacteria detection on tooth enamel, *Nature Commun.*, Vol. 3, Article 763.

Chapter 3

Properties and Applications of Modified Carbon Nanotubes

Nada F. Atta and Shereen M. Azab

3.1. Structure of Carbon Nanotubes

Carbon nanotubes (CNTs) are allotropes of carbon with a cylindrical nanostructure Nanotubes have been constructed with length-to-diameter ratio of up to 132,000,000:1 [1] significantly larger than for any other material. These cylindrical carbon molecules have unusual properties, which are valuable for nanotechnology, electronics, optics and other fields of materials science and technology. In particular, owing to their extraordinary thermal conductivity and mechanical and electrical properties, carbon nanotubes find applications as additives to various structural materials. The chemical bonding of nanotubes is composed entirely of sp^2 bonds, similar to those of graphite. These bonds, which are stronger than the sp^3 bonds found in alkanes and diamond, provide nanotubes with their unique strength.

Carbon nanotubes are composed of just one element, Carbon, and are easily produced by several techniques. A nanotube can bend easily but still is very robust. The nanotubes can be manipulated and contacted to external electrodes. Their diameter is in the nanometer range, whereas their length may exceed several micrometers, if not several millimeters. In diameter, the nanotubes behave like molecules with quantized energy levels, while in length; they behave like a crystal with a continuous distribution of momenta. The nanotubes can carry a large electric current; they are also good thermal conductors. It is not surprising, then, that many applications have been proposed for the nanotubes. One of their most promising applications is their ability to emit electrons when subjected to an external electric field. Carbon nanotubes can do so in normal vacuum conditions with a reasonablevoltage threshold, which make them suitable for cold-cathode devices. Nanotubes are also good candidates for the design of composite materials. They can increase the conductivity, either electrical or thermal, of polymer matrices which they are embedded in at a few weight percents, while improving the mechanical resistance of the materials. The small dimensions, high mechanical strength, high surface area, high chemical stability and the remarkable physical properties of these structures make them a very unique material with a whole range of promising applications.

3.2. Types of Carbon Nanotubes

Carbon nanotubes have attracted the fancy of many scientists world-wide.since CNTs significantly change their properties depending on their shapes, it is very important to selectively synthesize those that have the same properties, or to establish the technology which allows selection. CNTs are graphene sheets that are cylindrically curled and are mainly divided into four types such as; (1) Single-walled carbon nanotubes (SWCNTs) which consist of one-layer graphene wall, (2) Double-walled carbon nanotubes (DWCNTs) which consist of two-layer graphene walls, (3) Triple-walled carbon nanotubes (TWCNTs) which consist of three-layer graphene walls and (4) Multi-walled carbon nanotubes (MWCNTs) which consist of more than three-layer graphene walls (Fig. 3.1a). The electronic properties significantly vary depending on the chirality (spiral) and the number of graphene walls. SWCNTs show both metallic and semiconducting properties depending on their chiralities.Carbon nanotubes are considered to be composed of only one material; however, CNTs can be present in different forms and therefore their properties and applications can be diverse Apart from variations in diameter or length, individual nanotubes naturally align themselves into "ropes" held together by van der Waals forces, more specifically, pi-stacking. On the other hand, electronic properties depend on the structure of SWCNTs, mainly diameter and chirality. SWCNTs can be classified as metallic (arm chair) or semiconducting (zigzag or chiral) [2] (Fig. 3.1 b). The way the graphene sheet is wrapped is represented by a pair of indices (n, m). The integers$n$ and m denote the number of unit vectors along two directions in the honeycomb crystal lattice of graphene. If $m = 0$, the nanotubes are called zigzag nanotubes, and if $n = m$, the nanotubes are called armchair nanotubes. Otherwise, they are called chiral. The diameter of an ideal nanotube can be calculated from its (n,m) indices as follows:

$$d = \frac{a}{\pi}\sqrt{\left(n^2 + nm + m^2\right)},$$

where, $a = 0.246$ nm.

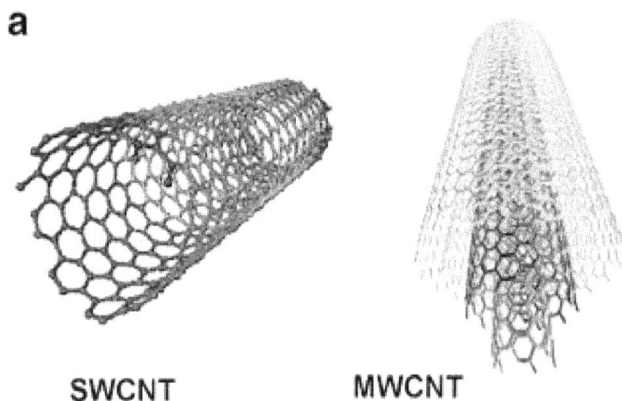

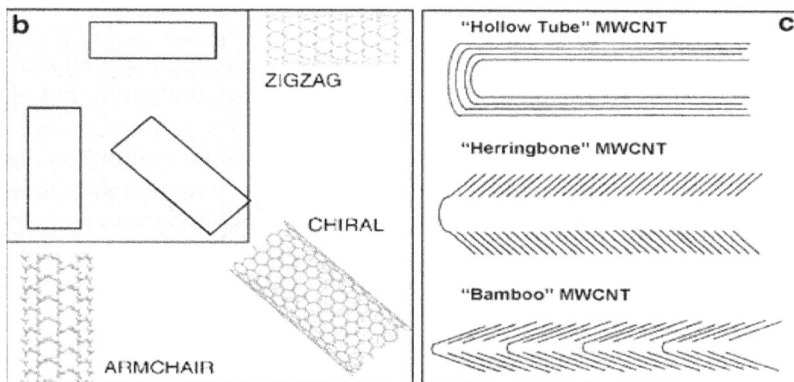

Fig. 3.1. (a) Single- (SWCNT) and multiple- (MWCNT) wall configurations; (b) SWCNT categories, and c) schematic cross section through different MWCNTs showing the orientation of the graphene sheets within the tube- Reproduce by permission of The RoyalSociety of Chemistry [3].

In the case of MWCNTs, "hollow-tube", "herringbone" or "bamboo" morphological variations can be found (Fig. 3.1 c). (DWCNT) form a special class of nanotubes because their morphology and properties are similar to those of SWNT but their resistance to chemicals is significantly improved. This is especially important when functionalization is required to add new properties to the CNT. In the case of SWCNT, covalent functionalization will break some C=C double bonds, leaving "holes" in the structure on the nanotube and, thus, modifying both its mechanical and electrical properties. In the case of DWCNT, only the outer wall is modified. Moreover, closed or open-ended CNTs can be found. Chemically functionalized carbon nanotube with groups such as –COOH, –OH, –SH or –NH$_2$ are available, other functionalization's are also possible. Thus, even when CNTs have a very simple chemical composition and atomic bond configuration, they can exhibit extreme diversity in structure and in turn in properties and behavior.

MWCNTs are now considered important modifiers due to their ability to promote electron transfer in electrochemical reactions, improve sensitivity and chemical inertness, other modifier and different types of surfactants are also used for electrochemical studies.

3.3. Solubilisation

Most applications of CNTs rely on the modification of working electrodes. Abrasive methodology in which CNTs are attached by gently rubbing a polished electrode on a paper containing nanotubes or a CNT-ionic liquid gel can be employed. Inclusion in the paste of a carbon paste electrode or in the ink of a screen–printed electrode is also a possibility for the fabrication of electrochemical sensors. However, most of the applications involve the use of a CNT suspension that is dropped onto the electrode or this is dipped into the CNTs solution for further washing or drying in both cases. Disaggregation into individual tubes is then required. Different solvents can be employed but it has to be taken into account that not all solvents are valid for all type of nanotubes. The mass /volume ratio has to be considered as well as the procedure, commonly sonication and centrifugation.

Aqueous solutions with surfactants are adequate media for dispersing CNTs because the hydropobic tails interact with the hydrophobic nanotubes while the hydrophilic part improves water solubilitydue to their cationic/anionic or hydrophobic nature. Non-covalent weak interactions can be used to attach small or big molecules including polymer chains that can help solubilisation. In fact, wrapping with polymers has provided a supramolecular approach to solubilise CNTs and also to prepare composite materials. A special case is the use of the polyanionic polymer Nafion that possesses a polar side chain and produces CNT solubilisation. However, one must take into account that this process forms a membrane that can act as a diffusion barrier on the surface of the electrode as well as acting as an ion exchanger. Similarly, dispersion can be obtained with chitosan, which in this case is a polycation polymer. Also DNA can be employed for solubilisationpurposes. It has been reported that single stranded DNA (ssDNA) interacts strongly with CNTs to form a stable DNA–CNT hybrid that effectively disperses CNTs in aqueous solution [4]; π-stacking interactions with the side-wall of carbon nanotubes and the positioning of the hydrophilic sugar – phosphate backbone to the exterior are responsible for this.

3.4. Improvements of the Electrochemical Behaviour

In the case of nanotubes, their special geometry and unique electronic, mechanical, chemical and thermal properties make them tremendously attractive for the development of electrochemical devices.since the first modification of a paste electrode in 1996, it has been proven that they have an outstanding ability to mediate fast electron-transfer kinetics for a wide range of electroactive species. Reducing overpotentials is important because the use of high potentials is not necessary and possible interferences are there by avoided. Moreover, such systems allow the resolution of the overlapping response of several analytes. The enhancement in the reversibility of the reactions is also documented as well as the resistance to surface fouling. These advantages imply that better analytical characteristics are obtained. In fact the importance of oxygenated species at the ends of carbon nanotubes has been reported for their favourable electrochemical properties [5]. Improvement in the electrochemical behaviour is also the goal of the co-immobilisation of nanoparticles and nanotubes. Platinum or gold nanoparticles can be employed and they can be generated previously or electrochemically deposited on a CNT-modified electrode. Electrical contact of the electrode through the CNT enables the whole structure to be used as an electrode. Most of the studies have been performed on conventional electrodes, commonly glassy carbon. Screen-printed electrodes (SPEs)posses both characteristics and have been demonstrated to be perfectly compatible with CNTs.

3.5. Nanotubular Electrodes

Miniaturisation has for some time been an important goal in electroanalytical research. In the case of nanotubular electrodes, their small dimensions and conductivity means that they can be considered as the smallest possible electrodes. Among all the properties their small size, which can be an inconvenience for other techniques, could be an advantage when dealing with electrochemical techniques. The nanoelectrode has a large effective surface, increased mass transport rate and decreased susceptibility to the solution resistance. The smallest nanoneedle-type biosensor was fabricated by attaching an MWCNT to a tungsten tip using a nanomanipulator [6].

Multiple CNTs can also be constructed on a conducting layer in certain architectures to fabricate nanoelectrode arrays (NEA s) or ensembles (NEEs). They can produce a much higher current than a single nanoelectrode and improve the signal to noise ratio. The interspacing of the individual electrodes should be much larger than the radius of each electrode: other wise it will behave similar to a macro-electrode. Usually the CNT array is grown on nickel catalyst film in such a way that site density of vertically aligned nanotubes is controlled. In some cases, bottom- up and top-down miniaturisation approaches are joined together because e.g.lithographically patterned silicon pillars (top-down) with transferred catalyst materials on to their tops form the basis for CNT growth (bottom-up). On the other hand, if a dielectric is encapsulated on a forest-like vertically aligned array, leaving only the very end of the CNTs exposed, an inlaid NEA with diminished background noise is formed.

3.6. Electrochemical Biosensors

It is usually the case that the transduction efficiency determines many of the analytical characteristics of the biosensors.The synthesis of nanomaterials has provided a new field for development of novel transduction matrices. However, electroanalytical methodologies take advantage of both electronic and sorption properties (actually adsorptive stripping voltammetry on CNT-modified electrodes takes advantage ofsorption phenomena. These are very important in the development of biosensors where apart from covalent, non-covalent interactions can facilitate the adsorption of analytes.

It has to be considered that nanotubes posses a large active surface that when properly functionalised, favour adsorptive processes. On the other hand, the performance of a biosensing electrode is strongly dependent on the type of raw nanotube material used, the transduction electrode and the modification/ immobilisation procedure. Enzymes (oxidase, dehydrogenase, peroxidase and catalase classes) are the most common biorecognition molecules employed, and direct electron transfer between CNT and enzymes has been demonstrated. Usually, the redox centre of an enzyme is electrically insulated by a protein shell and therefore the whole enzyme cannot suffer redox processes at any potential. CNTs allow molecular wiring and electron transport, thereby acting as electrical wires or conductive nanoneedles [7]. Immuno and DNAdetection through CNT-modified electrodes is also possiblesince non-covalent or covalent attachment between anti-bodies/DNA and CNTs is feasible.

3.7. Properties of Carbon Nanotubes

3.7.1. Mechanical Properties

Carbon nanotubes are the strongest and stiffest materials yet discovered in terms of tensile strength and elastic modulus respectively. This strength results from the covalent sp^2 bonds formed between the individual carbon atoms. Since carbon nanotubes have a low density for a solid of 1.3 to 1.4 g/cm^3 [8] its specific strength is the best of known materials. Under excessive tensile strain, the tubes will undergo plastic deformation which means the deformation is permanent. Although the strength of individual CNT shells is extremely high, weak shear interactions between adjacent shells and tubes leads to significant reductions in the effective strength of multi-walled carbon nanotubes and carbon nanotube bundles down to only a few GPa's [9]. This limitation has been recently addressed by applying high-energy electron irradiation, which crosslinks inner shells and tubes, and effectively increases the strength of these materials to ~60 GPa for multi-walled carbon nanotubes and ~17 GPa for double-walled carbon nanotube bundles.

3.7.2. Kinetic Properties

Multi-walled nanotubes are multiple concentric nanotubes precisely nested within one another. These exhibit a striking telescoping property whereby an inner nanotube core may slide, almost without friction, within its outer nanotube shell, thus creating an atomically perfect linear or rotational bearing. This is one of the first true examples of molecular nanotechnology, the precise positioning of atoms to create useful machines. Already, this property has been utilized to create the world's smallest rotational motor. Future applications such as a gigahertz mechanical oscillator are also envisaged.

3.7.3. Electrical Properties

Because of the symmetry and unique electronic structure of graphene, the structure of a nanotube strongly affects its electrical properties. For a given (n,m) nanotube, if $n = m$, the nanotube is metallic; if $n - m$ is a multiple of 3, then the nanotube is semiconducting with a very small band gap, otherwise the nanotube is a moderate semiconductor. Thus all armchair $(n = m)$ nanotubes are metallic, and nanotubes (6,4), (9,1), etc. are semiconducting [10]. However, this rule has exceptions, because curvature effects in small diameter carbon nanotubes can strongly influence electrical properties. Thus, a (5,0) SWCNT that should be semiconducting in fact is metallic according to the calculations. Likewise, *vice versa*-- zigzag and chiral SWCNTs with small diameters that should be metallic have finite gap (armchair nanotubes remain metallic).

3.7.4. Optical Properties

One of the more recently researched properties of MWCNTs is their wave absorption characteristics, specifically microwave absorption. Interest in this research is due to the current military push for radar absorbing materials (RAM) to better the stealth characteristics of aircraft and other military vehicles. There has been some research on filling MWCNTs with metals, such as Fe, Ni, Co, etc., to increase the absorption effectiveness of MWCNTs in the microwave regime. Thus far, this research has shown improvements in both maximum absorption and bandwidth of adequate absorption.

3.7.5. Thermal Properties

All nanotubes are expected to be very good thermal conductors along the tube, exhibiting a property known as "ballistic conduction", but good insulators laterally to the tube axis. Measurements show that a SWCNT has a room-temperature thermal conductivity along its axis of about 3500 W m^{-1} K^{-1}[11], compare this to copper, a metal well known for its good thermal conductivity, which transmits 385 W m^{-1} K^{-1} A SWCNT has a room-temperature thermal conductivity across its axis (in the radial direction) of about 1.52 W m^{-1} K^{-1} [12], which is about as thermally conductive as soil. The temperature stability of carbon nanotubes is estimated to be up to 2800 °C in vacuum and about 750 °C in air [13].

3.8. Surface Characterization of MWCNTs

The morphological studies of the MWCNTs were carried out by scanning electron microscopy (SEM). MWCNTs could be seen in the form of tubes some of which twisted together (Fig. 3.2). It is obvious that the MWCNTs were distributed uniformly on the surface of the electrode. The spaghetti-like MWCNTs formed a porousstructure. The entangled cross-linked fibrils offered high accessiblesurface area [14]. The modification of solid electrodes using multi-walled carbon nanotubes not only enlarges the ratio surface area of the electrode surface but also improves the electron transfer rate between the electrode surface and the bulk solution.

3.9. Carbon Nanotube's Applications

The uniqueness of the nanotube arises from its structure and the inherent subtlety in the structure, which is the helicity in the arrangement of the carbon atoms in hexagonal arrays on their surface honeycomb lattices. The other factor of importance in what determines the uniqueness in physical properties is topology, or the closed nature of individual nanotube shells; when individual layers are closed on to themselves, certain aspects of the anisotropic properties of graphite disappear, making the structure remarkably different from graphite.

Since their discovery, several demonstrations have suggested potential applications of nanotubes. These include the use of nanotubes as electron field emitters for vacuum microelectronic devices, individual

MWCNTs and SWCNTs attached to the end of an Atomic Force Microscope (AFM) tip for use as nanoprobe, MWCNTs as efficient supports in heterogeneous catalysis and as microelectrodes in electrochemical reactions, and SWCNTs as good media for lithium and hydrogen storage. Some of these could become real marketable applications in the near future, but others need further modification and optimization. Areas where predicted or tested nanotube properties appear to be exceptionally promising are mechanical reinforcing and electronic device applications.

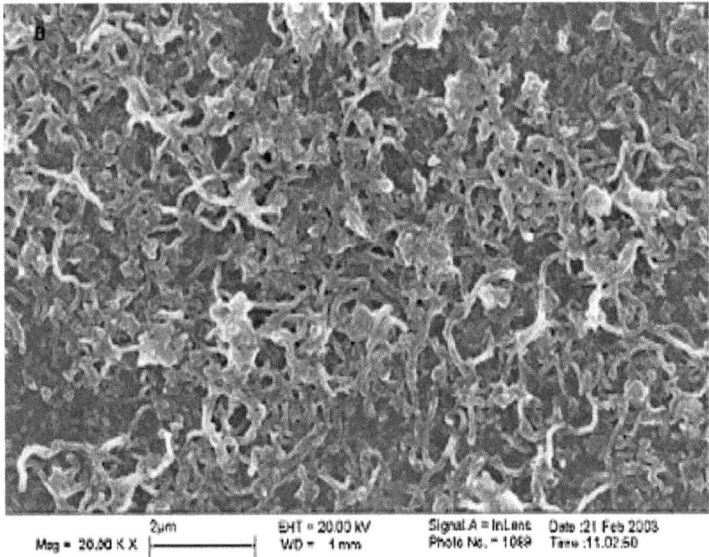

Fig. 3.2. SEM image of modified electrode by MWCNTs.

3.9.1. Current Applications

Current use and application of nanotubes has mostly been limited to the use of bulk nanotubes, which is a mass of rather unorganized fragments of nanotubes. Bulk nanotube materials may never achieve a tensile strength similar to that of individual tubes, but such composites may, nevertheless, yield strengths sufficient for many applications. Bulk carbon nanotubes have already been used as composite fibers in polymers to improve the mechanical, thermal and electrical properties of the bulk product.

Other current applications include:

• Tips for atomic force microscope probes;

• In tissue engineering, carbon nanotubes can act as scaffolding for bone growth.

3.9.2. Structural Applications

Because of the carbon nanotube's superior mechanical properties, many structures have been proposed ranging from everyday items like clothes and sports gear to combat jackets and space elevators. However, the space elevator will require further efforts in refining carbon nanotube technology, as the practical tensile strength of carbon nanotubes can still be greatly improved.Carbon nanotubes are also a promising material as building blocks in bio-mimetic hierarchical composite materials given their exceptional mechanical properties. Initial attempts to incorporate CNTs into hierarchical structures led to mechanical properties that were significantly lower than these achievable limits. Because of the high mechanical strength of carbon nanotubes, research is being made into weaving them into clothes to create stab-proof and bulletproof clothing. The nanotubes would effectively stop the bullet from penetrating the body, although the bullet's kinetic energy would likely cause broken bones and internal bleeding.

3.9.3. In Electrical Circuits

One major obstacle to realization of nanotubes has been the lack of technology for mass production. In 2001 IBM researchers demonstrated how metallic nanotubes can be destroyed, leaving semiconducting ones behind for use as transistors. Their process is called "constructive destruction", which includes the automatic destruction of defective nanotubes on the wafer [15]. This process, however, only gives control over the electrical properties on a statistical scale.

The first nanotube integrated memory circuit was made in 2004. One of the main challenges has been regulating the conductivity of nanotubes. Depending on subtle surface features a nanotube may act as a plain conductor or as a semiconductor.Large structures of carbon nanotubes can be used for thermal management of electronic circuits. An approximately 1 mm–thick carbon nanotube layer was used as a special material to fabricate coolers, this materials has very low density, ~20 times lower weight than a similar copper structure.

3.9.4. As Electrical Cables and Wires

Wires for carrying electrical current may be fabricated from pure nanotubes and nanotube-polymer composites. Recently small wires have been fabricated with specific conductivity exceeding copper and aluminum [16]; these cables are the highest conductivity carbon nanotube and also highest conductivity non-metal cables.

3.9.5. As Paper Batteries

A paper battery is a battery engineered to use a paper-thin sheet of cellulose infused with aligned carbon nanotubes [17]. The nanotubes act as electrodes; allowing the storage devices to conduct electricity. The battery, which functions as both a lithium-ion battery and a supercapacitor, can provide a long, steady power output comparable to a conventional battery, as well as a supercapacitor's quick burst of high energy - and while a conventional battery contains a number of separate components, the paperbattery integrates all of the battery components in a single structure, making it more energy efficient.

3.9.6. Solar Cells

One of the promising applications of SWNTs is their use in solar panels, due to their strong UV/Vis-NIR absorption characteristics. Research has shown that they can provide a sizeable increase in efficiency, even at their current unoptimized state. Solar cells developed by mixing of carbon nanotubes and carbon buckyballs (known as fullerenes) to form snake-like structures. Buckyballs trap electrons, but they can't make electrons flow. Add sunlight to excite the polymers, and the buckyballs will grab the electrons. Nanotubes, behaving like copper wires, will then be able to make the electrons or current flow.

3.9.7. Hydrogen Storage

In addition to being able to store electrical energy, there has been some research in using carbon nanotubes to store hydrogen to be used as a fuel source. By taking advantage of the capillary effects of the small carbon nanotubes, it is possible to condense gasses in high density inside single-walled nanotubes. This allows for gasses, most notably hydrogen (H_2), to be stored at high densities without being condensed into a liquid. Potentially, this storage method could be used on vehicles in place of gas fuel tanks for a hydrogen-powered car. Storage using SWCNTs would allow one to keep the H_2 in its gaseous state, thereby increasing the storage effciency. This method allows for a volume to energy ratio slightly smaller to that of current gas powered vehicles, allowing for a slightly lower but comparable range [18].

The effectiveness of hydrogen storage is integral to its use as a primary fuel source since hydrogen only contains about one fourth the energy per unit volume as gasoline. Extraordinarily high and reversible hydrogen adsorption in SWCNT-containing materials and graphite nanofibers (GNFs) has attracted considerable interest in both academia and industry. Table 3.1 summarizes the gravimetric hydrogen storage capacity reported by various groups [19]. However, many of these reports have not been independently verified. There is also a lack of understanding of the basic mechanism(s) of hydrogen storage in these materials.

Table 3.1. Summary of reported gravimetric storage of H_2
in various carbon materials ([19]).

Material	Max. wt% H_2	T (K)	P (MPa)
SWNTs(low purity)	5–10	133	0.040
SWNTs(high purity)	~ 4	300	0.040
GNFs(tubular)	11.26	298	11.35
GNFs(herringbone)	67.55	298	11.35
GNS(platelet)	53.68	298	11.35
Graphite	4.52	298	11.35
GNFs	0.4	298-773	0.101
Li-GNFs	20	473-673	0.101
Li-Graphites	14	473-674	0.101
K-GNFs	14	< 313	0.101
K-Graphite	5.0	< 313	0.101
SWNTs(high purity)	8.25	80	7.18
SWNTs(~ 50 % pure)	4.2	300	10.1

In addition to hydrogen, carbon nanotubes readily absorb other gaseous species under ambient conditions which often leads to drastic changes in their electronic properties.

3.9.8. Medical Uses

In the Kanzius cancer therapy, single-walled carbon nanotubes are inserted around cancerous cells, then excited with radio waves, which causes them to heat up and kill the surrounding cells. Researchers have shown that carbon nanotubes and their polymer nanocomposites are suitable scaffold materials for bone cell proliferation and bone formation [20, 21].

3.9.9. In Vacuum Microelectronics

Field emission is an attractive source for electrons compared to thermionic emission. It is a quantum effect. When subject to a sufficiently high electric field, electrons near the Fermi level can overcome the energy barrier to escape to the vacuum level. Electron field emission materials have been investigated extensively for technological applications, such as flat panel displays, electron guns in electron microscopes, microwave amplifiers [22]. For technological applications, electron emissive materials should have low threshold emission fields and should be stable at high current density.Compared to conventional emitters, carbon nanotubes exhibit a lower threshold electric field. The low threshold field for electron emission observed in carbon nanotubes is a direct result of the large field enhancement factor rather than a reduced electron work function.

3.9.10. Prototype Electron Emission Devices Based on Carbon Nanotubes

3.9.10.1. Cathode-Ray Lighting Elements

Cathode ray lighting elements with carbon nanotube materials as the field emitters is illustrated in Fig. 3.3, these nanotube-based lighting elements have a triode-type design. In the early models, cylindrical rods containing MWCNTs, formed as a deposit by the arc-discharge method, were cut into thin disks and were glued to stainless steel plates by silver paste. In later models, nanotubes are now screen-printed onto the metal plates. A phosphor screen is printed on the inner surfaces of a glass plate. Different colors are obtained by using different fluorescent materials.

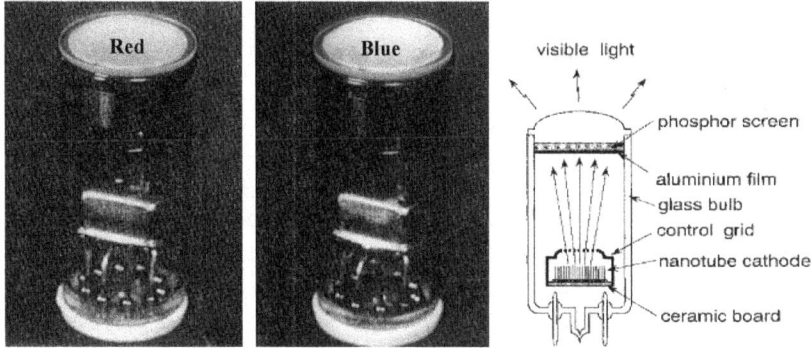

Fig. 3.3. Demonstration field emission light source using carbon nanotubes as the cathodes [23].

3.9.10.2. Flat Panel Display

Prototype matrix-addressable diode flat panel displays have been fabricated using carbon nanotubes as the electron emission source [24]. Recently, a 4.5 inch diode-type field emission display has been fabricated by Samsung, with SWCNT stripes on the cathode and phosphor-coated ITO stripes on the anode running orthogonally to the cathode stripes [25].

3.9.10.3. Gas-Discharge Tubes in Telecom Networks

Gas discharge tube (GDT) protectors, consisting of two electrodes parallel to each other in a sealed ceramic case filled with a mixture of noble gasesis one of the oldest methods used to protect against transient over-voltages in a circuit [26]. They are widely used in telecom network interface device boxes and central office switching gear to provide protection from lightning and ac power cross faults on the telecom network. They are designed to be insulating under normal voltage and current flow. These devices are robust, moderately inexpensive, and have a relatively small shunt capacitance, so they do not limit the bandwidth of high- frequency circuits as much as other nonlinear shunt components. Prototype GDT devices using carbon nanotube coated electrodes have recently been fabricated and tested by a group from UNC and Raychem Co [27]. The enhanced performance shows that nanotube-based GDTs are attractive over-voltage protection units in advanced telecom networks such as an Asymmetric-Digital-Signal-Line (ADSL), where the tolerance is narrower than what can be provided by the current commercial GDTs.

3.9.11. Energy Storage

Carbon nanotubes are being considered for energy production and storage. Graphite, carbonaceous materials and carbon fiber electrodes have been used for decades in fuel cells, battery and several other electrochemical applications. Nanotubes are special because they have small dimensions, a smooth surface topology, and perfect surface specificity, since only the basal graphite planes are exposed in their structure. The rate of electron transfer at carbon electrodes ultimately determines the efficiency of fuel cells and this depends on various factors, such as the structure and morphology of the carbon material used in the electrodes. Nanotube microelectrodes have been constructed using a binder and have been successfully used in bioelectrochemical reactions. Their performance has been found to be superior to other carbon electrodes in terms of reaction rates and reversibility.

Pure MWCNTs and MWCNTs deposited with metal catalysts (Pd, Pt, Ag) have been used to electro-catalyze an oxygen reduction reaction, which is important for fuel cells [28, 29]. Ru-supported nanotubes were found to be superior to the same metal on graphite and on other carbons in the liquid phase hydrogenation reaction of cinnamaldehyde. The properties of catalytically grown carbon nanofibers have been found to be desirable for high power electrochemical capacitors.

3.9.12. Electrochemical Intercalation of Carbon Nanotubes with Lithium

The basic working mechanism of rechargeable lithium batteries is electrochemical intercalation and de-intercalation of lithium between two working electrodes. Current state-of-art lithium batteries use transition metal oxides (i.e., $Li_x CoO_2$ or $Li_x Mn_2O_4$) as the cathodes and carbon materials as the anodes [30]. It is desirable to have batteries with a high energy capacity, fast charging time and long cycle time. The energy capacity is determined by the saturation lithium concentration of the electrode materials. It has been speculated that a high Li capacity may be obtained in carbon nanotubes if all the interstitial sites (inter-shell van der Waals spaces, inter-tube channels, and inner cores) are accessible for Li intercalation. Structural studies [31, 32] have shown that alkali metals can be intercalated into the inter-shell spaces within the individual MWCNTs through defect sites. The SWCNTs are also found to perform well under high current rates.

3.9.13. Nanoprobes and Sensors

The small and uniform dimensions of the nanotubes produce some interesting applications. With extremely small sizes, high conductivity, high mechanical strength and flexibility (ability to easily bend elastically), nanotubes may ultimately become indispensable in their use as nanoprobes. One could think of such probes as being used in a variety of applications, such as high resolution imaging, nano-lithography, nanoelectrodes, drug delivery, sensors and field emitters. The advantage of the nanotube tip is its slenderness and the possibility to image features (such as very small, deep surface cracks), which are almost impossible to probe using the larger, blunter etched Si or metal tips.

Biological molecules, such as DNA can also be imaged with higher resolution using nanotube tips [33]. In addition, due to the high elasticity of the nanotubes, the tips do not suffer from crashes on contact with the substrates. Any impact will cause buckling of the nanotube, which generally is reversible on retraction of the tip from the substrate. Attaching individual nanotubes to the conventional tips of scanning probe microscopes has been the real challenge. Bundles of nanotubes are typically pasted on to AFM tips and the ends are cleaved to expose individual nanotubes to measure binding forces between protein-ligand pairs and for imaging chemically patterned substrates (Fig. 3.4). Nanotubes can also be used as molecular probes, with potential applications in chemistry and biology.

Open nanotubes with the attachment of acidic functionalities have been used for chemical and biological discrimination on surfaces. These experiments open up a whole range of applications, for example, as probes for drug delivery, molecular recognition, chemically sensitive imaging, and local chemical patterning, based on nanotube tips that can be chemically modified in a variety of ways. In addition, the small dimensions and high surface area offer special advantages for nanotube sensors, which could be operated at room temperature or at higher temperatures for sensing applications.

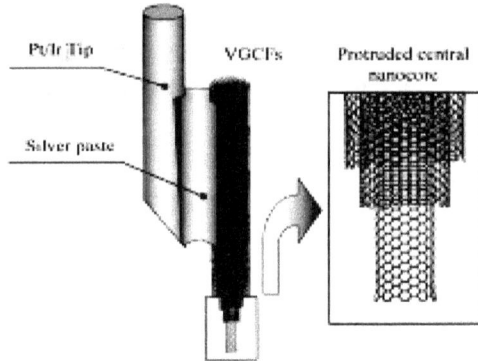

Fig. 3.4. Use of a MWCNT as an AFM tip.

3.9.14. Templates

Since nanotubes have relatively straight and narrow channels in their cores, it was speculated from the beginning that it might be possible to fill these cavities with foreign materials to fabricate one-dimensional nanowires. Early calculations suggested that strong capillary forces exist in nanotubes, strong enough to hold gases and fluids inside them [34]. Nanotubes have been used as templates to create nanowires of various compositions and structures (Fig. 3.5).

The critical issue in the filling of nanotubes is the wetting characteristics of nanotubes, which seem to be quite different from that of planar graphite, because of the curvature of the tubes. Wetting of low melting alloys and solvents occurs quite readily in the internal high curvature pores of MWCNTs and SWCNTs. In the latter, since the pore sizes are very small, filling is moredifficult and can be done only for a selected few compounds. Liquids such as organic solvents wet nanotubes easily and it has been proposed that interesting chemical reactions could be performed inside nanotube cavities [34]. The opening of nanotubes by oxidation can be achieved by heating nanotubes in air (above 600 °C) or in oxidizing solutions (e.g., acids). Filled nanotubes can also be synthesized in situ, during the growth of nanotubes in an electric arc or by laser ablation.

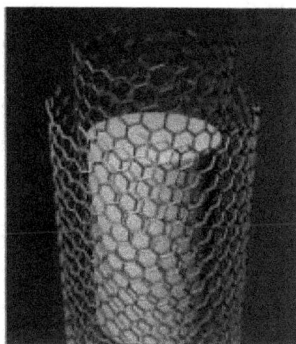

Fig. 3.5. A schematic shows the filling of one-dimensional hollow core of nanotubes with foreign substances.

Nanocomposite structures based on carbon nanotubes can also be built by coating nanotubes uniformly with organic or inorganic structures. These unique composites are expected to have interesting mechanical and electrical properties due to a combination of dimensional effects and interface properties. The interface formed between nanotubes and the layered oxide is atomically flat due to the absence of covalent bonds across the interface. It has been demonstrated that after the coating is made, the nanotubes can be removed by oxidation leaving behind freely-standing nanotubes made of oxides, with nanoscale wall thickness. These novel ceramic tubules, made using nanotubes as templates, could have interesting applications in catalysis. Recently, researchers have also found that nanotubes can be used as templates for the self-assembly of protein molecules [35].

3.10. Carbon Nanotubes Modified Electrode

3.10.1. CNT Modified Electrode Used in Analysis of Neurotransmitters

Dopamine (DA) is an important catecholamine neurotransmitter. CNT modified electrode used in electrochemical determination of DA can ameliorate the problems of high overpotential, slow electrode reaction, and low sensitivity, which occurs at conventional electrodes. In addition, this modified electrode can be applied for the determination of DA and coexisting ascorbic acid (AA) and uric acid (UA) selectively. In addition, DA could be determined in the presence of plenty of AA, for the sensitivity of DA was higher than that of AA. The separation mechanism of differentiating DA from AA was mainly attributed to the porous interfacial layer on the modified electrode. This kind of electrode can also be used for the simultaneous determination of DA and 5-HT with no interference from AA.

The electrochemical behaviors of nicotinamide adenine dinucleotide (NADH), epinephrine (EP), and norepinephrine (NE) at CNT and graphite powder modified electrodes were compared [36]. CNT electrode showed excellent electrocatalytic activity to NADH, EP, and NE and reduced peak-to-peak separations in the voltammetry in comparison with naked pyrolytic graphite electrode. Graphite powder modified electrodes showed similar results. CNT coated electrode and CNT embedded polymer electrode have attracted more attention and research because of the diversity of dispersants. The electrooxidation of DA was investigated on a GCE modified with the hybridization adduct of ferrocene (Fc)-SWNTs [37]. The electrode showed a dramatically enhanced electrocatalytic property toward DA due to the accelerated electron transfer prompted by the stable and strong π-π stacking interaction between Fc and SWNTs. A modified carbon-paste electrode (CPE) was prepared by incorporating thionine-nafion supported on MWNT. DA and AA could be determined simultaneously using this modified electrode. The prepared modified electrode thus shows a good selectivity for voltammetric response to AA and DA [38].

3.10.2. CNT Modified Electrode Used in the Analysis of Proteins

Direct electrochemical process of proteins on electrodes has been proved to be very difficult to carry out because of the deeply buried redox center. Therefore, decreasing the diffuse distance between the redox center and the surface of electrode can promote the kinetics of proteins. CNT electrode can improve the direct electron transfer (DET) between protein and electrode because of the promoting effect of CNT. Cytochrome c (Cyt c) is an important redox protein. CNT is of importance for the formation of stable monolayers and the enhancement of the DET between electrode surface and Cyt c [39]. CNT hybrid with other nanomaterials provides a synergical effect leading to the improvement in the response property of modified electrodes. Cai and co-workers [40] immobilized hemoglobin (Hb) on the surface of CNT electrode with surfactant cetyltrimethylammonium bromide (CTAB) as disperser.

The DET reaction of Hb was a one-electron transfer coupled with a one-proton transfer reaction process. Hb was coupled with CNT through ethylene dichloride (EDC) and assembled as Hb-CNT composites coated on GCE [41]. The DET of Hb could be effectively accelerated in the Hb-CNT assembly by EDC on GCE. Hb and Mb were adsorbed on MWNTs mainly by electrostatic interaction and the protein-MWNT films were coated on pyrolytic graphite electrodes. The protein-MWNT films exhibited a pair of well-defined, quasi-reversible cyclic voltammetric peaks, characteristic of hemeFe(III)/Fe(II) redox couples [42]. Protein immobilized on this kind of modified electrodes retained the bioelectrocatalytic activity to its substrate. Ferredoxin (Fd) is the protein with redox center of Fe-S; was immobilized on the surface of CNT modified with CTAB through electrostatic effect [43], then the modified CNT was coated onto the surface of GCE using Nafion. The electrochemical oxidation of NADH, a kind of coenzyme, at low overpotential is very important in making biosensors. CNT electrode with H_2SO_4 as disperser was used for electrochemical determination of NADH [44].

3.10.3. CNT Modified Electrode Used in Analysis of Nucleic Acids

Immuno and DNA detection through CNT-modified electrodes is also possible since non-covalent or covalent attachment between anti-bodies/DNA and CNTs is feasible. An elegant amplification strategy for immuno and DNA sensors involves application of CNTs with two different uses: a CNT-modified electrode for immobilising the primary antibody/DNA probe and a CNT for immobilising both the secondary antibody/DNA probe and multiple enzyme labels [45]. The nucleic acid is an important biomacromolecule that has typical π electronic structure and shows special electric and electrochemical features. On the basis of these features, the concentration and hybridization of nucleic acids can be detected using an electrochemical method. Tang et al. [46] reported that a synthesized single stranded ssDNA-1 (5'-XATGGGTATTCAACATTTCCG, X = $-NH_2$) was covalently immobilized on CNT modified gold electrode with coupling agents ethylene dichloride (EDC) and N-hydroxysuccinimide (NHS). Nucleic

125

acid hybridization and sequence-specific DNAs can be detected on this electrode by enhancing the response signals of DNAs.

The Atomic force microscope (AFM) using carbon nanotubes as probes, has been applied to investigate DNA or RNA structures and their various functional aspects [47]. Also a series of multi-walled carbon nanotube (MWNT) nano-electrode arrays prepared by embedding MWNT in a SiO_2 matrix have been developed for ultrasensitive determination of DNA/RNA [48]. The CNT can amplify DNA or protein recognition and transduction events, which may be used as an ultrasensitive method for electrical biosensing of DNA or proteins. Recently, a label-free or indicator-free DNA hybridization detection has been achieved by impedance and chronopotentiometric measurements based on MWNT modified glassy carbon electrode (GCE).

A direct amperometric detection of ssDNA or RNA concentration has been performed at mercury and electrochemically modified GCE. As presented in some studies the presence of MWNTs greatly amplified the amperometric signal of guanine or adenine residues by combining the advantages of screen printed carbon electrode with the properties of MWNTs. The proposed sensors could be prepared in batch and used for DNA or RNA concentration detection in new commercial products. Alsothe detection of trace levels of oligonucleotides and polynucleotides were performed using modified CNT's electrodes [49]. SWNT compound poly (acridine orange) modified GCE was used for the simultaneous determination of bases in DNA and showed excellent electrocatalytic activity toward electrooxidation of adenine, guanine, cytosine, and thymine and the oxidation peaks of purines and pyrimidines could be entirely separated [50].

3.10.4. CNT Modified Electrode Used in Analysis of Micromolecules

The CNT modified electrode was applied to study purines and its metabolic products, alkaloids, pharmaceuticals, and amino acids, etc. Uric acid is a metabolic product of purine in human body. The determination of UA concentration in urine and serum is helpful for understanding the metabolic state of purine. The simultaneous determination of UA and xanthine using MWNT modified GCE [51] in human serum without any preliminary treatment was performed.

The voltammetric behavior of UA at a MWNT-ionic liquid (BMIMPF6) paste coated GCE and the use of carbon nanotube-modified carbon fiber microelectrodes for in vivo voltammetric measurement of ascorbic acid in rat brain [52] are a proof of a marked electrocatalytic activity of the modified CNT's electrode. Also the overpotential of homocysteine at CNTPE was decreased by approximately 120 mV compared with that of traditional CPE [53]. The other amino acids, such as cysteine, glutathione, and n-acetylcysteine, showed similar results at CNTPE [54]. MWNT paste electrode was used for the determination of quercetin and rutin in urine and blood serum [55, 56]. The electrocatalytic oxidation of deferiprone, an anti-HIV replication drug, was investigated on a GC/CNT electrode. CNT can promote the rate of oxidation of deferiprone and the proposed method was applied to the direct assays of spiked human serum and urine fluids [57].

It is clear that CNT modified electrode used in bioelectrochemical analysis is the forefront research nowadays that involves the combination of nanomaterials, life sciences, and analysis. Its applications and prospects are summarized as follows:

(1) As far as the modified electrodes are concerned, the CNTPE and CNT intercalated electrode are not widely used in the determination of biomacromolecules and micromolecules than the CNT coated and polymer embedded electrodes because of the various kinds of dispersants. Therefore, to develop and use the materials with good and environmental friendly properties as dispersants is an important field.

(2) CNTs intermingled with other nanomaterials (such as nano Au particles, nano ZrO_2, etc) were modified on conventional electrodes. These materials provide a synergical effect that leads to the improvement in the response property of modified electrodes.

(3) CNT modified electrode applied to the study of protein and nucleic acid showed that not only was the DET realized between biomacromolecules and electrode but also the bioactivity of biomaromolecules was maintained. Therefore, the use of CNT to fabricate biosensor (such as enzyme biosensor, DNA biosensor, etc).

(4) Thus, further research on the mechanism of the electrochemical reaction between biomolecules and modified electrode is a very important field in relation to CNT modified electrode.

3.11. Gas Sensing

Recent research has also shown that nanotubes can be used as advanced miniaturized chemical sensors to detect low concentrations of gases such as nitrogen oxides, ammonia, hydrogen , carbon monoxide and some organic gases. By monitoring the change in the conductance of nanotubes, the presence of gases could be precisely monitored. Both CNT-based and CNT-doped gas sensors have been explored because of their specific properties such as nanometer hollow geometry, high specific surface area, high electron mobility, surface modification, and functionalization. Electrical resistance change of CNTs induced by gas molecules adsorption on their surface via van der Waals force has found its applications to develop CNT-based gas sensors for various gases from indoor air.

3.11.1. Formaldehyde Detection

Formaldehyde emitted from construction and decorative materials in our daily life, which is deleterious for our health with potentially carcinogenic, high rate of cancer, respiratory disease, and baby abnormalities . The World Health Organization (WHO) has derived an air quality guideline of 80 ppb averaged over 30 min [58, 59]. Therefore, detection of these indoor air pollutants is of particular concern because of their long-term effects on our health. So far, many methods have been investigated for detection of indoor formaldehyde, including spectrophotometry [60], potentiometric [61],amperometric [62], electrochemical biosensors [63], optical methods [64] and filter color testing methods [65]. Although these methods provide safe detection of formaldehyde, they still have some limitations and challenges such as low relative resistance changes and selectivity, long-term instability and so on. MWCNTs modified with amino groups is proposed to enhance chemical reaction between amino groups of the MWCNTs and formaldehyde molecules. Effects of content of amino groups of the MWCNTs on sensing responses are investigated. Relative resistance changes, selectivity and repeatability of the sensors are demonstrated by various measurements.Gas sensors with (MWCNTs) modified with amino-groups on interdigitated electrodes (IDE) were fabricated (Fig. 3.6) to detect low concentration of formaldehyde at room temperature.

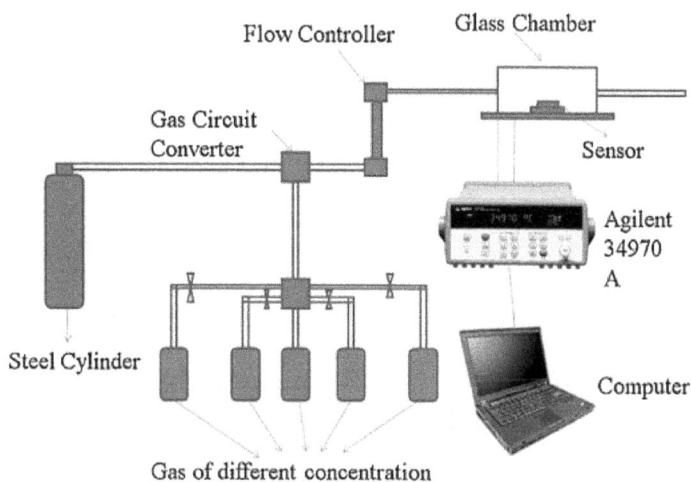

Fig. 3.6. Schematic diagram of experimental setup.

Effects of content of amino groups on sensing responses against various Interfering circumstances and low concentration of formaldehyde were investigated. The sensor behaved high relative resistance changes to formaldehyde and lower response to interfering gases such as acetone, carbon dioxide, ammonia, toluene and ethanol. When the concentration of formaldehyde was 20 ppb, the relative resistance changes of the sensor modified with 18.19 % amino-group reached 1.73 % (Fig. 3.7). The sensor displayed high chemical selectivity, fast response and good reproducibility to low concentration of formaldehyde, which was attributed to the properties of MWNTs and the interaction between the surface of MWCNTs and amino-group [66].

Three different gas sensors based on raw MWCNTs/Nafion film (ECNT), MWNTs-N-1/Nafion film (-18%amino) and MWNTs-N-2/Nafion film (ECNT-5%amino) were obtained. The responses of sensor ECNT-18%amino to methanol, ethanol, acetone, ammonia, carbon dioxide, and formaldehyde gas with the same concentration of 200 ppb was monitered. From Fig. 3.7, sensor ECNT-18%amino demonstrates a good relative resistance change with 5.54 % formaldehyde, while relative resistance changes for methanol, ethanol, ammonia and carbon dioxide are 0.71 %, 0.41 %, 0.71 %, 0.89 %, respectively.When concentration of target gases dropped to 20 ppb, the relative resistance change of formaldehyde gas reached 1.73 %, while for other five gases the response was very low. The reproducibility of the analysis data in response is almost the same [66].

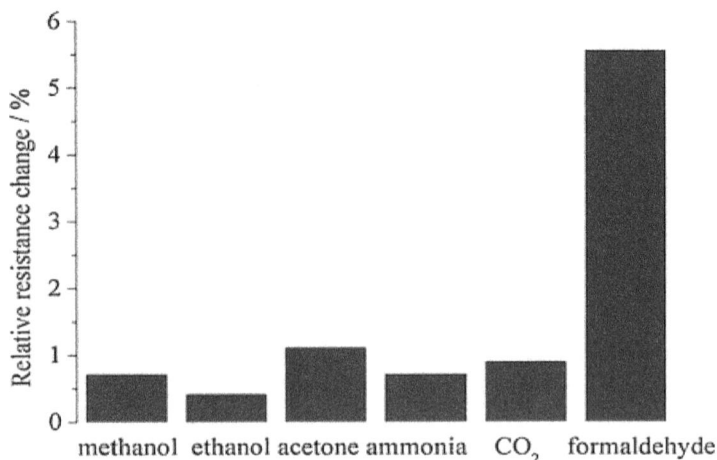

Fig 7. Selectivity measurements of sensor ECNT-18%amino.

3.12. Determination of Metal Ions

The determination of metal ions at trace level is very important in the content of environmental protection, food and agricultural chemistry as well as high purity materials. However, the direct determination of metal ions in complex matrices is limited due to their usually low concentrations and matrix interferences.

In trace analysis, therefore, a preconcentration and/or separation are necessary to improve sensitivity and selectivity of determination. Recently, Solid-phase extraction (SPE) is the most common technique used for preconcentration of analytes in environmental water because of its advantages of high enrichment factor, high recovery, rapid phase separation, low cost, low consumption of organic solvents and the ability of combination with different detection techniques in the form of on-line or off-line mode [67, 68]. In SPE procedure, the choice of appropriate adsorbent is a critical factor to obtain full recovery and high enrichment factor. MWCNTs and oxidized-MWCNTs have been used as adsorbent for preconcentration of many kinds of pollutants such as bisphenol A, 4-n-nonylphenol and 4-tert-octylphenol, pesticides, sulfonamides, phenols, herbicides, diazinon, dichlorodiphenyltrichloroethane, tetracyclines, organometals and metal ions. However, the selectivity of the raw or oxidized-MCWNTs for SPE is quite limited, especially for metal ions.

3.12.1. Detection of Sulfide Ion

The detection of sulfide ion has been paid much attentionas a consequence of the toxicity of hydrogen sulfide and thecorresponding risk associated with exposure in a number ofoccupational settings [69-71]. Thus, the determination of sulfideion was particularly important from industrial, environmental,and biological point of view. For this purpose, many analytical schemes have been developed [72-75]. Among these methods reported for sulfide ion, the chemiluminescence (CL)methods had been paid considerable attention since this analytical technique promised high sensitivity, wide linear range and requirement of simple instrumentation [76]. Although these methods presented straightforward approach to high sensitivity, some powerful or unstable oxidants used in these systems led to poor selectivity. Thus, a highly selective combined with excellently sensitive CL method for the determination of sulfide ion is strongly desirable. As reviewing the powerful CL systems, it was obviously found that most of the useful CL reactions were relatively fast, which led to better selectivity based on the CL speed-resolution power. Based on this advantage, two highly selective CL methods for urea or the ammonia were developed [77, 78]. Thus, to obtain high CL reaction speed, the powerful CL reaction systems often concerned the addition of some catalysts and the use of instable reagents, such as $OH-$, $BrO-$, $ClO-$,Mn^{3+},$Co^{3+}$and the superoxide ion was paid much more attention [79,80]. The main reason was as followings: firstly, superoxide ion, relative to hydrogen peroxide, presented not only the faster CL reaction speed, but also the higher oxidizing potential from the thermodynamic and kinetic points; secondly, due to the much lower absorption ability in UV–vis region, it nearly did not present the inner filter effect for CL emitter when it used as the CL reaction oxidant. In addition, no interference species was brought in by its reductive product, H_2O. Electrochemical reduction of dioxygen to superoxide ion depends on the reaction medium and electrode material.

For instance, a reversible, one-electron reduction occurs to yield the superoxide ion in aprotic media; Taylor and Humffray [81] suggested that dioxygen was reduced to superoxide ion on glass-carbon electrode first and Appleby and Marie [82] concluded the mechanism of dioxygen reduction on carbon black electrode in alkaline solution, in which superoxide ion was produced first and further oxidized to H_2O_2. On the other hand, both single-wall and multi-wall CNTs, have attracted increasing attention because of its remarkable nanostructures.

One promising application is based on the promotion ability of electron-transfer reactions of target molecules when used as an electrode material in electrochemical reactions. In addition, it has been also applied to dioxygen reduction [83, 84].

MWNTs and cobalt prophyrin-modified glass-carbon (GC) electrode exhibited electrocatalytic activity to the reduction of oxygen. It was found that the instable CL oxidant, super-oxide ion, could be rapidly generated at the surface of MWNTs-modified graphite electrode and further reacted selectively with sulfide ion, due to speed-resolution, producing a weak but fast electrogeneratedchemiluminescence (ECL) (Fig. 3.8) signal [85]. The electrolytic cell utilized a conventional three-electrode setup and was arranged as shown in Fig. 3.8. The working electrode was the MWNTs-modified graphite electrode; a Pt flake and Ag/AgCl (saturated KCl solution) were used as the auxiliary electrode and pseudo-reference electrode, respectively. The ECL intensity was transformed into an electrical signal by an R456 photomultiplier (PMT), which was operated at -800 V and the ECL cell was placed in front of the PMT.

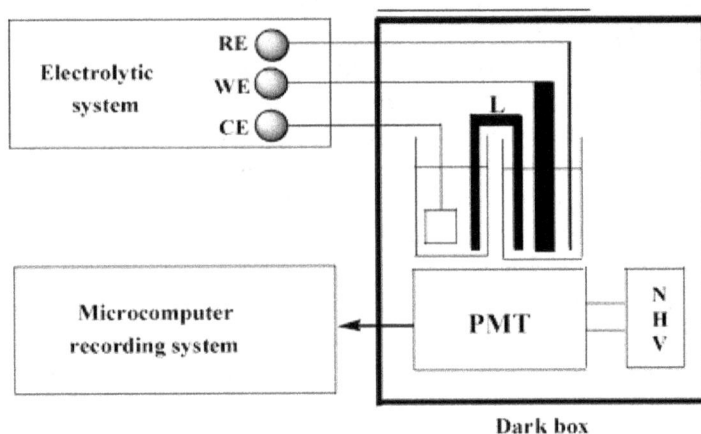

Fig. 3.8. The block diagram of ECL detection system. WE, working electrode; RE, Ag/AgCl reference electrode; CE, counter electrode (Pt); L, KNO_3 salt bridge; PMT, photomultiplier; NHV, negative high voltage supply.

At the same time, it was also found that this ECL signal could be strongly enhanced in the presence of the oxidative products of rhodamine B (OPRB) based on an energy transfer CL mechanism. Based on these observations, a highly selective and excellently

132

sensitive ECL method for sulfide ion detection at MWNTs-modified graphite electrode was applied to the determination of sulfide ion in environmental water samples.

3.12.2. Determination of Sulfite

Sulfites are used as preservatives to prevent oxidation, inhibit bacterial growth, and control enzymatic and non-enzymatic reactions with stabilizing and conditioning functions. Despite these useful advantages, sulfite should be applied in strictly limited amounts due to its potential toxicity. The existence of accurate methods for the determination of sulfites is necessary fundamentally for the food industry to ensure product quality. Metal hexacyanoferrates (MHCF) have been a study of wide interest due to their electroactive properties as excellent electron transfer mediators [86].

MHCF have been fabricated using various transition metal cations such as iron, cobalt, tin, indium, silver, zinc, chromium, and copper. Copper (II) hexacyanoferrate (CuHCF), have been previously investigated to be a good mediator in the catalytic oxidation of sulfite, has been used due to the CuHCF mediator's ease of preparation as well as integration into modified carbon paste working electrodes for the determination of sulfite [87]. CuHCF-CNT carbon paste electrode has been utilized for sulfite determination [88]. Flow injection analysis of sulfite, coupled with electrochemical detection, has become a method of choice due to the direct electrochemical oxidation of sulfite.

An on-line pervaporation-flow injection (PFI) method, using a highly sensitive CuHCF-CNT electrochemical sensor, was developed for the determination of sulfite in food products where the different aggregation and physical properties of the sample matrices always are a major problem and require sophisticated sample pretreatment prior to analysis. A pervaporation unit incorporated in the flow system was expected to improve the selectivity while the CuHCF-CNT-modified electrode in an amperometric flow-through cell enhanced the sensitivity of the flow injection system.

3.12.3. Determination of Cyanide

Monitoring of cyanide in industrial wastes is highly demanded for environmental control, especially in electroplating, precious metal refining and metal cleaning industries. Cyanide is an extremely toxic ion readily absorbed by living organisms by the inhalation, oral and dermal routes of exposure. Lethal levels of cyanide in blood are assumed to be 11.5 mM [89]. In the case of free cyanide, a concentration of 0.77 µM has been reported to be lethal for certain species of sea animals, whereas a concentration of 7.7 µM is allowable for drinking water supplies [89, 90]. Considering high toxicity of cyanide for humans, even at very low concentration levels, the development or improvement of analytical methods for the determination of low levels of cyanide ion in aqueous environments is thus of great importance. There are numerous methods for determination of cyanide in the water and wastewater samples. Among these methods, nanoparticles seem to be especially suitable for determination of very low levels of cyanide [91-93].

Nanomolar quantities of free cyanide ions can be electrochemically determined using silver hexacyanoferrate nanoparticles (SHFNPs) immobilized on MWCNT modified GC electrode (GC/MWCNT-SHFNPs) [94]. The GC/MWCNT-SHFNPs were prepared to detect cyanide based on specific reaction of cyanide ions with silver ions by square wave voltammetry technique. The electrochemical system was applicable for analysis of CN⁻ in industrial real samples. Fig. 3.9 shows the square wave voltammetry (SWV) obtained for GC/MWCNT-SHCFNPs modified electrodes. When CN⁻ was added to the solution, a new cathodic peak (in curve 2) was generated in half potential ($E_{1/2}$) −0.034 V (vs. SCE).

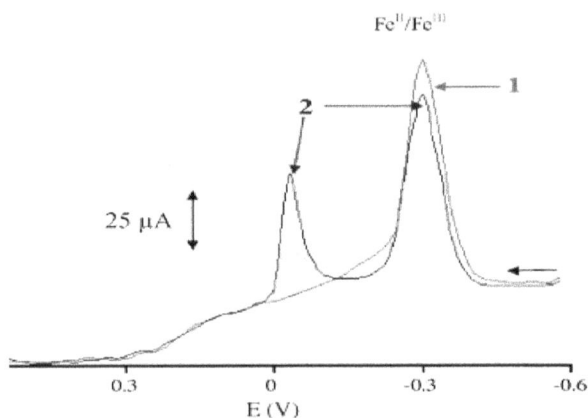

Fig. 3.9. square wave voltammograms of a GC/MWCNT-SHCFNPs in the (1) absence, and (2) presence of cyanide (5.0 µM), scan rate; 100 mV s^{-1}.

The corresponding reactions of these voltammograms can be explained as follows:

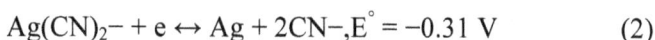

$$Ag_4[Fe^{II}(CN)_6] + 2CN^- - e \leftrightarrow Ag_3[Fe^{II}(CN)_6] + Ag(CN)_{2-} \quad (1)$$

$$Ag(CN)_2- + e \leftrightarrow Ag + 2CN-, E° = -0.31 \text{ V} \quad (2)$$

Cyanide can react with SHCFNPs (reaction (1)) [95] and produces $Ag(CN)^{2-}$ complex (Ag^I) which can be reduced to Ag in an electrochemical reaction.

Significantly lower detection limit, greater analytical sensitivity and stability response of this modified electrode compare favorably to other modified electrodes employed as CN^- sensors.

3.13. Conclusion

This chapter has described several possible applications of carbon nanotubes, with emphasis on materials science-based applications. Hints are made to the electronic applications of nanotubes. The overwhelming message we would like to convey through this chapter is that the unique structure, topology and dimensions of carbon nanotubes have created a superball-carbon material, which can be considered as the most perfect fiber that has ever been fabricated. The remarkable physical properties of nanotubes create a host of application possibilities, some derived as an extension of traditional carbon fiber applications, but many are new possibilities, based on the novel electronic and mechanical behavior of nanotubes. It needs to be said that the excitement in this field arises due to the versatility of this material and the possibility to predict properties based on its well-defined perfect crystal lattice. Nanotubes truly bridge the gap between the molecular realm and the macro-world, and are destined to be a star in future technology.

References

[1]. X. Wang, Q. Li, J. Xie, Z. Jin, J. Wang, Y. Li, K. Jiang, S. Fan, Fabrication of Ultralong and Electrically Uniform Single-Walled Carbon Nanotubes on Clean Substrates, *Nano Letters*, Vol. 9, Issue 9, 2009, pp. 3137–3141.

[2]. H. Dai, Carbon nanotubes: opportunities and challenges, *Surface Science*, Vol. 500, Issue 1-3, 2002, pp. 218-241.

[3]. C. E. Banks, T. J. Davies, G. G. Wildgoose, R. G. Compton, Electrocatalysis at Graphite and Carbon Nanotube Modified Electrodes: Edge-Plane Sites and Tube Ends Are the Reactive Sites, *Chemical Communications*, Vol. 36, Issue 18, 2005, pp. 829- 841.

[4]. M. Zheng, A. Jagota, E. D. Semke, B. A. Diner, R. S. Mclean, S. R. Lustig, R. E. Richardson, and N. G. Tassi, DNA-assisted dispersion and separation of carbon nanotubes, *Nature Material*, Vol. 2, 2003, pp. 338- 342.

[5]. A. Chou, T. Böcking, N. K. Singh, J. J. Gooding, Demonstration of the importance of oxygenated species at the ends of carbon nanotubes for their favourable electrochemical properties, *Chemical Communications*, Vol. 7, 2005, pp. 842-844.

[6]. H. Boo, R.-A. Jeong, S. Park, K. S. Kim, K. H. An, Y. H. Lee, A. H. Han, H. C. Kim, T. D. Chung, Nanoneedle Biosensor based on Single Carbon Nanotube, *Analytical Chemistry*, Vol. 78, 2006, pp. 617-620.

[7]. F. Patolsky, Y. Weizmann, I. Willner, Long-Range Electrical Contacting of Redox Enzymes by SWCNT Connectors, *Angewandte Chemie International Edition*, Vol. 43, Issue 16, 2004, pp. 2113-2117.

[8]. P. G. Collins, Nanotubes for Electronics, *Scientific American*, 2000.

[9]. T. Filleter, R. Bernal, S. Li, H. D. Espinosa, Ultrahigh Strength and Stiffness in Cross-Linked Hierarchical Carbon Nanotube Bundles, *Advanced Materials*, Vol. 23, Issue 25, 2011, pp. 2855-2860.

[10]. X. Lu, Z. Chen, Curved Pi-Conjugation, Aromaticity, and the Related Chemistry of Small Fullerenes (C_{60}) and Single-Walled Carbon Nanotubes, *Chemical Reviews*, Vol. 105, Issue 10, 2005, pp. 3643–3696.

[11]. E. Pop, D. Mann, Q. Wang, K. Goodson, H. Dai, Thermal conductance of an individual single-wall carbon nanotube above room temperature, *Nano Letters*, Vol. 6, Issue 1, 2005, pp. 96–100.

[12]. S. Sinha, S. Barjami, G. Iannacchione, A. Schwab, G. Muench, Off-axis thermal properties of carbon nanotube films, *Journal of Nanoparticle Research*, Vol. 7, Issue 6, 2005, pp. 651–657.

[13]. E. Thostenson, C. Li, T. Chou, Nanocomposites in context, *Composites Science and Technology*, Vol. 65, Issue 3–4, 2005, pp. 491–516.

[14]. Z. Wang, Q. Liang, Y. Wang, G. Luo, Carbon nanotube-intercalated graphite electrodes for simultaneous determination of dopamine and serotonin in the presence of ascorbic acid, *Electroanalytical Chemistry*, Vol. 540, 2003, pp. 129–134.

[15]. P. G. Collins, M. S. Arnol, P. Avouris, P, Engineering Carbon Nanotubes and Nanotube Circuits Using Electrical Breakdown, *Science*, Vol. 292, Issue 5517, 2001, pp. 706–709.

[16]. Y. Zhao, J. Wei, R. Vajtai, P. M. Ajayan, E. V. Barrera, Iodine doped carbon nanotube cables exceeding specific electrical conductivity of metals, *Scientific Reports (Nature)*, Vol. 83, 2011, pp. 1-5.

[17]. Beyond Batteries: Storing Power in a Sheet of Paper. Eurekalert. org. August 13, 2007. http://www.eurekalert.org/pub_releases/2007-08/rpi-bbs080907.php Retrieved 2008-09-15.

[18]. A. C. Dillon, K. M. Jones, T. A. Bekkedahl, C. H. Klang, D. S. Bethune, M. J. Heben, Storage of hydrogen in single-walled carbon nanotubes, *Nature,* Vol. 386, Issue 6623, 1997, pp. 377–379.

[19]. M. S. Dresselhaus, K. A. Williams, P. C. Eklund, Hydrogen Adsorption in Carbon Materials, *MRS. Bulletin,* Vol. 24, Issue 11, 1999, pp. 45-50.

[20]. X. Shi, B. Sitharaman, Q. P. Pham, F. Liang, K. Wu, B. W. Edward, L. J. Wilson, A. G. Mikos, Fabrication of porous ultra-short single-walled carbon nanotubenanocomposite scaffolds for bone tIssue engineering, *Biomaterials,* Vol. 28, Issue 28, 2007, pp. 4078–4090.

[21]. B. Sitharaman, X. Shi, X. Walboomers, L. Frank, C. Hongbing, W. Vincent, J. M. Lon, G. J. Antonios, A. John, In vivo biocompatibility of ultra-short single-walled carbon nanotube/biodegradable polymer nanocomposites for bone tIssue engineering, *Bone,* Vol. 43, Issue 2, 2008, pp. 362–370.

[22]. I. Brodie, C. Spindt, Vacuum Microelectronics, *Advanced Electronics Electron Physics,* Vol. 83, Issue 1, 1992, pp. 1-106.

[23]. Y. Saito, S. Uemura, K. Hamaguchi, Cathode Ray Tube Lighting Elements with Carbon Nanotube Field Emitters, *Japanese Journal of Applied Physics,* Vol. 37, 1998, pp. L346-L348.

[24]. Q. H. Wang, A. A. Setlur, J. M. Lauerhaas, J. Y. Dai, E. W. Seelig, R. H. Chang, A nanotube-based field-emission flat panel display, *Applied Physics Letters,* Vol. 72, Issue 22, 1998, pp. 2912-2913.

[25]. W. B. Choi, D. S. Chung, J. H. Kang, H. Y. Kim, Y. W. Jin, I. T. Han, Y. H. Lee, J. E. Jung, N. S. Lee, G. S. Park, J. M. Kim, A nanotube-based field-emission flat panel display, *Applied Physics Letters,* Vol. 75, Issue 20, 1999, pp. 3129-3131.

[26]. R. Standler, Protection of Electronic Circuits from Over-voltages, *Wiley,* New York, 1989.

[27]. R. Rosen, W. Simendinger, C. Debbault, H. Shimoda, L. Fleming, B. Stoner, O. Zhou, Cumulative Auther index, *Applied Physics Letters,* Vol. 76, Issue 13, 2000, pp. 1197.

[28]. P. J. Britto, K. S. V. Santhanam, A. Rubio, A. Alonso, P. M. Ajayan, Improved Charge Transfer at Carbon Nanotube Electrodes, *Advanced Materials,* Vol. 11, Issue 2, 1999, pp. 154-157.

[29]. G. Che, B. B. Lakshmi, E. R. Fisher, C. R. Martin, Carbon Nanotubule Membranes for Electrochemical Energy. Storage and Production, *Nature,* Vol. 393, Issue 6683, 1998, pp. 346-349.

[30]. M. Whittingham (Ed.), Recent Advances in Rechargeeable Li Batteries, *Solid State Ionics,* Vol. 69, Issue 3, 4, 1994, pp. 402.

[31]. O. Zhou, R. M. Fleming, D. W. Murphy, C. T. Chen, R. C. Haddon, A. P. Ramirez, S. H. Glarum, Defects in Carbon Nanostructures, *Science,* Vol. 263, Issue 5154, 1994, pp. 1744-1747.

[32]. S. Suzuki, M. Tomita, Observations of Potassium-Intercalated carbon nanotubes and their valence-band excitation Spectra, *Applied Physics,* Vol. 79, Issue 402, 1996, pp. 3739–3743.

[33]. S. S. Wong, J. D. Harper, P. T. Lansbury C. M. Lieber, Carbon nanotube tips: high resolution probes for imaging biological systems, *American Chemical Society,* Vol. 120, Issue 413, 1998, pp. 603-604.

[34]. E. Dujardin, T. W. Ebbesen, T. Hiura, K. Tanigaki, Capillarity and Wetting of Carbon Nanotubes, *Science,* Vol. 265, Issue 5180, 1994, pp. 1850-1852.

[35]. F. Balavoine, P. Schultz, C. Richard, V. Mallouh, T. W. Ebbesen, C. Mioskowsk, Helicale Kristallisation von Proteinen auf Kohlenstoffnanoröhren: ein erster Schritt zur Entwicklung neuer Biosensoren, *Angew. Chem.,* Vol. 111, Issue 13-14, 1999, pp. 2036-2039.

[36]. R. R. Moore, C. E. Banks, R. G. Compton, Basal plane pyrolytic graphite modified electrodes, *Analytical Chemistry,* Vol. 76, Issue 10, 2004, pp. 2677–2682.

[37]. S. Jiao, M. Li, C. Wang, D. Chen, B. Fang, Fabrication of Fc-SWNTs modified glassy carbon electrode for selective and sensitive determination of dopamine in the presence of AA and UA, *Electrochimica Acta,* Vol. 52, Issue 19, 2007, pp. 5939–5944.

[38]. S. Shahrokhian, H. R. Zare-Mehrjardi, Application of thionine-nafion supported on multi-walled carbon nanotube for preparation of a modified electrode in simultaneous voltammetric detection of dopamine and ascorbic acid, *Electrochimica Acta,* Vol. 52, Issue 22, 2007, pp. 6310–6317.

[39]. J. Wang, M. Li, Z. Shi, N. Li, Z. Gu, Direct electrochemistry of cytochrome c at a glassy carbon electrode modified with single-wall carbon nanotubes, *Analytical Chemistry,* Vol. 74, Issue 9, 2002, pp. 1993–1997.

[40]. C. X. Cai, J. Chen, Direct electron transfer and bioelectrocatalysis of hemoglobin at a carbon nanotube electrode, *Analytical Biochemistry,* Vol. 325, Issue 2, 2004, pp. 285–292.

[41]. R. Zhang, X. Wang, K. Shui, Accelerated direct electrochemistry of hemoglobin based on hemoglobin–carbon nanotube (Hb–CNT) assembly, *Colloid and Interface Science,* Vol. 316, Issue 2, 2007, pp. 517–522.

[42]. L. Zhao, H. Liu, N. Hu, Electroactive films of heme protein-coated multiwalled carbon nanotubes, *Colloid and Interface Science,* Vol. 296, Issue 1, 2006, pp. 204–211.

[43]. Y. F. Lü, C. X. Cai, Immobilization, Characterization and Direct Electron Transfer Reac-tion of Ferredoxin on Multi-walled Carbon Nanotube, *Acta Chimica Sinica,* Vol. 64, 2006, pp. 2396–2402.

[44]. M. Musameh, J. Wang, A. Merkoci, Y. H. Lin, Solubilization of carbon Nanotubes By nafion toward the Preparation of Amperometric Biosensors, *Electrochemistry Communication,* Vol. 4, Issue 10, 2002, pp. 743–746.

[45]. J. Wang, G. Liu, M. R. Jan, Ultrasensitive electrical biosensing of proteins and DNA: carbon-nanotube derived amplification of the recognition and transduction events, *American Chemical Society*, Vol. 126, Issue 10, 2004, pp. 3010-3011.

[46]. T. Tang, T. Z. Peng, Q. C. Shi, Sequence determination of DNA pieces using carbon nanotube modified, gold electrodes, *Acta Chimica Sinica*, Vol. 63, Issue 22, 2005, pp. 2042–2046.

[47]. A. G. Wu, L. H. Yu, Z. A. Li, H. M. Yang, E. K. Wang, Atomic force microscope investigation of large-circle DNA molecules, *Analytical Biochemistry*, 325, Issue 2, 2004, pp. 293–300.

[48]. J. Koehne, J. Li, A. M. Cassell, H. Chen, Q. Ye, H. T. Ng, J. Han, M. Meyyappan, The fabrication and electrochemical character-ization of carbon nanotube nanoelectrode arrays, *Material Chemistry*, Vol. 14, 2004, pp. 676–684.

[49]. M. L. Pedano, G. A. Rivas, Adsorption and Electrooxidation of nucleic acids at Carbon Nanotubes Paste Electrodes, *Electrochemical Communications*, Vol. 6, 2004, pp. 10–16.

[50]. Y. R. Wang, H. U. Ping, L. Qiong-Lin, L. Guo-An, W. Yi-Ming, Application of Carbon Nanotube Modified Electrode in Bioelectroanalysis, *Chinese Journal of Analytical Chemistry*, Vol. 36, Issue 8, 2008, pp. 1011–1016.

[51]. Y. Y. Sun, J. J. Fei, K. B. Wu, S. S. Hu, Simultaneous electrochemi-cal determination of xanthine and uric acid at a nanoparticle film electrode, *Analytical and Bioanalytical Chemistry*, Vol. 375, Issue 4, 2003, pp. 544–549.

[52]. M. N. Zhang, K. Liu, L. Xiang, Y. Q. Lin, L. Su, L. Q. Mao, Carbon nanotube-modified carbon fiber microelectrodes for in vivo voltammetric measurement of ascorbic acid in rat brain, *Analytical Chemistry*, Vol. 79, Issue 17, 2007, pp. 6559–6565.

[53]. N. S. Lawrence, R. Deo, J. Wang, Detection of homocysteine at carbon nanotube paste electrodes, *Talanta*, Vol. 63, Issue 2, 2004, pp. 443–449.

[54]. W. Y. Qu, H. Wang, K. B. Wu, Electrocatalytic Activities of 1,2-Naphthoquinone Modified Carbon Nanotube to the Electrochemical Oxidation of β-Nicotinamide Adenine Dinucleotide, *Chinese Journal of Analytical Chemistry*, Vol. 34, Issue 12, 2006, pp. 1688-1693.

[55]. X. Q. Lin, J. B. He, Z. G. Zha, Detection of partial discharge in SF_6 gas using a carbon nanotube-based gas sensor, *Sensors and Actuators B: Chemical*, Vol. 119, Issue 2, 2006, pp. 608–614.

[56]. Y. R. Wang, M. Zhang, P. Hu, Q. L. Liang, Y. M. Wang, G. A. Luo, *Chinese Traditional Patent Medicine*, Vol. 23, 2007, pp. 1476–1478

[57]. H. Yadegari, A. Jabbari, H. Heli, A. A. Moosavi-Movahedi, K. Karimian, A. Khodadadi, Electrocatalytic oxidation of deferiprone and its determination on a carbon nanotube-modified glassy carbon electrode, *Electrochimica Acta*, Vol. 53, Issue 6, 2008, pp. 2907–2916.

[58]. WHO, Air Quality Guidelines for Europe, 2nd ed., European Series, Vol. 91, *World Health Organization Regional Publications*, 2000.

[59]. Guidelines for indoor chemicals, Tech. Rep., *Ministry of Health, Labor and Welfare,* 2002.

[60]. O. Bunkoed, F. Davis, P. Kanatharana, Sol–gel based sensor for selective formaldehyde determination, *Analytica Chimica Acta,* Vol. 659, Issue 1-2, 2010, pp. 251–257.

[61]. Y. I. Korpan, M. V. Gonchar, A. A. Sibirny, C. Martrlet, A. V. El'skaya, T. D. Gibson, A. P. Soldatkin, Development of highly selective and stable potentiometric sensors for formaldehyde determination, *Biosensors and Bioelectronics,* Vol. 15, Issue 1, 2000, pp. 77–83.

[62]. M. Hammerle, K. Hilgert, S. Achmann, R. Moos, Direct monitoring of organic vapours with amperometric enzyme gas sensors, *Biosensors and Bioelectronics,* Vol. 25, Issue 6, 2010, pp. 1521–1525.

[63]. M. Hammerle, E. A. H. Hall, N. Cade, D. Hodgins, Electrochemical enzyme sensor for formaldehyde operating in the gas phase, *Biosensors and Bioelectronics,* Vol. 11, Issue 3, 1996, pp. 239–246.

[64]. L. Bareket, A. Rephaeli, G. Berkovitch, A. Nudelman, J. Rishpon, Carbon nano-tubes based electrochemical biosensor for detection of formaldehyde released from a cancer cell line treated with formaldehyde-releasing anticancer products, *Bioelectrochemistry,* Vol. 77, Issue 2, 2010, pp. 94–99.

[65]. K. Kawamura, K. Kerman, M. Fujihara, N. Nagatani, T. Hashiba, E. Tamiya, Development of a novel hand-held formaldehyde gas sensor for the rapid detection of sick building syndrome, *Sensors and Actuators B: Chemical,* Vol. 105, Issue 2, 2005, pp. 495–501.

[66]. H. Xie, C. Sheng, X. Chen, X. Wang, Z. Li, J. Zhou, Multi-wall carbon nanotube gas sensors modified with amino-group to detect low concentration of formaldehyde, *Sensors and Actuators B: Chemical,* Vol. 168, 2012, pp. 34–38.

[67]. K. Pyrzynska, Functionalized cellulose sorbents for on-line preconcentration of trace metals for environmental analysis, Crit. Rev. *Analytical Chemistry,* Vol. 29, Issue 4, 1999, pp. 313–321.

[68]. Y. Q. Cai, G. B. Jiang, J. F. Liu, Q. X. Zhou, Multiwalled carbon nanotubes as a solid-phase extraction adsorbent for the determination of bisphenol A, 4-n-nonylphenol, and 4 tert-octylphenol, *Analytical Chemistry,* Vol. 75, Issue 10, 2003, pp. 2517–2521.

[69]. P. Patnaik, A Comprehensive Guide to the Hazardous Properties of Chemical Substances, 2nd ed., *Wiley,* New York, 1999.

[70]. R. P. Pohanish, S. A. Greene, Hazardous Materials Handbook, *Van Nostrand Reinhold,* New York, 1996.

[71]. W. Puacz, W. Szahun, Catalytic determina tion of sulphide in blood, *Analyst,* Vol. 120, Issue 3, 1995, pp. 939-941.

[72]. J. Kurzawa, Determination of sulphur(II) compounds by flow injection analysis with application of the induced iodine/azide reaction, *Analytica Chimica Acta,* Vol. 173, 1985, pp. 343-348.

[73]. D. R. Canterfold, Simultaneous determination of cyanide and sulfide with rapid direct current polarography, *Analytical Chemistry*, Vol. 47, 1975, Issue 2, pp. 88-92.

[74]. K. Han, W. F. Koch, Determination of sulfide at the parts-per-billion level by ion chromatography with electrochemical detection, *Analytical Chemistry*. Vol. 59, Issue 7, 1987, pp. 1016-1020.

[75]. H. Weisz, T. Lenz, Kontinuierliche katalytisch-kinetische bestimmung von silber und sulfid im p.p.m.-bereich unter verwendung einer durchflusszelle, *Analytica Chimica Acta*, Vol. 70, Issue 2, 1974, pp. 359-364.

[76]. S. R. Spurlin, E. S. Yeung, On-line chemiluminescence detector for hydrogen sulfide and methyl mercaptan, *Analytical Chemistry*, Vol. 54, Issue 2, 1982, pp. 318-320.

[77]. X. C. Hu, N. Takenaka, M. Kitano, H. Bandow, Y. Maeda, M. Hattori, Determination of trace amounts of urea by using flow injection with chemiluminescence detection, *Analyst*. Vol. 119, Issue 8, 1994, pp. 1829-1833.

[78]. X. Hu, N. Takenaka, S. Takasuna, Determination of ammonium ion in rainwater and fogwater by flow injection analysis with chemiluminescence detection, *Analytical Chemistry*, Vol. 65, 1992, pp. 3489-3492.

[79]. J. S. Sun, S. G. Schulman, J. H. Perrin, Chemiluminescence of β-lactam antibiotics following oxidation by potassium Superoxide, *Analytica Chimica Acta*, Vol. 338, Issue 1-2, 1997, pp. 1-2.

[80]. S. Kim, R. DiCosimo, J. S. Filippo, Spectrometric and chemical characterization of superoxide, *Analytical Chemistry*, Vol. 51, Issue 6, 1979, pp. 679-681.

[81]. R. J. Taylor, A. A. Humffray, Electrochemical studies on glassy carbon electrodes: II. Oxygen reduction in solutions of high pH (pH>10), *Electroanalytical Chemistry*, Vol. 54, Issue 1, 1975, pp. 63-84.

[82]. A. J. Appleby, J. Marie, AC impedance measurements in molten $PbCl_2$ – KCl: In(III) reduction and lead sulphide oxidation, *Electrochimica Acta*, Vol. 24, Issue 12, 1978, pp. 1243-1245.

[83]. J. Y. Qu, Y. Shen, X. H. Qu, S. J. Dong, Preparation Of Hybrid Thin film modified carbon nanotubes on Glassy Carbon Electrode, *Electroanlysis*, Vol. 16, Issue 17, 2004, pp. 1444-1450.

[84]. F. Wang, S. S. Hu, Electrochemical reduction of dioxygen on carbon nanotubes–dihexadecyl phosphate film electrode, *Electroanalytical Chemistry*, Vol. 580, Issue 1, 2005, pp. 68-77.

[85]. R. Huang, X. Zheng, Y. Qu, Highly selective electrogenerated chemiluminescence (ECL) for sulfide ion determination at multi-wall carbon nanotubes-modified graphite electrode, *Analytica Chimica Acta*, Vol. 582, Issue 2, 2007, pp. 267–274.

[86]. A. Abbaspour, A. Ghaffarinejad, Electrocatalytic oxidation of L-cysteine with a stable copper–cobalt hexacyanoferrate electrochemically modified

carbon paste electrode, *Electrochimica Acta*, Vol. 53, Issue 22, 2008, pp. 6643-6650.

[87]. D. R. Shankaran, S. S. Narayanan, Chemically modified sensor for amperometric determination of sulphur dioxide, *Sensors and Actuators B: Chemical*, Vol. 55, Issue 2-3, 1999, pp. 191-194.

[88]. L. S. T. Alamo, T. Tangkuaram, S. Satienperakul, Determination of sulfite by pervaporation-flow injection with amperometric detection using copper hexacyanoferrate-carbon nanotube modified carbon paste electrode, *Talanta*, Vol. 81, Issue 3, 2010, pp. 1793–1799.

[89]. T. I. Mudder, M. M. Botz, A. Smith, Chemistry and Treatment of Cyanidation Wastes, second ed., *Mining Journal Books*, London, UK, 2001.

[90]. L. N. Denefield, J. F. Judkins, B. L. Weand, Process Chemistry for Water and Wastewater Treatment, *Prentice-Hall Inc.*, Englewood Cliffs, 1982.

[91]. W. J. Jin, M. T. Fernández-Argüelles, J. M. Costa-Fernández, R. Pereiro, A. Sanz-Medel, Photoactivated luminescent CdSe quantum dots as sensitive cyanide probes in aqueous solutions, *Chemical Communications*, Vol. 7, 2005, pp. 883–885.

[92]. W. J. Jin, J. M. Costa-Fernández, R. Pereiro, A. Sanz-Medel, Surface-modified CdSe quantum dots as luminescent probes for cyanide determination, *Analytica Chimica Acta*, Vol. 522, Issue 1, 2004, pp. 1–8.

[93]. H. Sun, Y. Y. Zhang, S. H. Si, D. R. Zhu, Y. S. Fung, Piezoelectric quartz crystal (PQC) with photochemically deposited nano-sized Ag particles for determin-ing cyanide at trace levels in water, *Sensors and Actuators B: Chemical*, Vol. 108, 1-2, 2005, pp. 925–932.

[94]. M. Noroozifar, K. M. Mozhgan, T. Aboozar, Determination of cyanide in wastewaters using modified glassy carbon electrode with immobilized silver hexacyanoferrate nanoparticles on multiwall carbon nanotube, *Hazardous Materials,* Vol. 185, Issue 1, 2011, pp. 255–261.

[95]. G. Svehla, VOGEL's Textbook of Macro and Semimicro Qualitative Inorganic Chemistry, 5th ed., *Longman Group Limited*, London, 1979.

Chapter 4

Nanosensors Based on Surfactant Modified Electrodes

Nada F. Atta and Ekram H. El-Ads

4.1. Modified Electrodes

The development of various modified electrodes has stimulated the interest of most of scientists in the electrochemistry field over the last past several decades. Upon modification of the electrode, the reactivity of the surface can be altered owing to reducing the size and accessibility of reactive sites at the interface [1]. One of the most important goals of electrode modification is the control of heterogeneous kinetics and/or molecular recognition processes. From the large size of literature available on modified electrodes, it is clear that electrode modification is an important area of study in modern electrochemistry and any research carried out in this direction will be of interest, especially due to the several application possibilities of these electrodes [2-12]. Modified electrodes have several applications in electrochromic display devices, controlled release of drugs, electrosynthesis, corrosion protection, electroanalysis and analytical applications [13]. Special emphasis in this chapter will be laid on the potential applications of modified electrodes in electroanalysis. Modified electrodes can offer great advantages for electroanalysis applications, including acceleration of electron transfer rate, preferential accumulation and selective membrane permeation. The last mentioned steps can provide higher selectivity, sensitivity, good stability and preventing electrode fouling which may arise from interfering species or from the analyte itself or its electrode reaction products [14-16].

4.2. Modes for the Electrode Modification

Two modes can be used for electrode modification:

1. Surface modification: it can be established in several ways listed below:

a- **Chemical modification:** in which the electrode surface is activated by chemical reaction, then the activated electrode surface reacts with another chemical species that becomes immobilized on the surface. A part from organic compounds and enzymes can often be immobilized by such procedures following the reaction of the electrode surface with reagents such as carbodimide and silane.

b- **Adsorption:** in which the electrode surface is coated with the modifiers either by dipping or, more commonly, by spin-coating (application of a drop of solution on the electrode surface followed by spinning to evaporate the solvent). This mode is particularly used for modifying the electrode surface with soluble polymeric species or polymeric species which have a tendency for self-assembly, leading to self assemble monolayers on the electrode surface.

c- **Electroadsorption and electrodeposition:** in which the adsorption is carried out under the influence of applied potential resulting in thicker modified layers formation. Electropolymerization of monomers is also possible. Formation of conducting polymer films can be achieved via these procedures.

d- **Plasma:** in which the electrode surface is cleaned by plasma resulting in active surface with dangling bonds. As a result, the further adsorption of any species, such as amines or ethenes, in the vicinity is very fast. This method is actually used to clean the surface and can be combined to other modifying techniques.

2. Bulk modification: Bulk modification rather than surface modification procedures are sometimes used. Herein the modifier is mixed intimately with the electrode material, such as with carbon paste [17].

On the light of these different strategies for the electrode modifications, special attentions are given to the use of surfactants (modifiers) for the modification of the electrode surface and the study of their beneficial role in electroanalysis. Introduction of surfactants in the area of modified electrodes adds a new and useful dimension to electroanalysis applications.

4.3. Surfactants

4.3.1. Structures and Types

Surfactants, a kind of amphiphilic molecules with a hydrophilic polar head that is compatible with water on one side and a long hydrophobic non-polar tail that is compatible with oil on the other, have been widely applied in electrochemistry [18-21]. The polar head group contains heteroatoms such as O, S, P, or N included in functional groups such as alcohol, thiol, ether, ester, acid, sulfate, sulfonate, phosphate, amine, amide etc. On the other hand, the non-polar tail group is in general a hydrocarbon chain of the alkyl or alkylbenzene type, sometimes with halogen atoms and even a few non-ionized oxygen atoms. The polar-nonpolar duality nature endows the surfactants with their unique solution and interfacial characteristics improving the property of the electrode/solution interface [20]. In other words, surfactants are chemical compounds that have surface-active properties or the ability to affect the interfacial relationship between two dissimilar substances such as oil and water.

Most commonly, surfactants are classified according to their polar head group as shown in Fig. 4.1(A). Moreover, surfactants are divided into two categories; ionic and non-ionic as shown in Fig. 4.1(B). Non-ionic surfactants have no charged groups in its head and don't ionize in aqueous solution because their hydrophilic group is of a non-dissociable type, such as alcohol, phenol, ether, ester or amide.

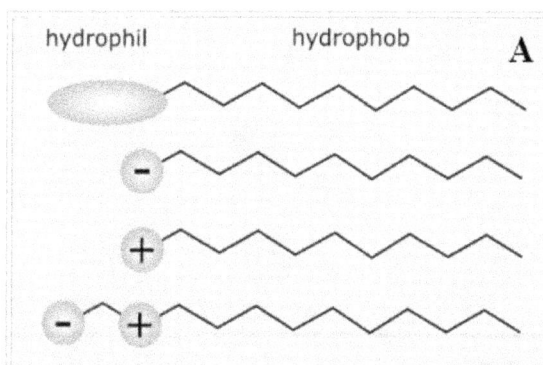

Fig. 4.1. (A) Surfactant classification according to the composition of their head: nonionic, anionic, cationic and amphoteric.

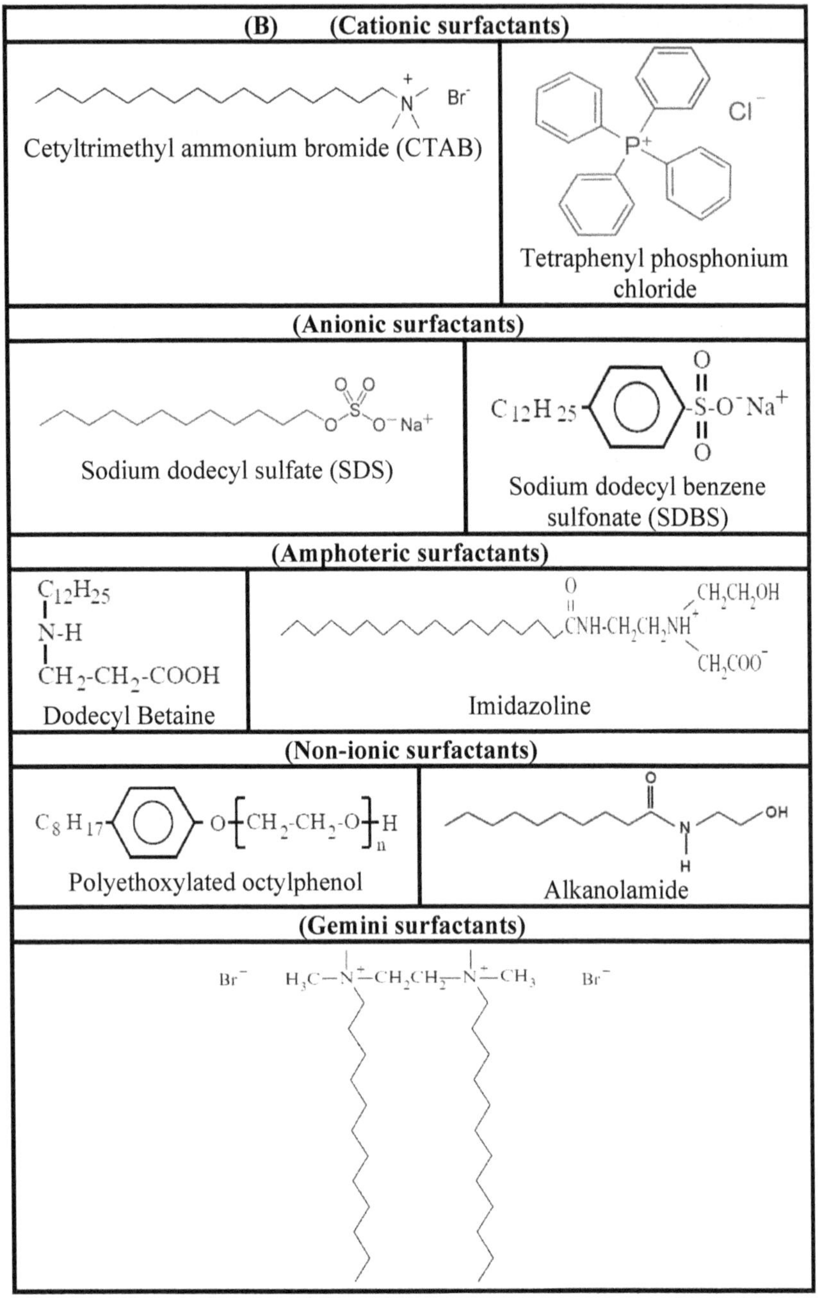

(B) (Cationic surfactants)
Br⁻
Cetyltrimethyl ammonium bromide (CTAB)
Cl⁻
P+
Tetraphenyl phosphonium chloride
(Anionic surfactants)
O⁻ Na+
Sodium dodecyl sulfate (SDS)
C12H25
-S-O⁻Na+
Sodium dodecyl benzene sulfonate (SDBS)
(Amphoteric surfactants)
C12H25
N-H
CH2-CH2-COOH
Dodecyl Betaine
CH2CH2OH
CNH-CH2CH2NH+
CH2COO⁻
Imidazoline
(Non-ionic surfactants)
C8H17
O[CH2-CH2-O]nH
Polyethoxylated octylphenol
OH
Alkanolamide
(Gemini surfactants)
Br⁻
H3C—N+—CH2CH2—N+—CH3
Br⁻

Fig. 4.1. (B) Examples for cationic, anionic, amphoteric, non-ionic and gemini surfactants.

On the other hand, ionic surfactants are subdivided into anionic and cationic surfactants depending on the type of their head molecular charge. If the charge is negative, the surfactant is more specifically called anionic; if the charge is positive, it is called cationic. In addition, amphoteric or zwitterionic surfactants, a special case, have oppositely charged centers in one molecule exhibiting both anionic and cationic functional groups [21, 22]. A new class of surfactants, the so-called gemini or dimeric surfactants, are composed of two hydrophobic or hydrophilic or neutral tails connected at the level of or close to head groups by a spacer group which may be also hydrophilic or hydrophobic or neutral and known to exhibit enhanced surface-active properties. Compared with those conventional single-chain (single head group) surfactants, gemini surfactants show quite unusual phase behaviors and other physicochemical properties [23].

Solubility of surfactants in water was found to be dependent on the length of the hydrocarbon chains due to the amphiphilic structure nature of the surfactant molecules. If the hydrophobic moiety is very large, the surfactant will not be soluble in water, but for smaller hydrophobic moieties, the surfactant is soluble. However, the contact between the hydrophobic group and the aqueous medium is energetically less favorable than the water-water contact. Thus, surfactant systems try to find alternative ways in order to reduce the contact between the hydrophobic group and the aqueous surrounding, hence reducing the free energy of these systems. Moreover, surfactants can reduce oil-water contact by the accumulation at various interfaces or the formation of different self-assembly structures in the solution [21, 22].

4.3.2. Self-assembly of Surfactants in Solution

Surfactants can reduce the oil-water contact via the formation of different self-assembly structures through which the hydrophobic domains of surfactant molecules can associate to form various structures achieving segregation of the hydrophobic parts from water. These various molecular structures may include micelles, micro-emulsions, liposomes, and a range of liquid crystalline phases [20, 21]. Formation of a specific self-assembly structure in solution depends on the type of surfactant (the size of the hydrophobic tail group, the nature, and size of the polar head group, ...) and the solution conditions (temperature, salt concentration, pH, ...).

4.3.3. Micelles in Aqueous Medium

Surfactants form true solutions at low concentrations. In these true solutions, the surfactant molecules are dispersed as individual molecules or ions through the solution and they don't associate themselves to form micelles (Fig. 4.2).

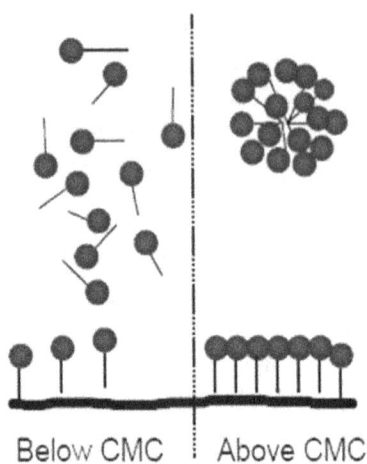

Below CMC   Above CMC

Fig. 4.2. The surfactant behavior below and above the CMC.

With increasing the concentration, however, the split personality structure nature of such amphiphilic substances leads their spontaneous self-association in solution resulting in the formation of micelles [22, 24]. The term micelles refers to aggregates of long-chain surfactant molecules or ions which are formed spontaneously in their solution at definite concentration. This concentration was found to be dependent on the size of the hydrophobic moiety, the nature of polar head group, the nature of counter ions (for charged surfactants), the salt concentration, pH, temperature, and presence of co-solutes [21, 24]. Micelles possess regions of hydrophilic and hydrophobic character. In water, the charged polar head groups oriented toward the water and the hydrocarbon chains oriented away from the water to face the interior of the micelles. Micelles are characterized by two important features; the aggregation number (N) and the critical micelle concentration (CMC). N is the number of molecules or monomers in the micelle determining the size and geometry of the micelle. Aggregation numbers for surfactants in aqueous solution generally range between 10 and 100

[24]. CMC is the concentration of surfactant at which a large number of micelles form in its solution that are in thermodynamic equilibrium with molecules or ions. The narrow range of concentration at which the micelles first become detectable is the CMC of the amphiphile. CMC also refers to that concentration where abrupt change in physical properties such as conductivity and surface tension occurs [21, 24]. The CMC is very characteristic for each surfactant, where dynamic aggregates are formed and it is necessary to know the CMC value for the commonly employed surfactants for quantitative understanding of experimental data. CMC values for commonly used surfactants range from 10^{-4} to 10^{-2} mol L^{-1} [24].

4.3.4. Self-assembly of Surfactants at Solid-liquid Interface

Upon reaching the CMC, the formation of micelles in solution starts to occur as mentioned before [22, 24]. At low surfactant concentration (less than the CMC), the cooperative hydrophobic forces are not sufficient to form micelles or any self-assembly structures in solution. As a result, surfactant molecules or ions pass out into the surface layer at the interface of the surfactant with the other phases (we will focus on solid-liquid interface), thus reducing the oil-water contact and hence lowering the free energy of the system [22] (Fig. 4.3). Soon, the surface layer becomes saturated and with further increase in the surfactant concentration, the system expels the hydrophobic chains from the water into the liquid "pseudo phase"- a micelle.

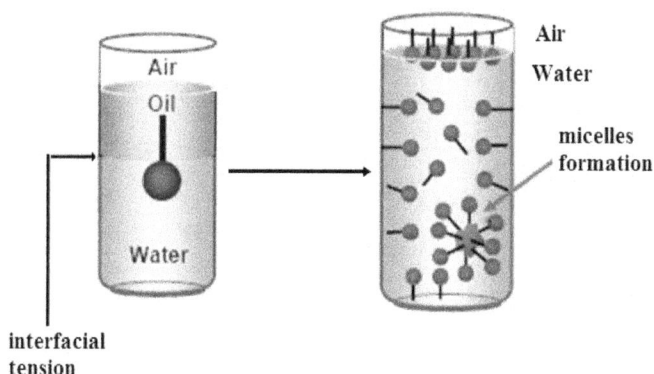

Fig. 4.3. A Schematic diagram showing how does surfactant molecules reduce the oil-water contact.

It is important to note that the adsorption of surfactants at solid substrate not only occurs before CMC, but also occurs above CMC. Different studies showed that ionic electro-active and electro-inactive surfactants above the CMC form full coverage aggregates on metal and carbon electrodes [25-29]. Different molecular structures have been reported for the adsorbed surfactant films on the surface of solid substrates in the form of monolayer [30-34], bilayer [35-37], multilayer (especially at extreme potentials of opposite sign for that of surfactant head group) [26, 27, 38-41], and surface micelles or hemimicelle (full sphere, full cylinder, or half cylinder) [36, 42-45].

The formation of specific self assembly structure on the solid substrate surface depends on the hydrophobicity and morphology of the substrate [33, 35, 36, 46-48], the nature of the electrolyte and counter ions [37, 46, 49, 50], the applied potential and the charge of the substrate [51-55], the surfactant concentration [56-61], the structure and type of surfactant (single or double chain hydrocarbon tail, length of the hydrocarbon chain, ionic or non-ionic head group) [33, 34, 36, 37], and the pH of the solution from which adsorption occurs [36, 44]. Typical surfactant aggregate structures are shown in Fig. 4.4 [20, 62].

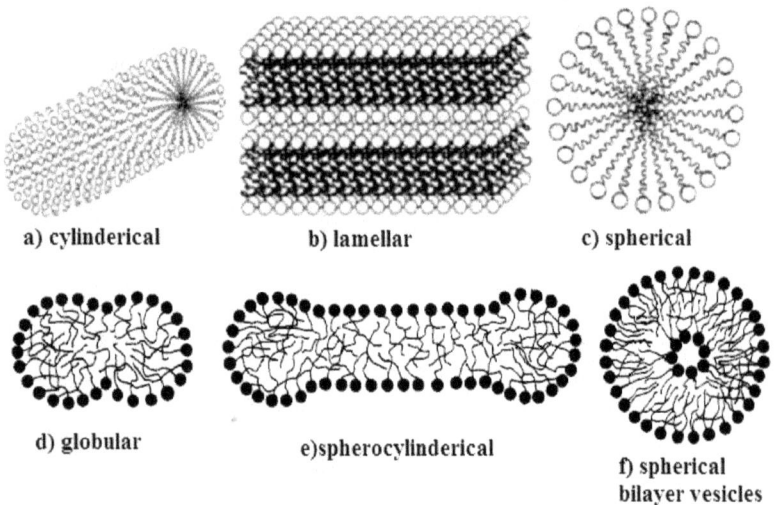

a) cylinderical   b) lamellar   c) spherical

d) globular   e)spherocylinderical   f) spherical bilayer vesicles

Fig. 4.4. Typical surfactant aggregates: (a) cylindrical; (b) lamellar; (c) spherical; (d) globular; (e) spherocylinderical micelles, and (f) spherical bilayer vesicles [20, 62].

4.4. Modification of the Electrode Surface by Surfactants

Two important modes by which one can use surfactants in order to modify the electrode surface:

1) Surface modification; in which spontaneous adsorption or self assembly of surfactants occurs on the electrode surface.

2) Bulk modification; in which surfactants (modifiers) are mixed intimately with the electrode material such as carbon paste.

4.4.1. Surface Modification by Surfactants

An elegant way for controlling the electrochemistry at the modified electrode surface is to modify it by amphiphilic compounds [63]. Different approaches have been reported for the adsorption of these amphiphilic compounds on the electrode surface [64-66]. One approach is based on the physico-chemisorption of highly ordered self-assembled monolayers (SAMs) [1]. Different self-assembly structures of surfactants were formed through the spontaneous physisorption of surfactant molecules at the electrode surface. This was achieved via the exposure of a clean electrode surface to a dilute solution of surfactants either by dipping or by the application of a drop of solution followed by spinning to evaporate the solvent (spin-coating). The focus in this chapter will be on the spontaneous physisorption of surfactants on the electrode surfaces.

4.4.1.1. Effect of the Surfactants on the Electrochemical Kinetics of the Electrode Reaction

In order to predict the effect of surfactants on the electrochemical kinetics of the electrode reaction, one must know dimensions, polarities, and molecular structures of surfactant aggregates on the electrode and the position of the electroactive species within them. Moreover, for a reactant molecule or ion in a micellar solution, such predictions may need to take into account the following influences [67]:

a) The distance between the electrode and the electron acceptor or donor moiety;

b) The environment surrounding the reactant at the time of electron transfer;

c) Structure and dynamics of surfactant aggregates on the electrode;

d) Dynamics of interaction of the reactant with surfactant structure on the electrode and with micelles.

To observe the influences of the adsorbed surfactant films on the discharge of electroactive solutes on the electrode surface, it is necessary to investigate the adsorption of surfactants from aqueous solution onto the electrode surface. The most important influences of the adsorbed surfactant films on the electron transfer rates include; (1) blocking by surfactants and (2) electrostatic interactions between electroactive solutes and the adsorbed surfactant films [67, 68]. The blocking effect is the common effect of these adsorbed films on the electrode process. The surfactant films may physically block the partial or full access of the electroactive species to the electrode surface inhibiting the electron transfer process [68]. The unfavorable entrance of hydrophilic species through the hydrophobic region of the adsorbed surfactant film or the coulombic repulsion between the charged head group of the surfactant and similar charged electroactive species was the origin of this blocking effect [69, 70]. As a result, the adsorbed surfactant film acts as a barrier between the electrode surface and the electroactive species.

On the other hand, mild kinetic enhancement for the electrode reaction may be observed due to the adsorbed surfactant film on the electrode. This enhancement depends on the pre-concentration process (the accumulation of the electroactive species through or into the adsorbed surfactant film). The coulombic attraction forces as well as the hydrophobic interactions between the electroactive species and the adsorbed surfactant layers are the main driving forces for the pre-concentration process [69-71].

Moreover, incorporation of electroactive species within the adsorbed surfactant films will introduce different distances between the electroactive species and the electrode surface. Thus, the rate of the electron transfer will decrease with increasing the distance between the electroactive species and the electrode surface within the adsorbed surfactant film [72-74]. This is in accordance with the electron transfer theory [75], which predicts the exponential decrease of the electron transfer rate with increasing the distance between the electroactive species and the electrode surface.

When electroactive species presents in micellar system, it may bind to the micelles present in the aqueous medium or partitioned between micelles and aqueous media [76]. Electroactive ions which are oppositely charged to the ionic micelles can bind at the micelle-water interface, whereas non-polar electroactive species can bind in hydrophobic regions of the micelles just below this interface [77]. Electroactive species solubilized in micelles or bonded to micelles can undergo electron transfer reactions in which the current is controlled by diffusional mass transport of micelles containing the electroactive species.

4.4.1.2. Surfactant Adsorption

One of the characteristic features of surfactants is their tendency to adsorb at the surface/interface mostly in an oriented fashion due to the specific amphiphilic structure [78, 79]. Surfactant adsorption is a process of transfer of surfactant molecules from bulk solution phase to the surface/interface. The adsorption of surfactant at the solid–liquid interface play an important role in many technological and industrial applications. There is a number of mechanisms by which surfactant molecules may adsorb onto the solid substrates from aqueous solution. In general, the adsorption of surfactants involves single ions rather than micelles [79].

a) **Ion exchange:** occurs via the replacement of counter ions adsorbed onto the substrate from the solution by similarly charged surfactant ions.

b) **Ion pairing:** occurs via adsorption of surfactant ions from solution onto oppositely charged sites unoccupied by counter ions.

c) **Hydrophobic bonding:** occurs when there is an attraction between a hydrophobic group of adsorbed molecule and a molecule present in the solution.

d) **Adsorption by polarization of π electrons:** occurs when the surfactant contains electron-rich aromatic nuclei and the solid adsorbent has strongly positive sites. Adsorption occurs via attraction between electron rich aromatic nuclei of the adsorbate and positive sites on the adsorbent.

e) **Adsorption by dispersion forces:** occurs via adsorption by London-van-der-Waals force between adsorbate (surfactant molecules) and adsorbent which increases with increasing molecular weight of the adsorbate [79].

4.4.1.3. Modes for Electron Transfer in Micellar Solutions

The surfactant aggregates on the electrode surface or inside the solution can control the electron transfer kinetics of electrode reaction process. Aggregates in the form of bilayers, cylinders or surface micelles have the tendency to be adsorbed onto the electrode surface in solutions with surfactant concentration above the CMC. On hydrophilic electrodes, head-down orientations of surfactants are preferred (Fig. 4.5 a, b) [67].

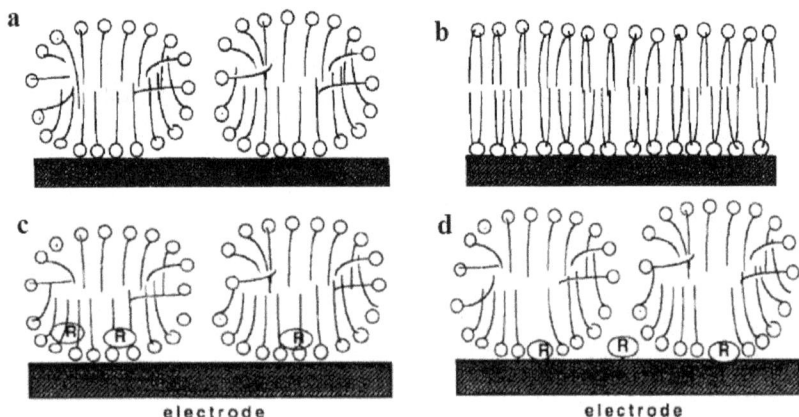

electrode   electrode

Fig. 4.5. Conceptual drawings of interfacial region on hydrophilic electrode in micellar solutions: (a) side view of surface micelles or cylinders; (b) side view of a bilayer, (c) position of reactants R which are dissolved in the surface aggregates; (d) position of reactants R which are specifically adsorbed to the electrode [67].

We will start by talking about the mechanism of discharge of electroactive species binding to the adsorbed dynamic surfactant film. The electron transfer process will take place when the electroactive species approaches to the vicinity of the electrode surface. Two main possibilities allow the charge transfer; the first is the displacement of the adsorbed surfactant by the reactant, thus, the reactant approaches the electrode closely (Fig. 4.5 c), and the second is the approach of the reactant to the electrode surface within one head group diameters of adsorbed surfactant moieties (Fig. 4.5 d). Different possibilities have been suggested for the molecular interpretation of electron transfer between a micelle-solubilized electroactive species and the electrode. One possibility is the dissociation of solute bound to the micelle followed by its entry into the adsorbed surfactant film, then its orientation near the surface, and finally electron transfer occurs. The entry into the films and orientation are expected to occur on a millisecond time scale. Another possibility is the join of the micelle in solution with the aggregates on the electrode surface, bringing the reactant close enough to the electrode for electron transfer [67].

4.4.1.4. The Importance of Surfactants to Electrochemistry

The importance of surfactants to electrochemistry is based on the following aspects [80]:

a) The presence of surfactants in aqueous solution will affect the double layer structure, the redox potential, the charge transfer, and the diffusion coefficient of the electrode processes [78, 80].

b) Surfactants can stabilize radicals or intermediate reaction products.

c) Surfactants can easily dissolve hydrophobic substances.

As known, surfactants can endow the electrode/solution interface with different electrical properties and change the electrochemical process. For example, the adsorption of ionic surfactants can form charged surfactant layers on the electrode surface, which have a strong accumulation capacity towards the oppositely charged analytes and an electrostatic repulsion ability towards the same charged analytes. Similarly, surfactants can form loose hydrophobic layers on the electrode surface by their hydrophobic tails exhibiting a strong accumulation capacity toward hydrophobic analytes. Moreover, the formation of surfactant layers on the electrode surfaces can also avoid the direct contact of the analyte with the electrode surface enhancing the antifouling capacity of the electrodes [81, 82].

On the other hand, surfactants play a very important role in electrode reactions, not only in solubilizing organic compounds but also by providing specific orientation of the molecules at the electrode interface. Surfactants are effective excipients in many drug formulations via improving dissolution rate and increasing drug solubility. The ability of surfactants to reduce interfacial tension and contact angle between solid particles and aqueous media leads to improving the drug wettability and increasing the surface availability for the drug dissolution [83].

Accordingly, surfactants were found to have several applications in electrochemistry, polarography, corrosion, batteries, fuel cells, electrometallurgy, electro-organic chemistry, photoelectrochemistry, electrocatalysis, and electroanalysis [84]. The focus in this chapter will be on surfactant applications in electroanalysis. Surfactants are very effective to be used in the electroanalysis of organic compounds, biologically important compounds, drugs, some important inorganic ions and metals [85-87]. Moreover, micelles have also been employed as selective masking agents to improve selectivity and sensitivity of electrochemical analysis [88].

4.4.1.5. Determination of Metal Ions at Modified Electrodes in Presence of Surfactants

The use of micellar system in the analysis of important inorganic ions and some heavy metals has been studied [85-87]. Kalcher et al utilized carbon paste electrode (CPE) for the determination of trace concentrations of molybdenum(VI) in the presence of cetyltrimethylammonium bromide (CTAB) using anodic stripping voltammetry. The preconcentration of Mo(VI) is performed by adsorption and reduction of ion-pairs of CTA and Mo(VI) oxalate at a potential of -0.4 V vs. saturated calomel electrode. The supporting electrolyte contains 0.01 M oxalic acid and 0.075 mM CTAB. The electrochemical redox reaction of Mo(VI) complexed with oxalate was more irreversible at bare CPE. In the presence of CTAB, the electrochemical reaction of Mo(VI) oxalate was facilitated due to the formation of ion-pairs at the electrode surface. Obviously, the lipophilic positive counter ion compensates the negative charge of the Mo-complexes and thus allows the aggregate to overcome the electrostatic repulsion more easily. Furthermore, the detection limit of Mo was 0.04 μg L^{-1} with an accumulation period of 10 min [89].

On the other hand, Deng developed a sensitive procedure for the determination of trace Mo(VI) based on the oxidation of Mo(VI)– 2 ', 3, 4 ', 5, 7- pentahydroxyflavone (morin) complex at a sodium dodecyl sulfate modified carbon paste electrode (SDS/CPE). The SDS/CPE was prepared as following: CPE was polished on a piece of tracing paper and then dipped into 1.0×10^{-2} M SDS aqueous solution for 1 min. Subsequently, the electrode was taken out from SDS solution and the solvent was left to evaporate under ambient conditions. 2 ', 3, 4 ', 5, 7- pentahydroxyflavone (morin) was used as the complexing ligand to determine Mo(VI). Compared to the poor response at the conventional CPE, the electrooxidation of Mo(VI)–morin complex at SDS/CPE was greatly improved via the significant enhancement of the peak current and the negative shift of the peak potentials. The hydrophobic and electrostatic interactions between morin molecules and SDS resulted in higher accumulation efficiency on the electrode surface thus facilitating the electron transfer between the electrode and the solution. In other words, the enhancement effect of SDS was mainly due to a 'synergistic adsorption' mechanism [90]. Furthermore, the influence of various surfactants on the oxidation of Mo(VI)–morin complex, such as anionic SDS, cationic CTAB and neutral Triton X-100, were investigated. Although all the three surfactants show an apparent enhancement effect on the oxidation of morin and Mo(VI)–morin complex, the highest oxidation peak current of Mo(VI)–morin complex occurs in the case of SDS instead of CTAB and Triton X-100. The difference in structure and the corresponding hydrophobic and electrostatic interactions with Mo(VI)–morin complex for various surfactants might explain their different enhancement effects. The sensitive response of Mo(VI)–morin complex at SDS/CPE and its successful application in the determination of trace Mo(VI) in real samples proved that this method is reliable sensitive, simple and cost effective [90].

Moreover, Ivan Švancara used CPE for the determination of three heavy platinum metals in the form of Pt(IV), Ir(III) and Os(IV) in the presence of cationic surfactants of the quaternary ammonium salt type using cathodic stripping differential pulse voltammetric mode. Platinum metals can be pre-concentrated in chloride-containing supporting media via $PtCl_6^{2-}$, $IrCl_6^{3-}$ and $OsCl_6^{2-}$ complex anions. The determination of platinum metals benefits from a "double" effect of the modifiers chosen; 1) Cationic surfactants of the $\dot{R}R_3N^+$ type (i.e. CTAB and 1-ethoxycarbonylpentadecyltrimethylammonium bromide (Septonex®)) or related compounds (cetylpyridinium bromide CPB) exhibit a strong ion-pairing affinity; 2) They are highly lipophilic in nature thus enabling to be anchored firmly onto the hydrophobic CPE. As a result, a very efficient pre-concentration of the respective anion with either tetravalent or trivalent platinum metal complexes was achieved. Moreover, to determine the effectiveness of the cationic surfactants toward Pt metals, three related quaternary ammonium salts were selected; CTAB, CPB, and Septonex® and tested. All of them exhibited the best performance at concentrations of about 1×10^{-5} mol l^{-1}. With respect to their selectivity towards the individual anions, Septonex® and CTAB were effective for $PtCl_6^{2-}$ and $IrCl_6^{3-}$, whereas $OsCl_6^{2-}$ interacted predominantly with Septonex®. On the other hand, CPB showed rather unsatisfactory performance in a majority of measurements due to some sterric effect of the pyridinium structure. The method elaborated was tested on real samples of industrial waste water showing satisfactory analytical performance [91].

In addition, Ivan Svancara described a procedure for the determination of chromium based on synergistic pre-concentration of the chromate anion at a CPE modified in situ with quarternary ammonium salts such as Septonex®, CTAB or CPB. The proper electrochemical detection utilizes the reduction Cr(VI) → Cr(III) performed in the differential pulse cathodic voltammetric mode. Considerable attention was paid to the accumulation mechanism at CPE in the presence of surfactants [92].

On the other hand, Kurt Kalcher described a method for the voltammetric determination of titanium(IV) using CPE modified in situ with CTAB. The cationic micellar surfactant adsorbed onto the electrode particularly at negative potentials facilitated the preconcentration of titanium(IV) as the oxalate complex simultaneously with reduction to titanium(III) [93].

4.4.1.6. Determination of some Neurotransmitters and Drugs at Different Modified Electrodes in Presence of Surfactants

It is well documented that the modification of the electrode surface by surfactant enhances the electron transfer rate between the electrode surface and analyte and also improves the detection limits [94]. On the other hand, there is a synergistic effect between the conducting substrate (conducting polymer, carbon paste, gold nanoparticles, self-assembly monolayer, etc.) and surfactant which enhances the use of surfactant modified electrodes as nanosensors with excellent reproducibility, high sensitivity, unique selectivity, and exceptional stability. Furthermore, the use of surfactants can be applied for the analysis of drugs with a direct analytical procedure in aqueous, drug formulations, and urine samples. This section will show the synergestic effect between the different conducting substrates and surfactants in details illustrating the different sensory applications of these modified electrodes.

4.4.1.6.1. Self-assembly Monolayer Modified Gold Nanoparticles Modified Electrodes

Atta et al constructed a novel electrochemical sensor (Au/Au$_{nano}$-Cys...SDS) via the formation of self-assembly monolayer (SAM) of cysteine on gold-nanoparticles modified gold electrode (Au/Au$_{nano}$-Cys) for the determination of dopamine (DA) [95] and epinephrine (EP) [18] in the presence of sodium dodecyl sulfate (SDS). Fig. 4.6 illustrates the schematic model of the proposed sensor. There is an electrostatic attraction between the cationic DA or EP and the anionic SDS which enhances the diffusion of the analyte through the positively charged cysteine SAM. Also, there is an interaction between the positively charged cysteine and anionic SDS which allows the reorganization of cysteine molecules on gold nanoparticles that results in the enhancement of the hydrogen bond formation between the cationic analyte and cysteine. As a result, increase of the oxidation current, faster electron transfer kinetics, decrease in the peak separation and hence better reversibility were observed [18, 95].

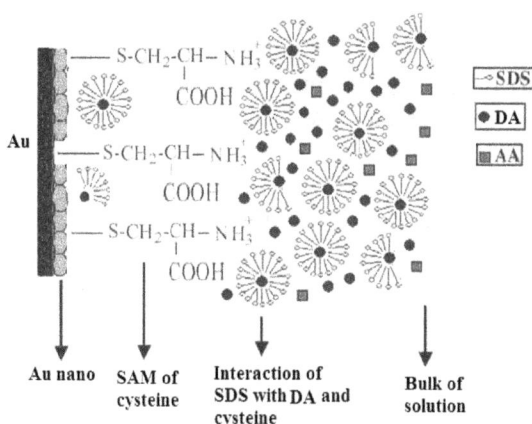

Fig. 4.6. Schematic model of Au/Au$_{nano}$-Cys...SDS modified electrode in presence of DA cations and AA [95].

Moreover, diffusion coefficient (D_{app}) values for DA and EP were 5.00×10^{-6} and 1.24×10^{-5} cm^2 s^{-1} at Au/Au$_{nano}$-Cys and 1.06×10^{-5} and 3.86×10^{-5} cm^2 s^{-1} at Au/Au$_{nano}$-Cys...SDS, respectively. The anionic surfactant SDS affects remarkably the diffusion component of the charge transfer at Au/Au$_{nano}$-Cys as indicated by the D_{app} values. The diffusion coefficient can be considered as an average value of the diffusion process in the bulk, within the surfactant aggregates in solution and the surfactant layer adsorbed at the surface of the electrode. The values of D_{app} show that the diffusion of DA and EP on Au/Au$_{nano}$-Cys is enhanced in presence of SDS rather than in absence of it [18, 95].

On the other hand, a highly selective and simultaneous determination of tertiary mixture of ascorbic acid (AA), DA [95] or EP [18], and acetaminophen (APAP) was explored at this modified electrode. Fig. 4.7 shows the DPVs of tertiary mixture of 1 mmol L^{-1} AA, 1 mmol L^{-1} DA and 1 mmol L^{-1} APAP in 0.1 mol L^{-1} PBS/pH 2.58 at Au/Au$_{nano}$-Cys with successive additions of 0–40 μL of 0.1 mol L^{-1} SDS. The inset of Fig. 4.7 shows the DPVs of the mixture in the absence and presence of SDS. Three well-defined oxidation peaks were obtained at Au/Au$_{nano}$-Cys at 300, 468, and 648 mV for AA, DA and APAP, respectively. Thus, Au/Au$_{nano}$-Cys can be used for the simultaneous determination of AA, DA or EP, and APAP in their mixture. By the successive additions of 10 μL of SDS in the mixture solution, the oxidation peak current of DA and APAP increases while the oxidation current of AA decreases until eventual complete disappearance. The addition of SDS will cause a formation of an amphiphilic surfactant film over Au/Au$_{nano}$-Cys. This film will align in a way where the head groups of surfactant molecules face the aqueous medium leaving the hydrophobic part in contact with each other and away from the aqueous medium. The negative charge of the adsorbed surfactant film as well as the hydrophobic character of the interior of this film will act to repel hydrophilic AA molecules or its AA$^-$ away from the electrode surface while enhancing the preconcentration/accumulation of hydrophobic cations of DA and APAP. In other words, the negatively charged SDS adsorbed onto the electrode surface controls the electrode reactions of AA, DA or EP, and APAP that differ in their net charge [18, 95].

Moreover, Au/Au$_{nano}$-Cys...SDS can be used for the simultaneous determination of binary mixture of EP and APAP prepared in PBS/pH 7.40 (inset of Fig. 4.8). Two resolved peaks at 222 mV and 431 mV were obtained at Au/Au$_{nano}$-Cys for EP and APAP, respectively. But, in presence of SDS, there is an observable increase in the oxidation peak current of EP due to the electrostatic interaction of the anionic surfactant with the protonated EP (pK$_a$ = 8.55), and a decrease in the oxidation current of APAP. The decrease in the anodic peak current of APAP may be due to its structure in which it behaves neutral in the pH of study (pK$_a$ = 9.5). As a result, its diffusion towards Au/Au$_{nano}$-Cys is slow in comparison with other cationic compounds and its interaction with the anionic SDS is retarded [18]. Furthermore, Au/Au$_{nano}$-Cys...SDS can selectively determine EP in the coexistence of a large amount of uric acid (UA) and glucose prepared in PBS/pH 7.40 (Fig. 4.8). There is no interference could be observed from glucose at the modified electrode in absence or in presence of SDS. At Au/Au$_{nano}$-Cys, two well separated oxidation peaks were obtained at 212 mV and 418 mV for EP and UA, respectively. By the addition of SDS, the oxidation peak current for EP increased due to its electrostatic interaction with the anionic surfactant, while the oxidation peak current for UA decreased due to its electrostatic repulsion with anionic surfactant SDS as UA (pK$_a$ = 5.4) is in the anionic form. Therefore, EP can be determined selectively in the presence of UA and glucose. Moreover, very low detection limit of 0.294 nmol L^{-1} and quantification limit of 0.981 nmol L^{-1} were obtained for EP at Au/Au$_{nano}$-Cys...SDS [18].

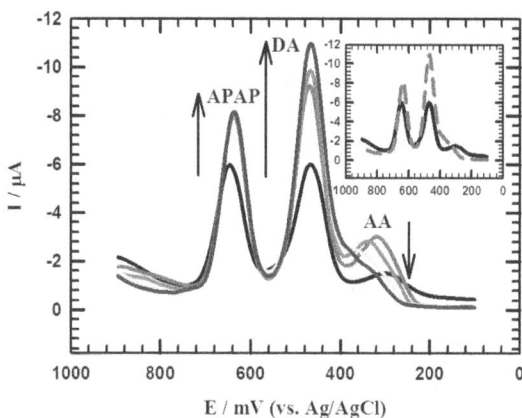

Fig. 4.7. Differential pulse voltammograms for 1 mmol L^{-1} AA, 1 mmol L^{-1} DA and 1 mmol L^{-1} APAP in PBS (0.1 mol L^{-1}) at Au/Au$_{nano}$-Cys with successive additions of (0–40 µL) of 0.1 mol L^{-1} SDS at pH 2.58, the inset represents the initial (in absence of SDS) and final (in presence of 40 µL SDS) DPVs [95].

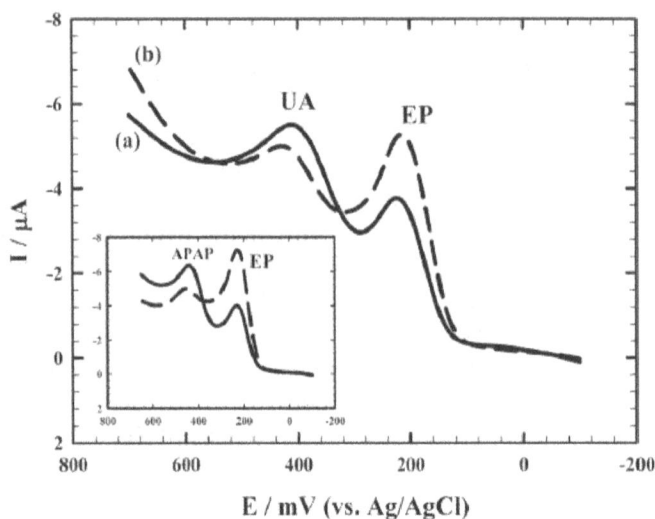

Fig. 4.8. Linear Sweep voltammograms (LSVs) of 0.5 mmol L^{-1} EP, 1 mmol L^{-1} UA in presence of 5 mmol L^{-1} glucose in 0.1 mol L^{-1} PBS/pH 7.40 at (a) Au/Au$_{nano}$-Cys, and (b) Au/Au$_{nano}$-Cys...SDS, inset; LSVs of 1 mmol L^{-1} EP, 1 mmol L^{-1} APAP/0.1 mol L^{-1} PBS/pH 7.40 at Au/Au$_{nano}$-Cys (solid line) and Au/Au$_{nano}$-Cys...SDS (dash line), scan rate 50 mV s^{-1} [18].

Furthermore, the surfactant film on the modified surface can be observed by the scanning electron microscopy (SEM). Fig. 4.9 (A) shows gold nanoparticles that are homogenously distributed and located at different elevations exhibiting a large surface area. Fig. 4.9 (B) shows gold nanoparticles modified with cysteine SAM and they are randomly distributed on the surface, have dendritic shape and sizes that are not homogenous. On the other hand, gold nanoparticles modified with cysteine SAM and further modified with SDS (Fig. 4.9 (C)) are homogenously distributed on the surface, better dispersed and highly packed. The interaction between the anionic SDS and cationic cysteine SAM allows the reorganization and redispersion of gold nanoparticles. Also, a spongy film is observed in Fig. 4.9 (C) due to the surfactant film on the surface which influences the conductivity level of the film and helps the attraction of the analytes to the electrode surface [18].

In conclusion, the synergistic effect between cysteine SAM modified gold nanoparticles and surfactant enhances the use of surfactant modified electrodes as nanosensors with excellent reproducibility, high sensitivity and unique selectivity.

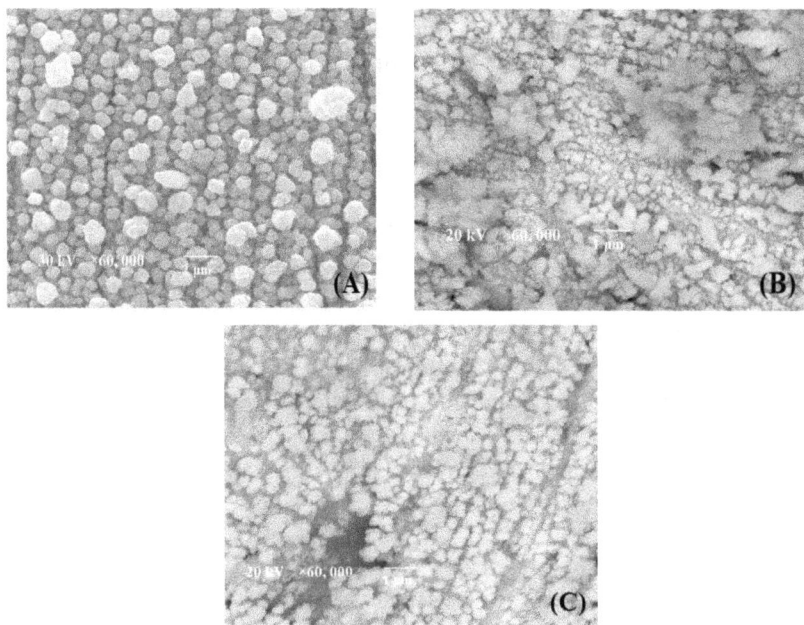

Fig. 4.9. SEM images of: (A) Au/Au$_{nano}$; (B) Au/Au$_{nano}$-Cys, and (C) Au/Au$_{nano}$-Cys...SDS electrodes [18].

4.4.1.6.2. Conducting Polymer Modified Electrodes

Atta et al fabricated a novel biosensor based on poly(3,4-ethylene-dioxythiophene) modified Pt electrode (Pt/PEDOT) for determination of DA in presence of SDS. The electrochemical response of DA was improved by SDS due to the electrostatic interactions of protonated DA with the negatively charged SDS resulting in the enhanced accumulation of protonated DA at the polymeric electrode surface and thus increasing the anodic peak current by one and half folds. The suggested mechanisms for the aggregation of surfactants on the modified electrode surface in the form of bilayers, cylinder, or surface micelles (in the case of relatively higher concentrations added of SDS) could explain the increase in the current in the presence of surfactants. The electron transfer process will occur when the electroactive species approaches the vicinity of the electrode surface. The facilitation of the charge transfer is due to the space of one to two head groups of adsorbed surfactant moieties that is extended from the electrode surface. Another possible mechanism suggested the formation of ion-pair of the charged surfactant and DA that tended to adhere to the modified surface through the lipophilic parts in both moieties [19]. Moreover, PEDOT/Pt in presence of SDS was employed for the selective and sensitive determination of DA in presence of AA and UA. The common overlapped oxidation peaks of AA, UA and DA can be resolved by using SDS as the DA current signal increases while the corresponding signals for AA and UA are quenched. The adsorption of the anionic surfactant SDS onto electrode surface may result in a negatively charged hydrophilic film with the polar head group points to the bulk of the solution. The micellar effect on the oxidation of DA is basically an electrostatic interaction between the surfactant film adsorbed on the electrode and the protonated DA which allows DA to reach the electrode vicinity faster. On the other hand, there is an electrostatic repulsion between the anionic SDS film and anionic species (AA and UA) at electrode surface thus decreasing their electron transfer rate [19].

Furthermore, PEDOT/Pt in presence of SDS was utilized for the determination of catecholamine compounds namely; epinephrine, L-norepinephrine, and L-DOPA, as well as serotonin (ST). The electrochemical data for the oxidation of catecholamines, serotonin, tryptophan, acetaminophen, and some interfering compounds such as UA and AA were collected from the cyclic voltammograms at PEDOT/Pt electrode in the absence and presence of 150 μL SDS. For cationic catecholamines, an increase in the anodic and cathodic peak current values was observed upon the addition of SDS due to the electrostatic interaction between the anionic surfactant film adsorbed on the electrode and the cationic catecholamine. For anionic compounds, the oxidation current response decreases in the presence of SDS. Tryptophan, AA and UA, which are in the anionic form at pH = 7.4, establish an electrostatic repulsion with anionic surfactant SDS resulting in a large decrease in the peak current value in micellar medium. L-DOPA and ACOP are neutral species at physiological pH (pH = 7.4) thus SDS would not affect the kinetic of these compounds. The change of the peak current values for cationic and anionic compounds after adding SDS can be assigned to the adsorption of the surfactant onto the electrode surface which may alter the overpotential of the electrode and influence the electron transfer rate. The formation of micellar aggregates may also influence the mass transport of electroactive species to the electrode [96].

On the other hand, the electrochemical response of ST in the presence of interference molecules such as UA, AA and glucose was investigated. The presence of more than 1000-fold excess of AA and 100-fold excess of glucose did not interfere with the response of ST. The presence of SDS in the medium plays a key role in the electrostatic attraction of ST toward the polymeric surface and causes repulsion toward the interfering species [96].

Furthermore, both DA and ST coexist in a biological system and they influence each other in their respective releasing. So, it was interesting to study the interaction of both compounds with SDS. At PEDOT/Pt electrode, DA and ST yielded two well-defined oxidation peaks at 0.20 and 0.35 V, respectively. The current response of ST increased while DA response decreased with successive additions of 150 μL SDS in 0.1 mol L^{-1} B-R pH = 7.4. This is attributed to the competitive interaction of DA and ST with the PEDOT/Pt film that is more pronounced in the case of ST. This is due to the large conjugated structure of ST which has the possibility to intercalate into the interior of PEDOT/Pt film. Another reason is the presence of $-NH_2$ and $-NH$ groups in ST which increases the positive charge density on ST enhancing its interaction with the anionic surfactant SDS thus facilitating its diffusion to the polymeric film [96].

On the other hand, drug analysis is an important branch of chemistry which plays an important role in drug quality control. Therefore, the development of sensitive, simple, rapid and reliable method for the determination of active ingredient is very important [97]. Surfactants have proven effective to be used in different occasions for the electroanalysis of drugs [85-87]. Improvement of drug analysis using surfactants was found to be concentrated in two important points:

1) The ability of surfactant systems to dissolve hydrophobic (insoluble) drugs.

2) The preferential accumulation of drug molecules on the electrode surface via electrostatic and hydrophobic interactions [98-103].

Atta et al employed PEDOT/Pt in presence of SDS for the determination of morphine (MO) which is often used to relieve severe pain in patients, especially those undergoing a surgical procedure. The anionic surfactant SDS facilitates reaching of MO to the electrode surface faster, improves its reaction rate and enhances greatly its anodic peak current. The lower oxidation potential and higher current response clearly indicate that PEDOT/Pt in presence of SDS has excellent electrocatalytic activity towards MO which is attributed to the presence of anionic SDS [104].

Moreover, DA usually interferes with MO analysis in urine or blood. By using PEDOT/Pt in presence of SDS, two well defined oxidation peaks were obtained for DA and MO at +0.22 V and +0.41 V, respectively. This illustrates that it is possible to discriminate MO from DA with good peak potential separation (ca. 190 mV) and relatively high oxidation current values. Also, the selective and simultaneous determination of MO and EP proved excellent. Thus, using the proposed biosensor for the selective determination of MO in presence of neurotransmitters is possible with high sensitivity and reproducibility. In addition, the electrochemical behavior of MO, UA and AA in a mixture solution was investigated. In presence of SDS, the oxidation peak current of MO increased, while the oxidation peaks of UA and AA disappeared. An electrostatic interaction of the anionic surfactant with the protonated MO in pH 7.4 was observed resulting in the high response for MO, but in case of anionic AA and UA, electrostatic repulsion takes place. Therefore, it is possible to determine MO selectively in presence of high concentration of AA and UA [104].

As well as, Atta et al utilized PEDOT/Pt electrode in presence of SDS for the selective determination of MO and codeine which is very similar to MO in structure and usually interferes with MO analysis in urine or blood. The presence of codeine may affect the detection of MO due to its competitive adsorption. Voltammetric response at the PEDOT/Pt electrode for solution containing equimolar concentration of MO and codeine was examined in presence of successive additions of 100 µL SDS in 0.1 mol L^{-1} B-R/pH 7.4. There is no interference from codeine was observed and an oxidation peak at 0.55 V for MO was obtained. Thus, PEDOT/Pt in presence of SDS can detect MO with no interference from the coexisting codeine. Since MO was the major component in opium poppy, therefore SDS can be used for the determination of MO in opium poppy in presence of codeine. It was possible to determine MO in presence of 10 fold concentration of codeine [104].

On the other hand, electrochemical impedance spectroscopy (EIS) was used to investigate the nature of MO interaction at PEDOT/Pt surface in presence and absence of SDS as EIS is a useful tool for studying the interface properties of surface-modified electrodes. In EIS, the semicircle diameter in the Nyquist plot represents the electron transfer resistance. Fig. 4.10 (A) shows the Nyquist plot of MO at PEDOT/Pt in the presence (a) and absence of SDS (b) at oxidation potential 0.42 V. The impedance responses of MO showed great difference after addition of SDS. In absence of SDS, the impedance spectra include a semicircle with a larger diameter. After addition of SDS, the diameter of semicircle diminishes markedly, the charge transfer resistance of electro-oxidation of MO decreases greatly, and the charge transfer rate is enhanced. Fig. 4.10 (B) represents the circuit used in the fitting where R_u is the solution resistance, R_p is the polarization resistance, CPE represents the predominant diffusion influence on the charge transfer process, and n is its corresponding exponent (n<1). C_f represents the capacitance of the double layer. Diffusion can create impedance known as the Warburg impedance (W).

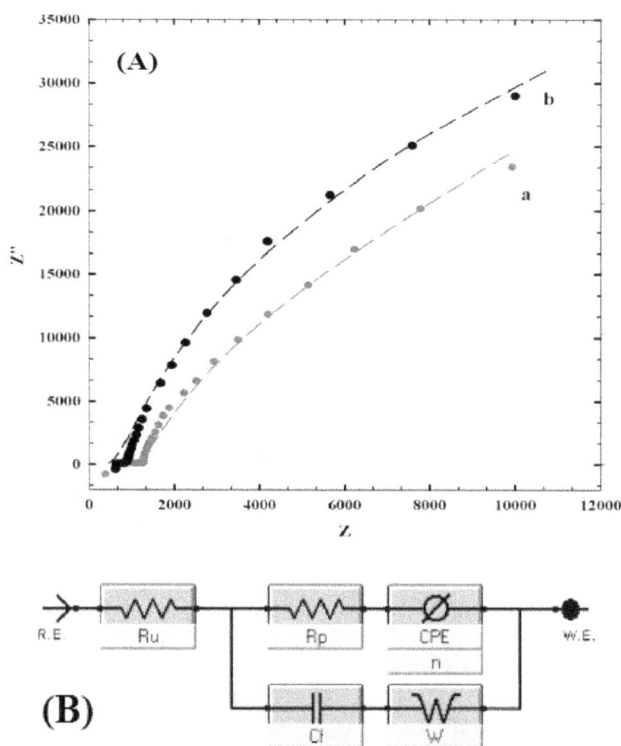

Fig. 4.10. (A) Nyquist diagrams (-Z$''$ vs. Z$'$) for the EIS measurements at PEDOT/ Pt at potential 420 mV for 5 mmol L^{-1} MO in 0.1 mol L^{-1} B-R pH 7.4, (a) in presence of SDS and (b) in absence of SDS. Amplitude: 5 mV, frequency range: 0.1–100 000 Hz, (B) The equivalent circuit [104].

167

Table 4.1 lists the fitting values calculated from the equivalent circuit for the impedance data. PEDOT/ Pt in presence of SDS showed increased values of the capacitive component compared to the case of absence of SDS due to the more conducting character of the surface regarding to ionic adsorption at the electrode surface and the charge transfer process. Also, the decrease in the interfacial electron transfer resistance is attributed to the selective interaction between SDS and MO which accelerates the electron transfer between the electrode and MO [104].

Table 4.1. Summary of the data obtained from EIS in the determination of MO using PEDOT/Pt electrode in absence and presence of SDS at the oxidation potential [104].

Electrode PEDOT/Pt	E (mV)	R_p (kΩ cm²)	R_u (kΩ cm²)	C_f (µF cm⁻²)	W (KΩ⁻¹ cm⁻²)	C_{CPE} (µF cm⁻²)	n
In absence of SDS	420	122	0.39	45	2.49	75	0.88
In presence of SDS	420	52	0.5	50	2.38	279.8	0.7

Also, Atta et al studied the voltammetric behavior of isoniazid (an antituberculous drug, INH) at PEDOT/Pt in the presence and absence of SDS and CTAB. At bare Pt electrode, INH showed a very poor electrochemical response which could be greatly enhanced at PEDOT/Pt in presence of SDS enabling a sensitive electrochemical determination of INH in pH 2.3 (INH is protonated in this pH). At PEDOT/Pt in presence of SDS, two well-defined irreversible anodic peaks of INH at +0.63 V and +0.82 V are observed. These anodic peaks are attributed to the irreversible oxidation reaction of INH. PEDOT/Pt did not show any voltammetric peaks for INH at this pH value due to the electrostatic repulsion between the PEDOT which has positive charge density and the cationic INH. The anionic SDS enhances greatly the anodic peak current of INH, facilitates reaching of cationic INH to the electrode surface faster, thus the reaction becomes easier. The lower oxidation potential and higher current response clearly indicate that PEDOT/Pt has excellent electrocatalytic activity towards INH in the presence of anionic SDS [105]. On the other hand, upon the addition of the cationic CTAB to INH solution/pH 2.3, a weak broad peak at PEDOT/Pt is observed which is attributed to the oxidation of INH and the anodic peak current decreases by the addition of CTAB. This is attributed to the electrostatic repulsion between the positively charged hydrophilic CTAB film and the cationic INH repelling the INH molecules from the electrode surface and hence retarding the diffusion process.

Moreover, the electrochemical response of INH at pH 7.4 was studied at PEDOT/Pt in presence of SDS and CTAB. At pH 7.4, the positive charge density of INH decreases according to its pK_a and becomes neutral, which affects its interaction with the polymeric layer and the surfactant molecules. The diffusion rate of INH towards the polymeric layer changed remarkably. One anodic peak of INH at +0.26 V is observed at PEDOT/Pt with anodic peak current 38 μA which did not appear in pH 2.3. In this case, INH is neutral and its diffusion towards the polymeric cationic film becomes easier. By successive additions of CTAB, the anodic peak current increases to 55 μA then becomes almost stable. The cationic CTAB enhances greatly the anodic peak current of INH at pH 7.4 due to the electrostatic force between CTAB and INH facilitating reaching of INH to the electrode surface faster, hence the reaction becomes easier. On the other hand, the opposite behavior was obtained in case of SDS. By addition of the anionic SDS, the anodic peak current of INH decreases due to the weaker electrostatic force between SDS and INH thus, the diffusion to the electrode surface becomes difficult and depends on the hydrophobic interaction. The adsorption of the surfactant molecules on the electrode surface that could be followed by the formation of micelle aggregates as the distance from the electrode surface increases. The possibility of aggregation of the drug with SDS in this pH can only be attributed to hydrophobic interactions and leads to reduced aggregation as compared to SDS in case of pH 2.3. The strength of interaction and binding between the drug and the surfactant should result in the observed distinct behavior and should also partially affect the transport of their corresponding aggregates in solution [105].

Furthermore, it is important to investigate the electrochemical response of INH in the presence of high concentration of AA and UA as the main interferences in biological environments. Selective determination of INH in presence of high concentration of AA and UA at pH 2.3 was achieved at PEDOT/Pt in presence of SDS. The electrostatic interaction between the anionic SDS film and the protonated INH enhances its response and the electrostatic repulsion between the anionic film and anionic species (AA, UA) at electrode surface decreases their electron transfer. Moreover, the same study was achieved in pH 7.4 in presence of CTAB. Two anodic peaks are observed at +0.1 V and +0.486 V for AA and UA, respectively, due to the electrostatic attraction force between cationic CTAB and the anionic AA and UA. By successive addition of CTAB, anodic peak for INH is observed at +0.7 V and its current increases by increasing the CTAB concentration. This is attributed to the neutral structure of INH which facilitates its diffusion towards the polymeric layer by the cationic CTAB [105].

In conclusion, the presence of surfactant in the medium plays a key role in the electrostatic attraction and repulsion of INH towards the polymeric surface in different pH values.

4.4.1.6.3. Gold Nanoparticles Modified Electrodes

Chunya Li et al utilized gold nanoparticles modified glassy carbon electrode (Au-NPs/GCE) for the determination of tryptophan (Trp) in presence of sodium dodecylbenzene sulfonate (SDBS). The oxidation peak potential of Trp at Au-NPs/GCE showed a negative shift of 50 mV and the peak current was improved in presence of SDBS indicating that the electron transfer between the electrode and the bulk of solution was facilitated. Surfactants can be adsorbed on hydrophobic surface to form surfactant film altering the overvoltage of the electrode and influencing the rate of electron transfer [106, 107]. In presence of SDBS, a hydrophilic film with negative charge was formed through the interaction of Au-NPs and sulfonic group of SDBS. This hydrophilic layer improves the accumulation of Trp at the electrode surface through electrostatic interaction thus the electron transfer of Trp was facilitated, its oxidation overpotential decreased and its current response was enhanced [106]. Moreover, Gong-Jun Yang studied the electrochemistry of ethamsylate (ESL), a homeostatic agent that appears to maintain the stability of capillary walls and correct abnormal platelet adhesion, at hydrophobic gold nanoparticles modified GCE ($C_{18}NH_2$-capped Nano-Au/GCE) in the presence and absence of CTAB. A significant negative shift of the oxidation potential and increase of the anodic peak current were observed in the presence of cationic surfactant CTAB indicating the role of CTAB in the oxidation of ESL (Fig. 4.11). Sulphoacid group in ESL is ionized and negatively charged in the weak acid solution. In presence of CTAB, a positively charged hydrophilic film of CTAB was formed on the modified surface. The hydrophobic interaction between ESL and hydrophobic chain of CTAB was more dominant than the static interaction with the polar head group thus the overvoltage was reduced and the electron transfer rate was enhanced. Therefore, the electrochemical oxidation of ESL was facilitated in the presence of cationic surfactant CTAB [107].

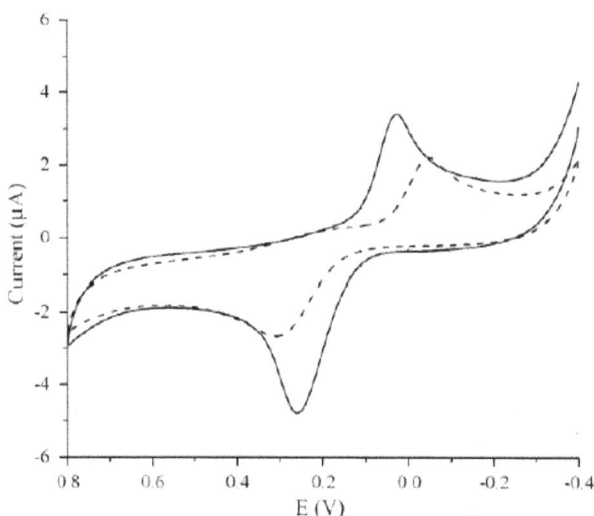

Fig. 4.11. Cyclic voltammograms of 1.0×10^{-4} M ESL in 0.01 M phosphate buffer solution (pH 6.5) at $C_{18}NH_2$-capped Nano-Au/GCE in the absence (dashed line) and in the presence (solid line) of 2.0×10^{-4} M CTAB, scan rate: 100 mV s^{-1} [107].

On the other hand, Atta et al studied the electrochemistry of DA at gold nanoparticles modified gold electrode in the presence of SDS (160 µL). The successive additions of 10 µL of 0.1 mol L^{-1} SDS to 1 mmol L^{-1} DA/0.1 mmol L^{-1} PBS/pH 2.58 leads to decreasing the oxidation current of DA from 5.6 µA in absence of SDS to 4.2 µA in presence of 160 µL of SDS. This indicates that there is an interaction between gold nanoparticles and SDS which leads to decreasing oxidation current of DA. SDS interacts with gold nanoparticles through its hydrophobic carbon chain, blocking the binding sites on gold nanoparticles and acting as an insulating layer rather than enhancing the charge transfer [19, 104]. Thus, the oxidation current of DA decreases in presence of SDS [95]. Moreover, the electrochemistry of DA at gold nanoparticles-coated poly(3,4-ethylenedioxythiophene) modified gold electrode was investigated in presence of SDS (Au/PEDOT-Au$_{nano}$...SDS). The voltammetric response of Au/PEDOT-Au$_{nano}$...SDS is highly selective toward DA compared to Au/PEDOT-Au$_{nano}$. The new composite electrode (Au/PEDOT-Au$_{nano}$...SDS) combines the properties of PEDOT to reduce the oxidation potential with the attractive electrocatalytic properties of gold nanoparticles and SDS to promote a fast electron transfer reaction. The addition of SDS enhances the preconcentration/accumulation of the hydrophobic DA$^+$ cations, enhances the diffusion of DA through the Au/PEDOT-Au$_{nano}$, accelerates the rate of electron transfer due to the electrostatic attraction of positively charged DA (pK$_a$ = 8.92) with the anionic surfactant SDS and improves the DA current signal at Au/PEDOT-Au$_{nano}$. The suggested mechanism for the aggregation of surfactants on the modified electrode surface was discussed before in section 4.4.1.6.2. Au/PEDOT-Au$_{nano}$...SDS explores the synergism between the PEDOT matrix, gold nanoparticles and SDS for the selective and sensitive determination of DA [108].

Furthermore, Au/PEDOT-Au$_{nano}$ electrode can be used for the simultaneous determination of AA, DA and APAP (prepared in 0.1 M PBS/ pH 2.58) in their mixture resulting in three well-defined oxidation peaks at 235, 499 and 678 mV for AA, DA and APAP, respectively. By the successive additions of SDS in the mixture solution, the oxidation peak current of DA increases while the oxidation current of AA decreases till its complete disappearance and the oxidation current of APAP decreases slightly. The negative charge of the adsorbed surfactant film as well as the hydrophobic character of the interior of this film will act to repel hydrophilic AA molecules or its AA$^-$ away from the electrode surface while enhancing the preconcentration/accumulation of the hydrophobic DA$^+$ cations. The slight decrease in the anodic peak current of APAP may be due to its structure in which it behaves neutral in the pH of study thus its diffusion towards Au/PEDOT-Au$_{nano}$ is slow in comparison with other cationic compounds and its interaction with the anionic SDS is retarded. The peak current of DA is enhanced at Au/PEDOT-Au$_{nano}$...SDS compared to that of other electrodes, even in the presence of AA and APAP. This clearly illustrates the synergism between the polymer matrix, gold nanoparticles and SDS. In addition, this modified electrode is used for the first time as a promising electrochemical sensor for DA in presence of AA and UA (prepared in 0.1 M PBS/ pH 7.40). The oxidation peaks are resolved at Au/PEDOT-Au$_{nano}$ with the peak potentials at 349 mV, 217 mV and 19 mV for UA, DA and AA, respectively. The large peak potential separation allows selective and simultaneous determination of UA, DA or AA in their mixture. In presence of SDS, the oxidation peak current for DA was enhanced and the oxidation peak current for both AA and UA were suppressed. The high response for DA was observed due to the electrostatic interaction of the anionic surfactant with the protonated DA (pK_a = 8.92) in pH 7.40, but in case of AA and UA, electrostatic repulsion takes place as AA (pK_a = 4.10) and UA (pK_a = 5.4) are in the anionic form. This confirms that the accumulation of SDS on the modified electrode forms a negative layer on the electrode surface thus DA can be determined selectively in the presence of AA and UA [108].

4.4.1.6.4. Carbon Paste Electrodes

It is well established that the interaction between surfactant aggregates and solutes in the solution phase is controlled by diffusion and takes place in the microsecond time scale. Electrode surfaces with hydrophobic characters such as carbon paste electrodes interact with surfactants through surface adsorption. Thus, carbon paste electrode modified with surfactants was proved to be useful for the determination of both inorganic species and biological compounds [109].

Atta et al studied the electrochemistry of 5×10^{-4} mol L^{-1} isoniazid (INH)/B–R buffer/pH 2.0 at bare CPE using three different types of surfactants; anionic, cationic and neutral surfactant (Fig. 4.12). In absence of surfactant, INH exhibited a well defined irreversible oxidation peak at nearly +970 mV at bare CPE (solid line). Upon the addition of one of the anionic surfactants (SDS) (the dashed line), the anodic current response of INH increased, the electron transfer was enhanced and the anodic peak potential was shifted to a lower value. While upon the addition of the cationic CTAB and neutral albumin surfactants, the anodic current responses decreased and the anodic peak potentials were shifted to higher values. INH is positively charged at low pH values thus upon the addition of SDS or SDBS, electrostatic attraction between the positively charged INH and the negatively charged adsorbed surfactant film as well as the hydrophobic interaction will act in a parallel way for the pre-concentration of INH on or into the adsorbed surfactant film. On the other hand, the oxidation peak current of INH decreased with incremental additions of sodium octyl sulfate (SOS), this may be attributed to the difference in the structure of the surfactants, the short chain length of SOS that contains 8 carbon atoms compared to SDS which contains 12 carbon atoms and SDBS which contains also 12 carbon atoms beside its aromatic ring. Furthermore, the oxidation peak current of INH decreased with incremental additions of the cationic surfactant CTAB and the electron transfer of INH was inhibited. Electrostatic repulsion between the positively charged adsorbed surfactant film and the positively charged INH was observed. Finally, the oxidation peak current of INH decreased with incremental additions of the non-ionic surfactants (NIS) Triton X-405, this may reflect a lowering of the diffusion rate considering the fact that the electroactive species exists away from the surface. In conclusion, it was revealed that aromaticity, charge, and the chain length of surfactants are all complementary factors that depend on each other and can't be neglected when we talk about surfactants [97].

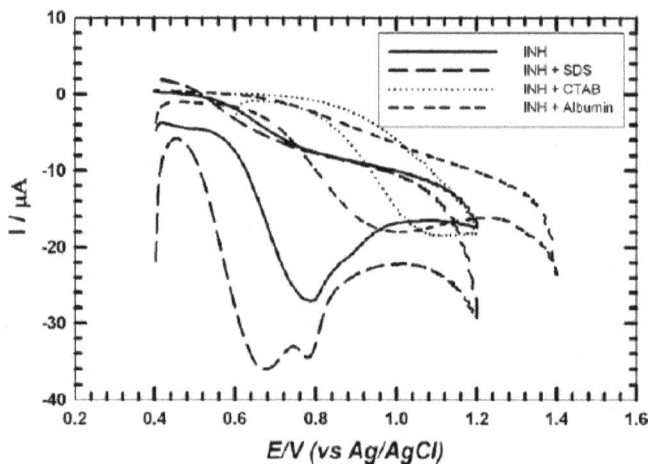

Fig. 4.12. Cyclic voltammograms of 5×10^{-4} mol L^{-1} (INH) at bare CPE (—) (in B–R buffer, pH 2.0) with scan rate 100 mV s^{-1}, using three different types of surfactants (i.e. SDS (– – –), CTAB (- - -) and albumin (. . .)) [97].

Moreover, Ramírez-Silva utilized CPE for the selective electrochemical determination of DA in the presence of AA using SDS micelles as masking agent. At bare CPE, an overlapped peak was observed for a mixture containing DA and AA (0.10 mM DA and 0.28 mM AA/ 0.1 M NaCl, pH 3). In the presence of 3 mM SDS, it is possible to selectively discriminate DA from AA and two resolved peaks were obtained at 450 mV and 688 mV for DA and AA, respectively. Two important features were observed upon the addition of SDS. Firstly; the individual components DA and AA were oxidized at a smaller potential by 143 mV and at a greater potential by 193 mV, respectively when compared with the case without SDS. Thus, SDS has exerted a separating effect where DA undergoes the oxidation process first relative to that of AA indicating that SDS prefers the oxidation of DA.

Secondly; an increase of the DA signal and decrease of the AA signal was observed when compared with the signal representing the sum of the separate signals. This may be attributed to the catalytic effect that appears in the process when DA and AA are in the same solution. In the presence of SDS, DA is oxidized first and the un-oxidized AA is able to reduce the oxidation product of DA thus provoking an increment in the DA signal as compared with the case when AA is absent in the same solution and diminishing that of AA since the amount that could be oxidized is smaller. Thus, SDS micelles promoted the strong adsorption of DA on the CPE and facilitated DA charge transfer reaction [110].

Moreover, Reza Hosseinzadeh studied the effect of cationic surfactant CTAB on DA determination in presence of AA at bare CPE and tin hexacyanoferrate (SnHCF) modified CPE [78]. In presence of 4 mM AA and 2 mM DA, two peaks were observed at 0.425 and 0.620 V vs. SCE, respectively (Fig. 4.13, curve a), that weren't separated clearly at bare CPE. Using the optimum concentration of CTAB (3 mM), AA peak was shifted to left more than DA peak due to the electrostatic effect between related ionic molecules and charged modified electrode (Fig. 4.13, curve b). On the other hand, a significant shift to less positive peak potentials (150 and 200 mV) and increase of peak currents were observed in presence of 3 mM CTAB at SnHCF modified CPE (Fig. 4.13, curves c and d). In presence of CTAB, the electrode acquired positive charge due to the arrangement of the cationic surfactant molecules on the negatively charged electrode via electrostatic interactions. As a result, AA (negatively charged) was adsorbed on the electrode surface but DA (positively charged) was repelled from the electrode surface. Therefore, DA oxidation potential was shifted to higher value and AA oxidation potential was shifted to lower value in presence of CTAB. As well as, the peak currents increased evidently enhancing the measurement ability in low concentrations for both DA and AA. CTAB not only exhibited a strong electrocatalytic activity towards the oxidation of AA but also resolved the overlapping anodic peak of DA and AA into two distinct peaks so that the AA content can be detected selectively in a mixture. In addition, one of the most important advantages of this double-modified electrode is the fact that AA is oxidized first at this surface so that DA concentration was carefully determined. At the surface of other electrodes, DA was oxidized first (due to the own redox potential of these compounds) and in presence of excess amount of AA it can be reduced again thus DA concentration can not be determined carefully because of its redox cycling. In conclusion, DA and AA concentrations can be carefully determined at the surface of the proposed electrode without any cycling interfering of DA–AA [78].

Palomar-Pardavé studied the influence of [SDS] on the electrochemical behavior of DA in aqueous solution. The voltammetric parameters gathered from the CV's recorded in the system CPE/ 0.1 M NaCl, 0.01 M HCl (pH 2.10), 2.5×10^{-4} M DA with different [SDS] were investigated. In absence of SDS, the DA electrochemical oxidation is a quasi-reversible process and mass transfer controlled that is coupled with a slow chemical reaction (EC mechanism). As the [SDS] increases in the system, ΔE drastically diminishes, the ($-i_{pc}/i_{pa}$) ratio tends to unity and the heterogeneous standard rate constant (k^0) drastically increases. When the [SDS] < critical micellar concentration (CMC) (0 – 0.5 mM), the DA electrochemical oxidation changes from a quasi-reversible system to a reversible one. When [SDS] reaches a value of 5 mM, which is higher than the corresponding CMC, the DA electro-oxidation reaction becomes adsorption controlled and the chemical reaction is avoided. The process is adsorption controlled confirmed from the effect of applying different potential scan rates [111].

In addition, Swamy et al developed a sensitive and selective electrochemical method for the electrochemical determination of DA using a hydroxy double salt/surfactant film modified CPE (HDS/SDS/CPE). The HDS/SDS/CPE showed an excellent electrocatalytic activity towards the oxidation of DA in 0.1 M phosphate buffer solution/pH 7.4 and the detection limit of DA was 1×10^{-7} M. Furthermore, the modified electrode exhibited an excellent selectivity towards DA in the presence of excess AA and UA. The separation of the oxidation peak potentials for DA–AA and DA–UA were about 187.8 mV and 110.7 mV, respectively which are large enough for the individual and simultaneous determination of AA, DA and UA [112].

Furthermore, a simple and selective method for the determination of DA using TX-100 modified CPE (TX-100 MCPE) was described. TX-100 MCPE showed excellent electrocatalytic activity towards the oxidation of DA. Interference of AA, UA and 5-HT was tested on the modified electrode achieving selective detection of DA. When AA and 5-HT were added as interferents in the solution containing 0.5×10^{-4} M DA, two additional oxidation peaks appeared at 157 mV and 410 mV for AA and 5-HT, respectively besides the peak of DA. The peak current of DA remained unaffected even when the concentration of AA was increased 10-folds. Moreover, when UA and AA were added as interferents to the solution containing DA, two oxidation peaks of AA and UA appeared at 161 mV and 402 mV, respectively along with the peak of DA. In conclusion, TX-100 MCPE showed very good selectivity, sensitivity and antifouling properties toward DA sensing [113].

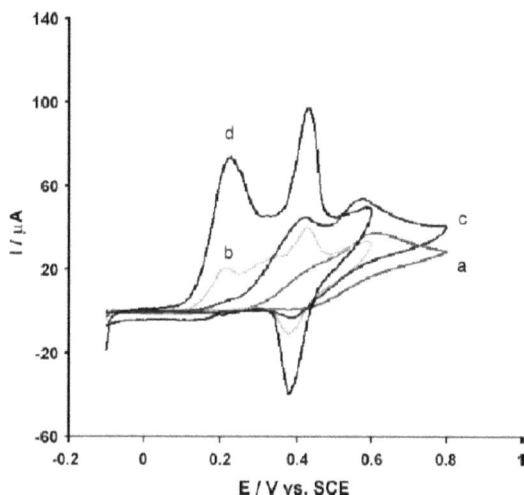

Fig. 4.13. Cyclic voltammograms of CPE: (a) in absence of 3 mM CTAB and presence of 4 mM AA and 2 mM DA in 0.5 M phosphate buffer/pH 7 as supporting electrolyte (b) in presence of CTAB and 4 mM AA and 2 mM DA in buffer solution. Cyclic voltammograms of SnHCF ME: (c) in absence of 3 mM CTAB and presence of 4 mM AA and 2 mM DA in buffer solution (d) in presence of CTAB and 4 mM AA and 2 mM DA in buffer solution [78].

Moreover, Kumara Swamy studied the electrochemical oxidation of DA at murexide modified CPE (Mu-MCPE). The results of comparative study of CPE, Mu-MCPE and TX-100/Mu-MCPE suggested that TX-100/Mu-modified CPE showed excellent electrocatalytic activity for detection of DA [114].

On the other hand, Jianbin Zheng utilized CPE modified with a monolayer film of SDS for detection of DA. Fig. 4.14 showed that CPE modified with SDS (SDS/CPE) exhibited higher sensitivity for DA than SDS/GCE. CPE was more suitable because the paraffin oil in CPE gave a more hydrophobic surface which was in favor of the SDS adsorption on the electrode surface. At SDS/CPE, an improved response was demonstrated suggesting the efficiency of CPE modification by SDS. SDS formed a monolayer on CPE with a high density of negatively charged end directed outside the electrode. As a result, SDS/CPE exerted discrimination against AA in physiological circumstance. Thus, it can selectively determine DA even in the presence of 220-folds AA. Furthermore, the effect of SDS concentration on the reduction (dot curve) and the oxidation (solid curve) peak currents of 4.0×10^{-5} mol L^{-1} DA was shown in Fig. 4.15.

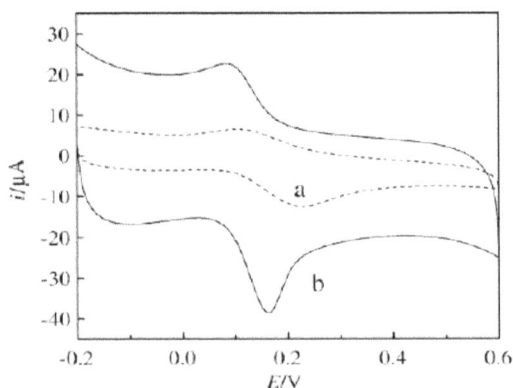

Fig. 4.14. Cyclic voltammograms of DA (0.15 mmol L^{-1}) recorded at SDS/GCE (a), and SDS/CPE (b) in pH 7.40 PBS, scan rate of 50 mV s^{-1} [115].

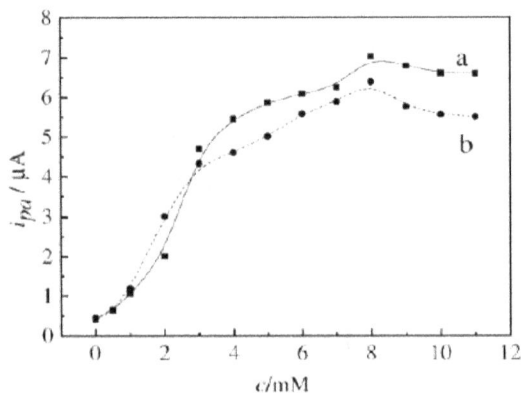

Fig. 4.15. Effects of SDS concentration on the redox peak current of 0.04 mmol L^{-1} DA. The solid curve with squares (a) represents anodic peak (i_{pa}) and dot curve with dots (b) represents cathodic peak current (i_{pc}). Accumulation time: 10 min [115].

178

When SDS concentration is lower than 8.0 mmol L^{-1} (CMC of SDS at room temperature), both i_{pa} and i_{pc} increased rapidly with the increase of SDS concentration. However, the signals of DA apparently decreased in the range of 8.0 – 11.0 mmol L^{-1}. Surfactant molecules formed micelles at high concentration (> CMC), causing the decrease of the available dissociative SDS molecules and the adsorbed SDS on the surface. The accumulation ability of SDS/CPE towards DA is directly proportional to SDS adsorption on electrode surface. Thus, CPE modified by 8.0 mmol L^{-1} SDS is covered by a compact SDS monolayer where the hydrophobic alkyl terminal group in the SDS molecule adsorbed at the hydrophobic electrode surface through van-der-waals interaction and the hydrophilic sulfate terminal group oriented outside. In conclusion, the synergistic effect between CPE and SDS, due to the hydrophobic interactions, was more obvious than that between GCE and SDS [115].

In addition, Swamy et al utilized flake-shaped CuO nanoparticles in the preparation of modified carbon-paste electrodes (MCPE) for the electrochemical detection of DA. Further modification of MCPE was achieved by the formation of a layer of polyglycine on the electrode surface then further modified with a layer of SDS or CTAB. Fig. 4.16 showed the electrochemical responses of 1×10^{-5} M DA in 0.2 M phosphate buffer solution/pH 6.0 at the MCPE prepared with different films of flake-shaped CuO nanoparticles. The MCPE prepared with SDS/polyglycine/flake-shaped CuO nanoparticles exhibited an enhanced current response with sharp redox peaks for DA compared to the MCPEs prepared with CTAB/polyglycine/flake-shaped CuO nanoparticles, polyglycine/flake-shaped CuO nanoparticles and flake-shaped CuO nanoparticles. SDS micelles promoted the strong adsorption of DA on the MCPE and facilitated DA charge transfer reaction due to the electrostatic attraction while there is an electrostatic repulsion between cationic CTAB and DA. Moreover, SDS/polyglycine/flake-shaped CuO nanoparticles exhibited a high selectivity ($E_{AA}-E_{DA}$ = 0.28 V) for the simultaneous investigation of DA and AA [116].

Furthermore, Brahman investigated the enhanced electrochemical response of flutamide, anticancer drug, at polymer modified CPE in presence of CTAB. At polymer modified CPE, a well defined and sharp cathodic peak is obtained and the reduction peak current increases in the presence of 1 % CTAB. The cathodic peak current increases steadily with the increase in CTAB concentration reaching its maximum value at 1 % and after that deceases continuously. It may be interpreted that at 1 % CTAB, the adsorption behavior changes from monomer adsorption to monolayer adsorption with increase in concentration of CTAB at the electrode surface. The decrease of the peak current with further increase in CTAB concentration may be due to the inhibition of electron transfer by micellar aggregates and the increase of hydrophobicity of CTAB micelles leading to decreasing the electron transfer rate constant and the peak current. Therefore, 1 % CTAB is chosen as the optimum concentration. In addition, the modified electrode exhibited enhanced electro-catalytic activity, high sensitivity and stability and is applicable over wide range of concentration for the determination of flutamide [94].

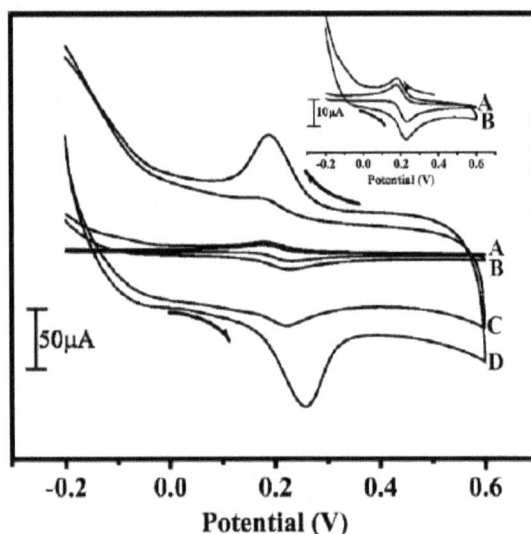

Fig. 4.16. Cyclic voltammograms of 1×10^{-5} M DA at (A) the MCPE prepared with flake-shaped CuO nanoparticles, (B) the MCPE prepared with polyglycine/flake-shaped CuO nanoparticles, (C) the MCPE prepared with CTAB/polyglycine/flake-shaped CuO nanoparticles, and (D) the MCPE prepared with SDS/polyglycine/flake-shaped CuO nanoparticles, scan rate of 0.050 V s^{-1} [116].

On the other hand, Pitre described a new electrochemical method for the determination of atropine at multi-wall carbon nanotube electrode (MWCNTE) based on the enhancement effect of the anionic surfactant SDBS. At MWCNTE, atropine yields a very weak oxidation peak and upon the addition of 0.4×10^{-4} M SDBS, the oxidation peak current of atropine greatly increased. The remarkable peak current enhancement revealed that SDBS facilitated the electron transfer of atropine. SDBS was adsorbed onto MWCNTE surface via hydrophobic interaction between C–H chain and graphite altering the structure and property of the electrode/solution interface. Furthermore, this method has excellent selectivity towards atropine as many other foreign substances, 300–400 fold concentration of AA, ofloxacin, norfloxacin, and gatifloxacin, do not interfere with the current response of 7.96 ng/ml atropine [117].

On the other hand, Srivastava developed a carbon nanotube paste electrode modified in situ with Triton X 100 (TX 100) for the individual and simultaneous determination of acetaminophen (ACOP), aspirin (ASA) and caffeine (CF). It was revealed that the oxidation of ACOP, ASA and CF is facilitated at an in situ surfactant-modified multiwalled carbon nanotube paste electrode (ISSMCNT-PE). On employing TX 100, the peak current for all three molecules increased drastically due to the strong adsorption of TX 100 at the surface of the carbon paste via hydrophobic interactions. Consequently, adsorbed TX 100 induced ACOP, ASA and CF adsorption on the electrode surface. Moreover, ACOP, ASA and CF were solubilized by TX 100 enhancing their residence time near the electrode (i.e., TX-100 forms a thin layer on the electrode surface into which the analyte molecules are preconcentrated). This phenomenon increased the probability of electron transfer between the electrode and all three molecules. As a result, the oxidation peak potentials were shifted to less positive values due to the induced and associated adsorption [118].

Moreover, Yuan showed that nonionic surfactants, such as Triton X-100 and Tween-20, improved the electrocatalytic activity of screen-printed carbon paste electrodes (SPCE). The electrochemical response of SPCE toward H_2O_2 was enhanced by 10 folds with nonionic surfactants modification. In addition, the glucose biosensors fabricated from nonionic surfactant-modified SPCE exhibited 6.4–8.6 folds higher response toward glucose than unmodified SPCE. A concentration effect is proposed for nonionic surfactant to bring neutral reactants to the electrode surface. Interestingly, the nonionic surfactant-modified SPCE exhibited an opposite effect to AA, which causes interference during clinical diagnosis, exhibiting a lower electrochemical response on the nonionic surfactant modified SPCE than on the unmodified SPCE. In other words, the differential responses of nonionic surfactant-modified SPCE toward H_2O_2 and AA suggested its potential in the development of biosensors for clinical diagnosis [119].

Moreover, a variety of single-chain surfactants with different charge properties and tail lengths can spontaneously be adsorbed on the hydrophobic surface of CPE and form stable monolayers on the electrode surface. The resulting surface confined Hb exhibited well-defined direct electron transfer behaviors in all positively, neutrally and negatively charged surfactant films. This is attributed to the hydrophobic interactions between Hb and surfactants which enhanced the adsorption of Hb on surfactant films. When the density of surfactant monolayers was controlled to be the same, Hb was found to possess a better direct electron transfer behavior on monolayers of cationic surfactants with a longer tail length. All the films were catalytically active for the reduction of H_2O_2 and NO [120].

4.4.1.6.5. Glassy Carbon Electrodes

Atta et al studied the effect of adding surface-active agents to electrolytes containing terazosin, an antihypertensive drug, on the voltammetric response of glassy carbon electrode (GCE). Addition of SDS to the terazosin-containing electrolyte was found to enhance the oxidation current signal while CTAB showed an opposite effect [109]. The presence of positive charge on the amino-group of terazosin at pH 5 and its hydrophobic character enhanced the aggregation of the latter with SDS, which possesses negatively charged polar groups. The possibility of aggregation of terazosin with CTAB can only be attributed to hydrophobic interactions leading to reduced aggregation as compared to the SDS case. The strength of interaction and binding between the drug and the surfactant should result in the observed distinct behavior and should also partially affect the transport of their corresponding aggregates in solution. The previous results were confirmed by diffusion coefficient values (D_0 cm^2 s^{-1}), which are 1.61×10^{-7} at GCE, 4.19×10^{-6} in presence of SDS and 9.61×10^{-8} in presence of CTAB. In other words, the diffusion is enhanced in presence of SDS and retarded in presence of CTAB compared to GCE [109]. Moreover, the interaction of anionic SDS or cationic CTAB and terazosin in aqueous buffer solution was followed by UV–vis spectroscopy. Fig. 4.17 (a) shows the effect of different concentrations of SDS on the absorption spectrum of terazosin. Basically, the anionic surfactant SDS showed no absorption background. The anionic character of SDS favors coulombic forces with the drug and should lead to the formation of aggregates in the solution phase. All the bands in the UV and visible regions at ca. 210 nm, 240 nm and 330–340 nm (broad), decreased with each SDS addition. It was mentioned previously that aggregation in aromatic systems could be also attributed to the formation of larger units (possibly due to the formation of longer repeat unit chains). This "oligomerization" was due to the London–Margenau attractive forces between the π-electrons that is counterbalanced by the coulombic and Lenard–Jones repulsive forces. This should be accompanied with a blue-shift or a red-shift in the corresponding spectra that was not observed in the present case for terazosin. This indicates that the charge interaction of the drug with SDS is the main contribution to the association that resulted in the decrease in the absorption spectra. CTAB is a cationic surfactant therefore, coulombic repulsions are expected to be significant and should result in the exclusion of terazosin. Thus, the successive additions of CTAB showed no significant effect on the absorption intensity bands at ca. 210 nm, 240 nm and 330–340 nm (broad). The repulsive coulombic forces between the positively charged amino-group of terazosin and the positively charged ammonium group of CTAB prevent the aggregation of the drug molecules in solution (Fig. 4.17 (b)). Therefore, the only existing attractive forces competing with the repulsive ones are the hydrophobic interactions [109].

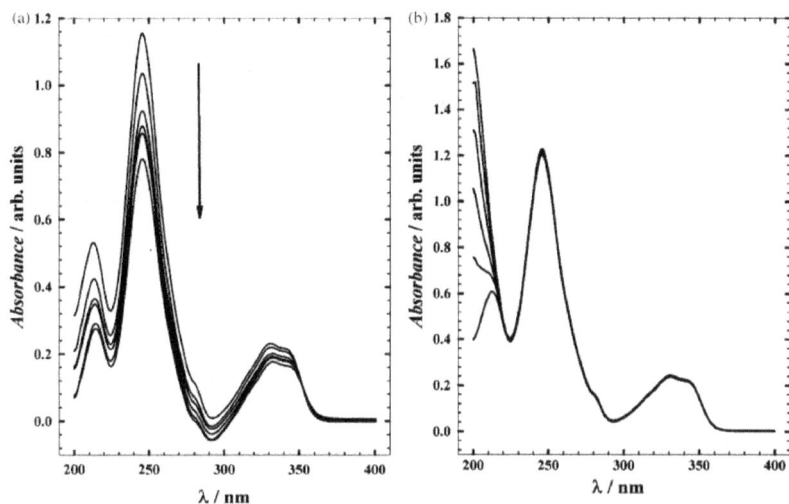

Fig. 4.17. (a) The effect of different concentrations of SDS surfactant on the absorption spectrum of 4.76×10^{-5} mol L^{-1} terazosin dye in aqueous universal buffer, pH 2.0; (b) The effect of different concentrations of CTAB surfactant on the absorption spectrum of 4.76×10^{-5} mol L^{-1} terazosin dye in aqueous universal buffer, pH 2.0 [109].

Furthermore, Mello developed a method for electrochemical detection of AA at GCE modified with a cationic surfactant; cetylpyridinium chloride (CPC). The presence of CPC affected the electrochemical behavior of AA by shifting its oxidation potential to less positive value (220 mV) compared to bare GCE and enhancement of its peak current. This may be attributed to a surface effect in which the adsorption of the surfactant onto the electrode surface resulted in the formation of a positively charged hydrophilic film with the polar head group oriented toward the solution bulk, which interacted with the hydrophilic anionic AA resulting in a great effect on its electron transfer rate. The proposed electrode was very sensitive toward AA and good recovery results were obtained contributing to the applicability of the proposed method for AA determination in complex samples [121].

Moreover, Zhong-Li Liu studied the micellar effect on the electrochemistry of DA and its selective detection in the presence of AA at a GCE in the presence of CTAB and SDS. The anodic peak potential (E_{pa}) and peak current (I_{pa}) were found to be remarkably dependent on the charge and the concentration of the surfactant. The cyclic voltammetric determination of a mixture of DA and AA (aqueous solution, pH 6.8) showed only a single overlapped peak at GCE as the anodic peak potentials of DA and AA are very close. However, taking advantage of the opposite micellar effect on the anodic oxidation of DA and AA, it should be possible to separate the two anodic peaks in ionic micelles. DA is positively charged and AA is negatively charged in neutral and acidic solutions. Thus, the addition of cationic surfactant CTAB to the mixture of DA and AA shifted the anodic peak of DA to positive values and that of AA to negative values, making the two anodic peaks well separated by 400 mV that is more than enough to allow selective detection of the two compounds. As a result, DA can be determined in the presence of 100 times excess of AA. A similar anodic peak separation should also be achieved in SDS micellar solution since the anionic surfactant decreases the anodic peak potential of DA, whilst it increases that of AA. Moreover, the I_{pa} of DA is greatly enhanced due to its catalytic oxidation enabling the quantitative determination of both compounds [122].

In addition, Shen-Ming Chen utilized a cationic surfactant; didodecyldimethylammonium bromide, modified GCE (DDAB/GCE) for simultaneous determination of various combinations of neurotransmitters (DA, norepinephrine and epinephrine) and AA. The DDAB-modified film is positively charged and neurotransmitters are positively charged species in the neutral solution whereas AA is a negatively charged one. Well separated voltammetric peaks (peak potential separation = 300 mV) were observed for DA and AA at DDAB/GCE. The oxidation potential of AA was shifted to a less positive value due to a positive electro-catalytic activity of the DDAB/GCE, while neurotransmitters were oxidized at a more positive potential due to electrostatic repulsion [123].

On the other hand, the electrochemical behavior of betahistine hydrochloride, anti-vertigo drug, at GCE was investigated by Rajeev Jain. Square-wave voltammetric response of betahistine hydrochloride in organic solvents (acetonitrile, DMF and 1,4-dioxane), water and surfactants (a neutral, a cationic and an anionic type) was recorded (Fig. 4.18). Addition of anionic surfactant (sodium lauryl sulfate SLS) to the betahistine hydrochloride solution enhanced the reduction current signal while neutral surfactant (Tween-20) and cationic surfactant CTAB showed an opposite effect [124].

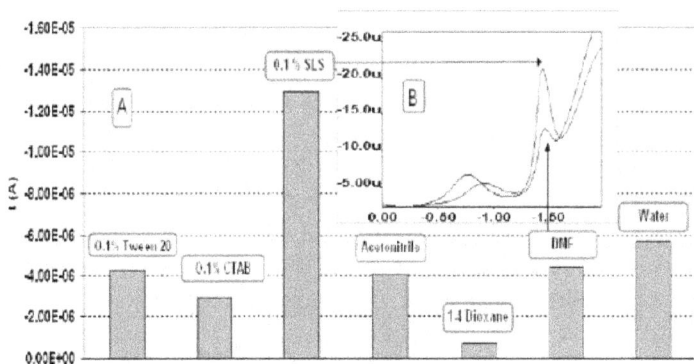

Fig. 4.18. (A) Comparison of cathodic peak current I (A) response of betahistine hydrochloride (700 µg L^{-1}) in different organic solvents (acetonitrile, DMF and 1,4-dioxane), water and surfactants (0.1 % Tween-20, 0.1 % CTAB and 0.1 % SLS), (B) Inset picture represents overlapped square-wave voltammograms of betahistine hydrochloride (µg L^{-1}) in 0.1 % SLS (maximum response in surfactants) and DMF (maximum response in organic solvents). X-axis shows potential U (V) and Y-axis represents cathodic peak current I (A) [124].

On the other hand, Shengshui Hu investigated the effect of CTAB on the electrochemical behavior of thyroxine, important hormone produced in the thyroid gland, at GCE modified with single-walled carbon nanotubes (SWNTs/GCE). At SWNTs/GCE, a well-defined oxidation peak of thyroxine at 0.78 V was obtained but its reduction peak was disappeared. When trace CTAB was added to thyroxine solution, the reduction current could be greatly enhanced and the oxidation current remained stable. The cationic surfactant CTAB may be adsorbed onto the electrode surface through the hydrophobic interaction with SWNTs endowed with strong adsorptive properties and large specific surface area and form a positively charged film. At the same time, CTAB can combine with thyroxine to form a special negatively charged complex by the interaction between the bromide ions in CTAB and iodine ions in thyroxine. This negatively charged complex can be strongly adsorbed to the surface of SWNTs/GCE via the electrostatic interaction with the positively charged CTAB film at the modified electrode surface. Thus, the concentration of thyroxine at the electrode surface is greatly accumulated enhancing the electron transfer rate and increasing the reduction current of thyroxine [125].

Moreover, Chunhai Yang examined the electrochemical response of estrone in the presence of CTAB. It was found that the oxidation peak current of estrone at GCE and congo red (CR) functionalized multi-walled carbon nanotubes (MWNT) modified GCE (MWNT-CR/GCE) remarkably increased in the presence of CTAB indicating the enhancement effect of CTAB towards estrone. CTAB can form a layer at the electrode surface via the hydrophobic interaction showing efficient accumulation towards estrone and improving its electrochemical response. Moreover, CTAB layer effectively improved the anti-fouling capacity of MWNT-CR/GCE via inhibiting the direct adsorption of estrone at the electrode surface and reducing the risk of fouling the active sites. Therefore, combining the excellent properties of MWNT-CR film and the obvious enhancement effect of CTAB improved the determination of estrone greatly [81].

Furthermore, Shengshui Hu investigated the electrochemical response of estradiol at GCE and MWNT-CR/GCE. The electrochemical response of estradiol at MWNT-CR/GCE was greatly enhanced which was further amplified by the addition of trace CTAB in solution along with the enhanced antifouling capacity of the modified electrode. The weak hydrophobic adsorption of CTAB on the hydrophobic and smooth surface of MWNTs was found to be the key for simultaneously improving the sensitivity and antifouling capacity of carbon nanotube-based electrochemical sensors by surfactants. In addition, the influence of surfactant type on the electrochemical response of estradiol at

MWNTs–CR/GCE was examined (Table 4.2). Clearly, the addition of negatively charged SDS and neutral Triton X-100 caused a slight decrease in the oxidation current and a positive shift in the oxidation potential. However, the addition of cationic surfactants generally resulted in enhanced increase in the oxidation current. This is attributed to the accumulation of partially ionized estradiol at the positively charged surfactant layers on the electrode surface in the weak basic electrolyte. Moreover, with the increase of the tail length of cationic surfactants, the oxidation current apparently increased except for STAB [82].

Table 4.2. The potential and current changes of the oxidation peak for 4.0×10^{-6} M estradiol at MWNTs-CR/GCE in the presence of 8.0×10^{-5} M various surfactants [82].

Surfactant	E_p (V)	ΔE_p (V)[a]	I_p (µA)	$\Delta I_p\%$ [a]
No surfactant	0.503	–	6.28	–
Triton X-100	0.539	0.036	5.24	−16.56
SDS	0.550	0.047	6.01	−4.30
LTAB	0.504	0.001	7.93	26.27
TPB	0.517	0.014	8.81	40.29
CTAB	0.507	0.004	13.68	117.83
STAB	0.517	0.014	9.43	50.16

[a] The change of the oxidation potential (ΔE_p) and the change percent of the oxidation current ($\Delta I_p\%$) with the addition of various surfactants were compared with those in the absence of surfactants. SDBS; sodium dodecylbenzene sulfonate, LTAB; lauryltrimethylammonium bromide, TPB; tetradecanepyridinium bromide, STAB; stearyltrimethylammonium bromide.

On the other hand, surfactants are effective in preventing electrode fouling from matrix constituents in the detection of heavy metals and bioorganic analytes. The stabilizing effect of surfactants in the electrochemical detection of phenylenediamine, p-nitrophenol and chlorpromazine, which formed electrode-fouling products, had been investigated illustrating the improved stability of the amperometric signals in the presence of surfactants [126]. Hoyer showed that surfactants are highly effective in stabilizing the voltammetric signal of serotonin (5- hydroxytryptamine, 5-HT) which is prone to severe decay due to the adsorption of its oxidation products at bare electrodes. Electrode fouling by oxidation products of 5-HT resulted in a decrease in its voltammetric signal by 66 % during 21 repeated scans. But, the decrease was reduced to 7 and 10 % in the presence of 5000 mg L^{-1} of the cationic surfactants; cetyltrimethyl ammonium chloride (CTAC) and cetylpyridinium chloride (CPC), respectively, (Fig. 4.19). Addition of only 1 mg L^{-1} of the cationic surfactants CTAC or CPC led to a significant improvement of the stability of 5-HT signal. The stabilization effect increased with the increase in the surfactant concentration. Moreover, CTAC produces a somewhat stronger stabilization effect than CPC for equal concentration of surfactants. On the other hand, the anionic surfactants sodium dodecyl sulfate and sodium decyl sulfonate also stabilized 5-HT signal, but less than the cationic surfactants while non ionic surfactants were least effective. Adsorbed oxidation products of 5-HT on the electrode surface during the first 21 scans are not removed by the surfactants. Thus, the surfactants must be added to the medium prior to any measurements of 5-HT in order to achieve maximum protection of the electrode surface against fouling. The experimental results are interpreted in terms of the surface activity and aggregation behavior of the surfactants [126].

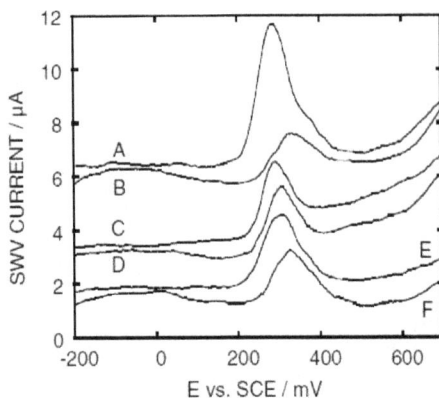

Fig. 4.19. Square wave voltammograms of 5-HT. The first and twenty-first scan recorded in the absence of surfactant (curves A and B) and in the presence of 5000 mg L^{-1} CTAC (C and D) and 5000 mg L^{-1} SDS (E and F) are shown [126].

On the other hand, direct electrochemistry of soluble fragment of the subunit II from cytochrome c oxidase containing the dinuclear Cu$_A$ (Cu$_A$ protein) was carried out on GCE with surfactants of different polarities. The protein showed highly reversible electrochemistry with equal anodic to cathodic peak currents in the case of the cationic surfactant, CTAB as the electrode modifier. The other surfactants such as; the neutral surfactant Triton X-100 and the negative surfactant SDS, showed only slightly better electrochemical response than bare GCE. The ratios of I_{pa} to I_{ca} were found to be <1 and the values of ΔE_p were found to be > 60 mV but <115 mV in the presence of SDS or TX-100 indicating that these surfactants were better than bare GCE but less efficient than CTAB in promoting direct electrochemistry of Cu_A protein. In the case of CTAB, there are two types of possible interactions between Cu_A and the surfactant. One of them is the electrostatic interaction between the positively charged head groups of the surfactant and the negatively charged amino acid residues. The other is the hydrophobic interaction between the surfactants and adjacent hydrophobic patches of the surface of Cu_A. Thus, the micelles in the diffusion layer possibly surround the protein surface in a manner so that the negatively charged surface of the protein that is close to the metal centers come close to the electrode as shown in Fig. 4.20.

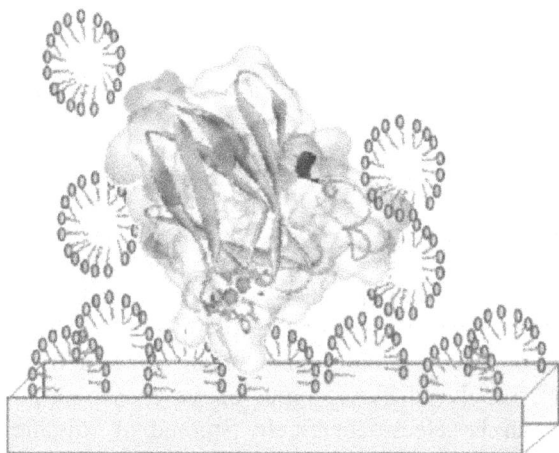

Fig. 4.20. Schematic representation for the possible interactions of the Cu_A with the CTAB micelles and orientation of the protein on the GCE surface [127].

This possibly protects the protein from irreversible adsorption and also helps to bring the protein at the electrode surface in the orientation that is favorable for efficient electron transfer to the metal ion center. In case of SDS modified electrode, the hydrophobicity of the surfactant may avoid irreversible adsorption of the protein, but the electrostatic repulsion between the micellar surface and the negative residues near the active site of Cu_A would not allow favorable electron transfer between the electrode and the protein. The neutral surfactant, TX-100 would only provide hydrophobic interaction but no electrostatic interaction between the electrode modified by the surfactant and the protein [127].

Moreover, modification of GCE with neutral surfactants resulted in an improved response in the direct electrochemistry of horse heart myoglobin and horseradish peroxidase (HRP) as revealed by Shyamalava Mazumdar. These proteins showed very slow electron transfer kinetics at bare GCE. Upon modification of the electrode surface with neutral surfactants, the electrochemical response of myoglobin as well as HRP was significantly improved exhibiting enhanced heterogeneous electron transfer rate. The electrode response was found to depend on the structure of the surfactants. The hydrophobicity of the surfactants rather than their charge was found to be crucial in promoting the electrode response of these proteins at GCE [128].

4.4.1.6.6. Mercury Electrodes

Recently, the enhancement of the oxidation peak currents of certain pharmaceuticals in the presence of surfactants have been reported by different studies [129]. The effect of surfactants on the peak height of poorly soluble drugs can be explained by the fact that; the amphiphilic structure of surfactants and their assembly in aqueous solution provided a multifunctional environment for the solubilization and partitioning of aqueous soluble and insoluble compounds. Micellar assemblies have the ability to dissolve different types of water insoluble redox active probes that can be electrochemically studied at suitable electrodes. Aqueous micellar solutions, which are surfactant based organized systems, can be used as less hazardous and versatile substitutes for organic solvent in voltammetry, HPLC separation and in catalysis [83].

Rajeev Jain studied the voltammetric behavior of ornidazole at dropping mercury electrode (DME) in different surfactant media; anionic, cationic and non-ionic surfactants over the pH range 2.5 to 12.0 in phosphate buffer solution. Ornidazole can be used in the treatment of hepatic and intestinal amoebiasis, giardiasis, trichomoniasis of the urogenital tract and bacterial vaginosis. It was found that addition of non-ionic surfactant (Tween 20) to the ornidazole containing electrolyte leads to maximum increase in the peak current arising from the occurrence of lateral interaction in the adsorbing species. Once the adsorbed ions reach a certain critical concentration at the interface, they begin to associate into two dimensional patches of ions "hemi-micelles". The ornidazole molecules which are essentially non polar observed by its low solubility in water, are attracted to non polar regions of these hemi-micelles, which are oriented towards the electrode surface. Thus, more ornidazole molecules reach to the electrode surface resulting in an increase in the peak current. On the other hand, the addition of the anionic surfactant sodium lauryl sulfate and cationic surfactant CTAB showed a small enhancement in the peak current [83].

On the other hand, the use of surfactants as drug carriers makes necessary the study of interaction of drugs with micellar systems implying the elucidation of the nature of these interactions. Rajeev Jain developed a voltammetric method for the determination of entacapone, used in the treatment of Parkinson's disease, at hanging mercury drop electrode (HMDE) based on the enhancement effect of Tween 20. Addition of neutral surfactant (Tween 20) to the entacapone containing electrolyte enhanced the reduction current signal and the detection limit was very low. On the other hand, anionic surfactant (sodium lauryl sulfate) and cationic surfactant (cetrimide) showed an opposite effect [130]. Moreover, Ferreira studied the voltammetric behavior of 4-methylbenzelidene camphor (MBC) at mercury electrode using square wave voltammetry (SWV). The highest reduction peak current of MBC was obtained in Britton-Robinson buffer and cationic surfactant; CTAB. The use of neutral surfactant; Triton X-100 and anionic surfactant; lauryl sulfate had not assisted in MBC reduction. Only the cationic surfactant CTAB provided the voltammetric reduction of MBC as CTAB is adsorbed onto the mercury electrode surface by hydrophobic and electrostatic attractions at negative potentials. At concentrations lower than CMC, the surfactants act as an ion-pairing agent and monomers are adsorbed onto the electrode surface to a very small extent. At concentrations higher than CMC, a strong adsorption is observed at negative potentials which can lead to an extraordinary enhancement in the sensitivity of voltammetric measurements [131].

On the other hand, Dar investigated the electrochemical behavior of quinine, antipyretic (fever reducing), antimalarial, analgesic and anti-inflammatory drug, at HMDE in presence of surfactants. The reduction peak current of quinine was remarkably improved, quasi-reversible electron transfer reaction and lower detection limit of quinine were found in presence of 1 % CTAB. SDBS and Tween-20 showed the same effect but the peak current was more enhanced in presence of CTAB (Fig. 4.21). The enhancement effect upon addition of surfactants can be rationalized by the adsorption of the surfactant at the electrode surface which may alter the electron transfer rate and by the formation of micellar aggregates which may influence the mass transport of drug to the electrode. The adsorption of CTAB at the electrode surface may form a hydrophilic film on the electrode with its polar head group directed towards the bulk water phase. Quinine molecules which are essentially non polar observed by its low solubility in water, are attracted to non polar regions of these hemi-micelles, which are oriented towards the electrode surface. Thus, more quinine molecules reach the electrode surface enhancing the peak current. Furthermore, the adsorption of the hydrophobic tail on the electrode surface helps to release the reduction product from the electrode surface thus reducing the overvoltage [88].

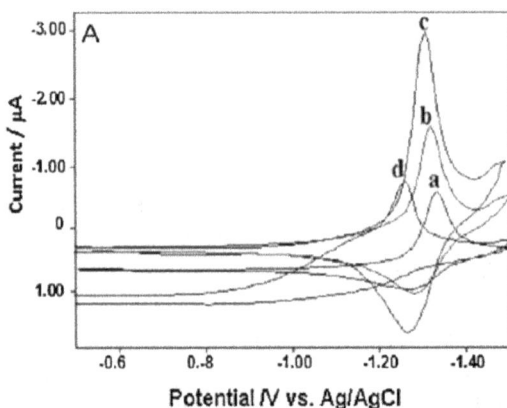

Fig. 4.21. Cyclic voltammograms of 120 ng mL^{-1} quinine at HMDE in different surfactant media and 0.02 M Britton–Robinson buffer (pH 10.38). Scan rate: 500 mV s^{-1}, $t_{accumulation}$ = 60 s. (a) Ethanol, (b) 1 % Tween-20, (c) 1 % CTAB, and (d) 1 % SDBS [88].

Jianbin Zheng studied the voltammetric behavior of 4′,7-dimethoxy-3′-isoflavone sulfonic sodium (DISS), effective scavengers of active oxygen radicals, at a static dropping mercury electrode. DISS gave two waves between pH 8.0 and 12.0. Above pH 8.0, the peak current of first wave P_{c1} of DISS was greatly enhanced in the presence of CTAB. On the other hand, polyvinyl alcohol (PVA) and SDS had little effect on the peak current when the pH value was in the range from 8.0 to 12.0. The adsorption of DISS on the mercury electrode was induced by CTAB which enhanced greatly the peak current of the P_{c1} wave of DISS [132]. Moreover, Jain and co-workers studied the effect of surface active agents on the voltammetric response of nortriptyline hydrochloride at HMDE. Addition of Tween 20 to the nortriptyline hydrochloride containing electrolyte enhanced the reduction current signal and improved the detection limit of the analyte [129].

4.4.1.6.7. Other Conducting Substrates

Surfactants have proved to be effective in the electroanalysis of biological compounds and drugs. It was recently shown that anionic surfactants could be used to improve the accumulation of some electroactive organic molecules such as ethopropazine at gold electrodes. In another study, the influence of micelles in the simultaneous determination of two components was also demonstrated, as in the case of catechol and hydroquinone. Jain and co-workers had studied the effect of changing the charge of the surfactant; anionic, neutral and cationic on the cefdinir peak current; the addition of cationic surfactant enhanced the reduction current signal [129].

The use of surfactant solutions as modifiers can improve the sensitivity and selectivity of the voltammetric measurements. Abdulkadir Levent studied the electrochemical oxidation of nicotine in aqueous as well as micellar media at a pencil graphite electrode (PGE). The compound was oxidized irreversibly at low positive potentials in one (in acidic and neutral media) or two (in alkaline media) oxidation steps. The effect of two ionic surfactants; CTAB and SDS, which were added to the electrolyte solution, on the peak current was investigated. Voltammograms obtained at pH 7.0 using phosphate buffer showed that CTAB did not play a significant role on the electrode process when compared with the unmodified PGE. On the other hand, the electron transfer rate was improved and the sensitivity was clearly enhanced with better defined signal at SDS modified electrode. The sensitivity of SDS modified electrode was about 1.6 times higher than that of bare electrode. In addition, the peak potential was shifted to less positive value by about 80 mV with an increase in surfactant concentration up to 10 mM, indicating that the oxidation of compound becomes easier in presence of micellar system. SDS forms a negatively charged hydrophilic film onto the electrode surface and oriented towards the water bulk phase. Based on this fact, positively charged nicotine has a tendency to accumulate on the negatively charged crown of anionic SDS micelles, which may decrease the overpotential of the electrode and increase the electron transfer rate [133].

Shengshui Hu investigated the electrochemical behavior of adrenaline at acetylene black electrode in presence of SDS (Fig. 4.22).

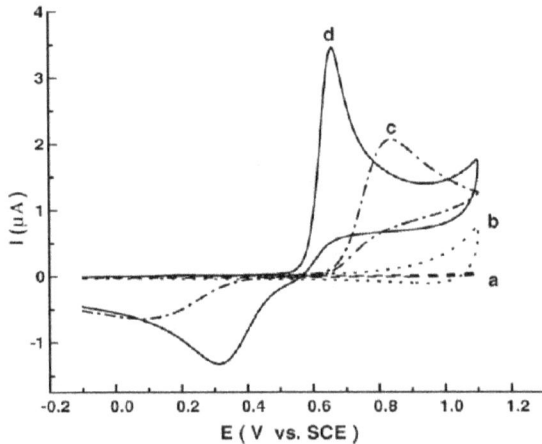

Fig. 4.22. Cyclic voltammorgrams in 0.5 mol/L sulfuric acid at an acetylene black electrode: (a) in the absence of adrenaline and SDS; (b) in the presence of SDS; (c) in the presence of adrenaline; (d) in the presence of adrenaline and SDS. Scan rate: 100 mV/s; accumulation time: 70 s; adrenaline: 1.0×10^{-4} mol/L; SDS: 1.0×10^{-4} mol/L [134].

In absence of SDS, adrenaline exhibited a weak anodic peak at 0.83 V and a broad cathodic peak at 0.10 V (curve c). After the addition of 1.0×10^{-4} mol/L SDS, the anodic and cathodic peak currents of adrenaline are both markedly enhanced. The anodic peak potential was shifted negatively to 0.65 V and the cathodic peak potential was shifted positively to 0.31 V (curve d). The electrochemical response of adrenaline was apparently improved by SDS due to the enhanced accumulation of protonated adrenaline via electrostatic interaction with negatively charged SDS at the hydrophobic electrode surface. Furthermore, the influence of different kinds of surfactants including anionic; SDS and SDBS, neutral Triton X-100 and cationic CTAB on

194

the electrochemical signals of adrenaline were depicted in Table 4.3. The peak currents of adrenaline were almost not improved in the presence of CTAB or Triton X-100. The increase percents of oxidation peak current were 4.17 % and 4.32 %, respectively. The interaction between cationic adrenaline and cationic CTAB or neutral nonionic Triton X-100 resulted in the weak adsorption onto the electrode surface. However, after the addition of anionic surfactants, a pair of greatly enhanced and well-defined peaks of adrenaline was obtained. In addition, the diverse degree in the peak current enhancement reflected by different anionic surfactants is due to their differences in structure, the hydrophobic interactions and the ion-exchange interactions [134].

Table 4.3. The changes of oxidation peak potential and current in the absence and presence of different kinds of surfactants [134].

	ΔE_{pa} (V)	%	ΔI_{pa} (μA)	%
SDS	−0.178	−21.3	1.455	74.2
SDBS	−0.152	−18.3	1.153	67.7
Triton X-100	0.082	9.62	0.071	4.32
CTAB	0.032	3.91	0.064	4.17

Moreover, Shou-Qing Liu fabricated a biosensor using graphene nano-sheets (GNS) as the component for signal amplification and CTAB as the discriminating agent with the aim of achieving selective and sensitive determination of DA in the presence of AA. An overlapped peak for DA and AA in coexisting solution was obtained at graphene nano-sheets paste (GNSP) electrode, while in the presence of CTAB, DA and AA can be simultaneously oxidized at 0.32 V and 0.11 V, respectively. The potential peak separation was 210 mV which is large enough to confirm specificity. Thus, selective determination of DA using the GNSP electrode is possible in an AA-DA mixed solution with the addition of a surfactant that exerts discriminating effects. Micelles were formed by the aggregation of the positively charged CTA on the surface of the negatively charged GNSP electrode. The structure of such micelles is shown in Fig. 4.23 considering the negative charges on the graphene nano-sheets and the positive charges on the hydrophilic heads and hydrophobic tails of the amphiphilic molecules [135].

Moreover, Orawon Chailapakul developed an electrochemical paper-based analytical device (ePAD) for the selective determination of DA in model serum sample. In absence of the anionic surfactant SDS, an overlapped oxidation peak of DA, AA and UA was obtained. With the addition of SDS, DA oxidation peak was shifted to more negative values and was clearly distinguishable from AA and UA. The oxidation potential shift was presumably due to the preferential electrostatic interactions between the cationic DA and the anionic SDS. Indeed, whilst the SDS-modified paper improved the DA current five-folds, the non-ionic Tween-20 and cationic tetradecyl trimethyl ammonium bromide surfactants had no effect or reduced the current, respectively. Furthermore, only the SDS-modified paper showed the selective shift in the oxidation potential of DA. The e-PAD seems suitable as a low cost, easy to-use and portable device for the selective determination of DA in human serum samples [136].

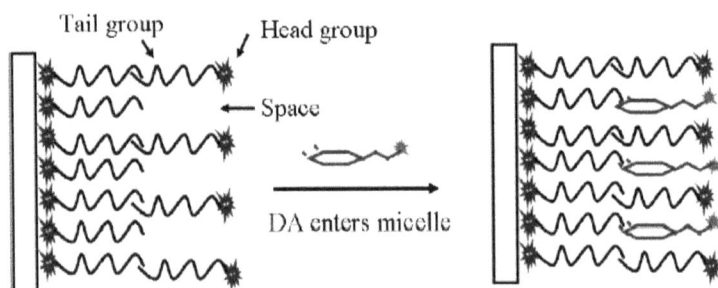

Fig. 4.23. Illustration of DA molecules entering a micelle on the interface between the graphene surface and solution [135].

Through out the previous studies, the effect of surfactants on the electrochemistry of the studied ions can be summarized as follows; in some cases synergistic preconcentration may occur for these ions on the electrode surface, in other cases surfactant may act as an antifouling and homogenizing agent.

4.4.2. Bulk Modification by Surfactant and its Electroanalytical Applications

Unlike surface modification, bulk modification represents the case in which surfactant (modifier) is mixed intimately with the electrode material such as carbon paste.

Shahrokhian modified CPE by incorporating cobalt-5-nitrosalophen (CoNSal), an effective redox mediator, and tetraoctylammonium bromide (TOAB), a cationic surfactant, for the simultaneous determination of AA and DA. The favorable ionic interaction (electrostatic repulsion) between the cationic DA and the cationic surfactant (TOA$^+$) caused a positive shift in the anodic peak potential of DA. By increasing the weight percent of TOAB in the matrix of the modified CPE, the potential peak separation (ΔE_p) for oxidation of AA and DA increased making it suitable for simultaneous voltammetric detection of AA and DA (Fig. 4.24). However, hydrophobicity of TOAB at the surface of the modified electrode showed a great influence on the sensitivity of the voltammetric responses especially for DA. Thus, an optimum 2 wt.% of TOAB together with 3 wt.% of CoNSal was used for construction of the modified electrode and two well-resolved anodic peaks for AA and DA were obtained. The very low detection limit (sub-micromolar), high sensitivity, selectivity, the very easy preparation, surface regeneration of the modified electrode and the reproducibility of the voltammetric response make the proposed modified system very useful in the construction of simple devices for the simultaneous determination of AA and DA in clinical and pharmaceutical preparations [137].

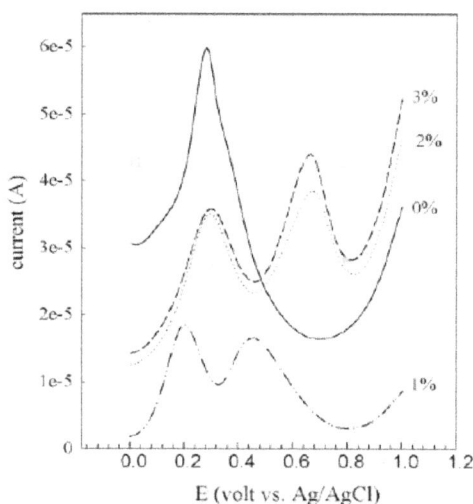

Fig. 4.24. Differential pulse voltammograms for a mixture of 1 mM AA and 1 mM DA in acetate solution of pH 5.0 at the surface of CoNSal-modified carbon paste electrodes containing various percents of TOAB, pulse amplitude was 50 mV [137].

Moreover, Swamy et al utilized a CPE modified by SDS for the detection of $K_3Fe(CN)_6$ and DA. Upon the formation of SDS monolayer, the electrode/solution interface was replaced by the SDS monolayer/solution interface. No peaks are observed at both bare CPE and SDS/CPE in the blank supporting electrolyte (KCl). The background current is greatly enhanced at SDS/CPE compared to that at bare CPE (more than 100 times), indicating that the surface property of the modified electrode has been significantly changed. In contrast to the poor response at bare CPE, the electrochemical signal of DA is enhanced at SDS/CPE via the improvement of both the shape of redox peaks and the magnitude of the peak currents (i_p). Furthermore, the same catalytic effect was observed for $K_3Fe(CN)_6$ [138].

In addition, Yaping Ding developed cetylpyridine bromide (CPB)/chitosan composite film-modified GCE for the simultaneous determination of DA and AA. In absence of CPB, only one peak was observed (Fig. 4.25). By introducing CPB in the matrix of the modified electrode, the peak potential changed abruptly at very low surfactant concentration and immediately reached a plateau, indicating the saturation of the electrode surface by the adsorbed surfactant. The peak potential of DA was shifted to more positive value while that of AA was shifted to less positive value. Obviously, CPB (cationic surfactant) plays a key role in separating the two oxidation peaks of DA and AA by preventing DA cations from approaching the electrode surface retarding its reaction and by attracting more AA anions to approach the electrode surface facilitating its reaction. For AA and DA in mixture, two well-defined oxidation peaks were obtained with 0.3 V potential difference, which is large enough to allow their simultaneous determination [139].

On the other hand, Batanero et al developed an electrochemical method for the determination of aceclofenac (AC), anti-inflammatory agent, at the ppb level using adsorptive stripping voltammetric techniques at conventional and surfactant (non-ionic; Triton X-100, and Triton X-405 and anionic SDS) modified carbon paste electrodes (CMCPE). It was revealed that the presence of certain surfactants in the CMCPE improves the electroanalytical determination of AC owing to the increase of the chemical affinity between the drug and the surfactant molecules. The drug can be adsorbed on the surfactant CMCPE, mainly when the modifier is Triton X-405. Therefore, a very sensitive response (a detection limit of 7 ppb) at Triton X-405 CMCPE was obtained. As well as, the method was applied for AC control in tablets and it was quite accurate in comparison with other reported methods for AC determination [140].

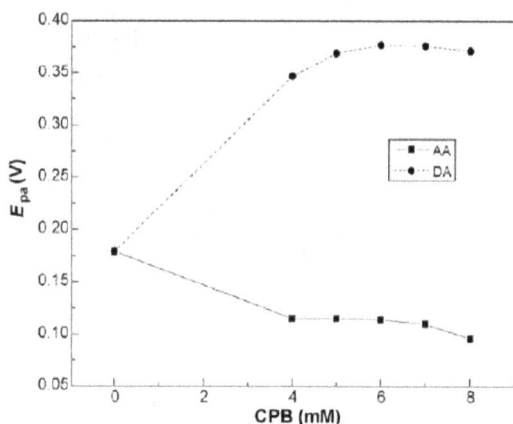

Fig. 4.25. Variation of oxidation potential (E_{pa}) of DA and AA at the chitosan modified electrode with various concentrations of CPB. In mixture of DA (1.2×10^{-4} M) and AA (2.5×10^{-4} M) in 0.10 M PBS (pH 4.7) [139].

Kawakami constructed a modified CPE in which the carbon particles were coated with a thin layer of a nonionic surfactant (NIS) with a pasting liquid containing ubiquinone (UQ, coenzyme Q_{10}) or menaquinone (MQ, vitamin K_2). A couple of peaks were observed at UQ and MQ-based electrodes in phosphate buffer solution/ pH 7.0 due to the oxidation and reduction of the respective quinones. The peak current showed a tendency to decrease when the NIS-coated particles were employed. This indicated that there was a decrease in the diffusion rate as the electroactive species existed away from the electrode surface and had to cross the NIS layer for electron transfer. The NIS layer acted not only as a diffusion barrier which prevented irreversible adsorption of quinone to occur on the electrode surface, but also as a matrix for the redox reaction of quinones at electrode surface. The effects of NIS layer on the electrochemical behavior of quinines depend on both the physico-chemical structure of the surfactant and the kind of quinines [141].

Owing to the cation exchange capacity of zeolites (their structures have high negative charge density) and the great affinity of the cationic surfactants to negative charge, the cationic surfactants are sorbed on the surface exchange sites of zeolite when a zeolite enters in a solution of cationic surfactant. This property was used to modify the external surface of the zeolite improving its anion exchange capacity [142]. Surfactant-modified zeolite (SMZ) was prepared according to the previously reported method [143]. The SMZ-modified carbon paste electrodes were prepared as following; appropriate amounts of the SMZ (5–25 wt% with respect to graphite) were mixed with 100 mg graphite powder and then Nujol was added. After through hand mixing in a mortar, the homogeneous paste was ready for using [143].

Recent studies indicate that SMZ are effective sorbents for multiple types of contaminants including some anionic and organic pollutants [142]. As a result, Nezamzadeh-Ejhieh proposed a nitrate-selective electrode based on surfactant (hexadecyl trimethyl ammonium bromide)-modified zeolite (SMZ) particles into carbon paste electrode (SMZ-CPE). The incorporation of the SMZ into a carbon paste makes it a suitable anion selective electrode. SMZ-CPE might be a useful analytical tool for the determination of nitrate in real samples. The proposed assembly was employed as an indicator electrode in the potentiometric titration of nitrate with solution of ferroammonium sulfate. The electrode showed relatively low detection limit, fast response time, reasonable selectivity, long term stability and good potential reproducibility [143]. Moreover, Nezamzadeh-Ejhieh demonstrated that the incorporation of cationic surfactant (hexadecyl trimethyl ammonium cation, HDTMA) onto zeolite surfaces creates suitable anionic exchangers. Combining these SMZs with polyvinyl chloride (PVC)-based membranes yields a suitable membrane electrode for the determination of anions such as sulfide. The sensor has a fast response time of 7–10 s, low detection limit 6.3×10^{-8} mol L^{-1} and can be used for at least 8 weeks without any deviation in potential. The proposed potentiometric PVC-membrane electrode can be used for the determination of sulfide anion in the presence of considerable concentrations of common interfering ions. Applicable pH range and potentiometric selectivity coefficients of the proposed electrode make it a superior device compared to other electrodes used for the determination of this ion in real samples [142].

Furthermore, Nezamzadeh-Ejhieh described a novel polyvinyl chloride (PVC) membrane electrode to be used as a potentiometric oxalate sensor. The sensor comprises a surfactant (hexadecyl trimethyl ammonium bromide (HDTMABr))-modified zeolite (SMZ) as a modifier, dioctyl phthalate (DOP) as a plasticizer, and PVC matrix. SMZ can be used for sorption of anionic species based on the conceptual model illustrated graphically in Fig. 4.26. HDTMA forms at low concentrations a monolayer on the zeolite surface with its hydrophobic region oriented toward solution. With increasing coverage, a second HDTMA layer is formed. This bilayer structure induces positively charged functional groups to orient toward the solution thus creating sorption sites for anions [144].

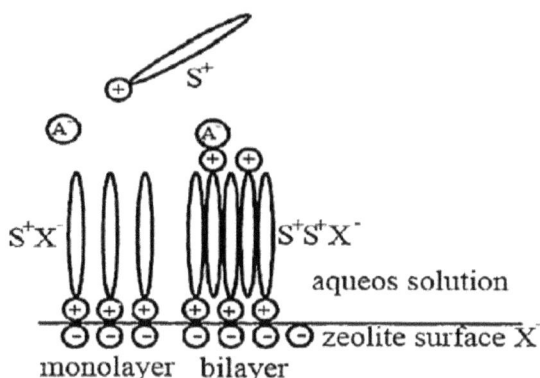

Fig. 4.26. A model for the adsorption of HDTMA onto the zeolite surface. The following parameters were used: A^-, anion; S^+, free surfactant molecule; S^+X^-, surfactant adsorbed in the monolayer and $S^+S^+X^-$, surfactant adsorbed in the bilayer [144].

On the other hand, Zeng et al fabricated a novel modified electrode comprised of hydrophobic ionic liquid (trihexyl tetradecyl phosphonium bis (trifluoromethylsulfonyl) imide, $[P_{6,6,6,14}][NTf_2]$), multiwalled carbon nanotubes (MWNTs) and cationic gemini surfactant ($C_{12}H_{25}N(CH_3)_2-C_4H_8-N(CH_3)_2C_{12}H_{25}Br_2$, $C_{12}-C_4-C_{12}$) to investigate the electrochemical behavior of Sudan I, found in food products. A small broad anodic peak is observed for Sudan I at bare GCE and the peak at $[P_{6,6,6,14}][NTf_2]$/GCE and MWNTs/GCE is also quite small (Fig. 4.27). On the other hand, $[P_{6,6,6,14}][NTf_2]$–MWNTs/GCE gives a well-defined anodic peak with higher anodic current which is attributed to the interaction of the hydrophobic ionic liquid and MWNTs. When gemini surfactant $C_{12}-C_4-C_{12}$ is present, the anodic peak of Sudan I occurs at lower potential and its anodic current becomes two times as high as that at $[P_{6,6,6,14}][NTf_2]$–MWNT/GCE. The gemini surfactant can facilitate the oxidation of Sudan I and improve its response due to the hydrophobic interaction between gemini surfactant and Sudan I. The components of the proposed sensor showed good synergistic interaction toward Sudan I sensing. As a result, the modified electrode presented good stability, repeatability and higher sensitivity. In addition, it was applied for the detection of Sudan I in hot chilli powder and ketchup samples with acceptable recovery results [145].

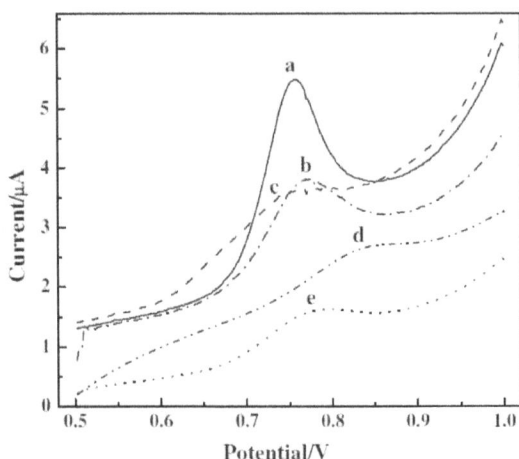

Fig. 4.27. LSVs of C_{12}–C_4–C_{12}–$[P_{6,6,6,14}][NTf_2]$–MWNT/GCE (a), $[P_{6,6,6,14}][NTf_2]$–MWNT/ GCE (b), MWNT/GCE (c), $[P_{6,6,6,14}][NTf_2]$/GCE (d), and bare GCE (e) in 5.0 μmol l^{-1} Sudan I. Scan rate: 0.1 V s^{-1}; supporting electrolyte: 0.05 mol l^{-1} potassium biphthalate buffer plus acetonitrile solution (v/v = 9:1, pH 4.5); accumulation time: 120 s (on open-circuit) [145].

On the other hand, Molina et al showed a new strategy for the construction of dehydrogenase amperometric biosensors using surfactant to enhance the sensitivity of diaphoase/ferrocene (DAP/FcH) modified CPE for the electrocatalytic oxidation of nicotinamide adenine dinuclotide (NADH). The study showed that the sensitivity and operation stability towards NADH in flow injection analysis (FIA) were greatly improved (5 times) upon adding Tween 20 into the electrode matrix. The improved performance was attributed to the more hydrophilic environment provided by the surfactant which enhanced the interactions between enzyme, cofactor and mediator. Consequently, this new approach represents decisive step for a versatile preparation method of amperometric biosensors [146].

Moreover, Zhang and Hu demonstrated that liquid phase deposition (LPD) technique provides a novel approach for the immobilization of hemoglobin (Hb) in TiO$_2$ film for studying the direct electron transfer

202

of Hb. Using the LPD process, a hybrid film composed of Hb, TiO_2 and sodium dodecylsulfonate (SDS) is successfully prepared on the electrode surface. The surface morphology of as-deposited $Hb/SDS/TiO_2$ film shows a flower-like structure which is attributed to the effect of the doped SDS molecules. $Hb/SDS/TiO_2$ hybrid LPD film can be utilized as H_2O_2 sensor showing high sensitive response. It was found that SDS could improve the electrocatalysis of Hb towards H_2O_2 after Hb was adsorbed on the hybrid film [147]. In addition, Hu et al investigated the direct electron transfer between myoglobin molecules and GCE without the aid of any electron mediator by immobilizing myoglobin in a new zwitterionic gemini surfactant cast on GCE surface. The electrocatalytic activities of myoglobin embedded in this film for the electroreduction of H_2O_2 was estimated in phosphate buffer solution/pH 7.4 and a higher sensitive electroanalytical method for H_2O_2 determination was developed. Proteins in the surfactant film on the electrode surface could retained their original conformation and high affinity and response sensitivity for H_2O_2 were obtained. This gemini surfactant with good water-solubility provided an efficient method for the achievement of direct electron transfer of protein with the electrode constructing a new potential platform for the development of biosensors for H_2O_2 or other biological molecules [23].

On the other hand, a simple and non-destructive method for solubilization of nanotube is based on non-covalent interactions of amphiphilic molecules (surfactants) with nanotube surfaces: hydrophilic parts of such molecules interact with the solvent and hydrophobic parts are adsorbed onto the nanotube surface. Thus, amphiphilic molecules can solubilize CNTs and prevent them from the aggregation into bundles and ropes [148-150]. Moreover, the use of surfactant would offer a quick and effective method to disperse graphene and prevent its tendency towards aggregation on the electrode surface [151]. Yuan-Di Zhao proposed a novel amperometric biosensor based on immobilization of hemoglobin (Hb), chitosan (CS), graphene (GR) and the cationic surfactant (CTAB) on a GCE to quantitatively measure nitric oxide (NO); a key signaling molecule in different physiological processes of plants. The use of CTAB would offer a quick and effective method to disperse GR and prevent the tendency towards GR aggregation on the surface of GCE presenting a well-dispersed GR suspension. The fabricated nanocomposite biosensor showed direct electrochemistry with a fast electron transfer rate and high electrocatalytic activity towards NO reduction. In addition, the biosensor had excellent long term stability, as well as reproducibility and selectivity and should have a potential application in monitoring NO in plant samples [151].

Furthermore, Gao utilized SDS as a useful dispersing agent for pristine and purified multiwall carbon nanotubes (MWCNTs) to prepare MWCNTs-modified electrodes. Voltammetric responses at MWCNTs–SDS modified GCE towards detecting H_2O_2 are observed to compare the electrochemical action of MWCNTs in different circumstances. The best electrochemical action of pristine MWCNTs toward H_2O_2 is at about 0.4 wt.% MWCNTs dispersed in 2 wt.% SDS aqueous solution modified electrode. In contrast to the properties of MWCNTs dispersed in distilled water, SDS is helpful for the dispersion and the electrochemical action of pristine and purified MWCNTs. Moreover, these MWCNTs–SDS modified electrodes might be used in biosensors after immobilizing some biomaterials [148]. Moreover, Xiaoquan Lu fabricated single-walled carbon nanotubes (SWCNTs) by SDS (f-SWCNTs) modified GCE (f-SWCNTs/GCE) for the simultaneous determination of AA, DA and UA. The f-SWCNTs/GCE; which is negatively charged, exhibited better electrocatalytic activities toward AA, DA and UA compared to bare GCE offering low electro-oxidation potentials. Moreover, the f-SWCNTs/GCE presents high sensitivity and low detection limit for selective determination of DA in the presence of AA and UA. The method was successfully applied for determination of DA in some biological fluids with good recovery results [149]. In addition, Ndiaye et al described the preparation of the CNTs-based gas sensors, which are achieved by the utilization of aqueous dispersions of carbon nanotubes (CNTs) using sodium dodecyl benzene sulfonate (NaDDBS) as a surfactant. The sensors are made of interdigitated electrodes (IDEs) on which the CNTs dispersion are drop-cast deposited. The presence of surfactant provided a better solubilization, dispersion and stabilization of the SWNTs in aqueous media. In the case of NaDDBs, the enhanced dispersion ability was attributed to the presence of the benzene group which contributes to an additional stabilization of the CNTs through п-п interactions [150].

On the other hand, acetylene black (AB) nanoparticles were easily dispersed into water in the presence of a special surfactant: dihexadecyl hydrogen phosphate (DHP) by Guohua Fan et al. After that, the surface of GCE was coated with AB-DHP composite film after evaporating water. The resulting AB-DHP film exhibited remarkable enhancement towards the oxidation of clenbuterol and greatly increased its oxidation peak current. Clenbuterol is a β_2-adrenergic agonist used as a drug for the treatment of pulmonary disease and asthma and recently applied illegally in livestock for diverting nutrients from fat deposition in animals to the production of muscle tissues. A novel electrochemical method was developed for the detection of clenbuterol with low detection limit 10 µg L^{-1}. The method was successfully used to detect the concentration of clenbuterol in pork and liver samples with excellent recovery ranging from 95.2 % to 106.8 % [152].

Hong Li applied a multiple sweep voltammetry effectively to the electrochemical assembly of polypyridyl cobalt(III) complexes [Co(phen)$_2$(tatp)]$^{3+}$ and [Co(phen)$_3$]$^{3+}$ (where phen = 1,10-phenanthroline and tatp = 1, 4, 8, 9 - tetra - aza - triphenylene) on a GCE modified with sodium dodecyl sulfate (SDS)-dispersed multi-walled carbon nanotubes (MWCNTs). The SDS-MWCNTs are found to impact the redox reactions of [Co(phen)$_2$L]$^{3+}$ (L = phen or tatp) adsorbed on the electrode surfaces. The anionic surfactant (SDS) facilitates the adsorption of [Co(phen)$_2$L]$^{3+}$ on the MWCNT surfaces via a fully-packed π–π stacking mode. Moreover, the SDS-MWCNT modified electrodes are suitable for the detection of 6-mercaptopurine (6-MP) using [Co(phen)$_2$(tatp)]$^{3+}$ as voltammetric probes. The proposed amperometric sensor exhibited a high selectivity toward 6-MP and a good linear response was obtained [153].

On the other hand, surfactants can be added during the polymerization to enhance the polymer growth rate. Ojani studied the electrocatalytic oxidation of some cephalosporins at poly(o-anisidine)/SDS/Ni modified CPE. At first, poly(o-anisidine) was formed by cyclic voltammetry in monomer solution containing SDS on CPE surface. The Addition of SDS to monomer solution resulted in increasing the polymer growth rate. Then, Ni(II) ions were incorporated into the electrode by the immersion of the polymer modified electrode having amine group in 0.1 mol L^{-1} Ni(II) ion solution. Cephalosporins were successfully oxidized on the surface of this nickel ions dispersed poly(o-anisidine) modified CPE. Moreover, the electrode can be used for simple, selective and precise voltammetric determination of cephalosporins in pharmaceutical preparations [154]. Furthermore, Qun Xu designed an amperometric sensor with a single working electrode for simultaneous determination of electro-inactive anions and cations as a detector in ion chromatography. The modification of the working golden electrode was based on the incorporation of dodecyl sulfate into polydiphenylamine by electropolymerization of diphenylamine in the presence of sodium dodecylsulfate. In ion-exclusion/cation-exchange chromatography, a set of well defined peaks of these anions and cations was obtained at the working potential, 11.35 V (vs. saturated calomel electrode) using citric acid solution as eluent. The common anions and cations in mineral water samples were determined using this ion-chromatographic system with satisfactory results [155].

Moreover, the amphiphilic surfactant molecules find numerous applications in electrochemistry signifying that the micro-heterogeneous-assemblies of surfactant aggregates are potential candidates for stabilizing inorganic complexes both in solution phase and on electrode surfaces for various applications [156]. Pillai constructed H_2O_2 amperometric sensor based on Prussian blue (PB)–CTAB composite film electrode. PB–CTAB was prepared in the presence of CTAB surfactant by electrochemical potential cycling method. The PB–CTAB composite has been advantageously employed as an amperometric sensor for H_2O_2 with extended linearity, improved catalytic activity and low detection limit compared to the unmodified PB film electrode. The PB–CTAB composite possesses enhanced film growth, efficient and rapid charge transfer, and also extremely high stability. The extraordinary stability of the composite film is attributed to the electrostatic stabilization offered by the cationic CTA^+ ion to the oppositely charged PB in the film thus preventing it from dissolution. The developed surfactant stabilized PB film electrode possesses high sensitivity towards H_2O_2 ca. 9.79 A M^{-1} cm^{-2} and also holds better qualities than the well-known H_2O_2 sensors [156]. Moreover, Salazar developed a highly selective and sensitive H_2O_2 sensor based on the electro-deposition of PB onto screen-printed carbon electrodes (SPCEs) modified by benzethonium chloride (BZTC) surfactant. This methodology provides a time-efficient method for producing stable films in the presence of BZTC. The presence of BZTC promotes the deposition of larger quantities of PB onto the SPCE surface. BZT^+ is adsorbed forming a bilayer between the negative electrode surface (unmodified electrode or PB film) and solution. This positive bilayer containing well oriented BZT^+ shows higher reactivity towards $Fe(CN)_6^{3-}$, which is the precursor in forming PB. Moreover, BZTC (2 mM)/PB operating at ~ 0 V vs. SCE displayed the highest H_2O_2 sensitivity (1.07 ± 0.03 A M^{-1} cm^{-2}, n = 5) reported in the literature to date for PB-modified SPCEs and showed excellent detection limit (<10^{-7} M). Finally, BZTC (2 mM)/PB-based sensor stored dry at room temperature over 4 months retained ~ 90 % of its initial response toward H_2O_2 [157]. In addition, Salazar reported a comparison of the beneficial effects of different cationic surfactants; cetyl trimethyl ammonium bromide (CTAB), benzethonium chloride (BZT) and cetylpyridinium chloride (CPC), for the electrochemical synthesis of Prussian Blue (PB) films on screen-printed carbon electrodes (SPCEs). All surfactant enhanced PB-modified SPCEs displayed a significant improvement in their electrochemical properties compared to PB-modified SPCEs formed in the absence of surfactants. Surfactant-modified electrodes displayed a consistently higher PB surface concentration value indicating that PB deposition efficiency was improved by 2–3 folds. The H_2O_2 sensitivity of surfactant-modified PB films is higher than those reported previously for SPCEs by other authors [158].

On the other hand, Zhou and Liu et al synthesized gold nanoparticle (AuNPs) stabilized by gemini surfactant (GEM16-3-16-Au) in aqueous solution of 1,3-bis (cetyldimethyl ammonium) propane dibromide (GEM16-3-16) by the reduction of $HAuCl_4$ with UV irradiation. GEM16-3-16 played a crucial role in the synthesis of the AuNPs. Without gemini surfactant, AuNPs can not be synthesized in $HAuCl_4$ solution with UV irradiation for 4 h. On the other hand, GEM16-3-16 was used as stabilizer to protect AuNPs in situ. The obtained GEM16-3-16-Au nanocomposite is stable in water for at least 6 months. The direct electrochemistry of hemoglobin (Hb) can be facilely achieved by incorporation into the GEM16-3-16-Au film on GCE. The electron transfer between Hb and the electrode surface can be explained by the following hypothesis illustrated in Fig. 4.28. GEM16-3-16-Au nanocomposite combined two kinds of promoter; gemini surfactant and AuNPs that improve the electron transfer between the electroactive centers of Hb and electrode surface. There are two weak interactions between gemini surfactant and Hb molecules (Fig. 4.28A). The binding of Hb to the charged nanocomposite was dominated by electrostatic interactions between the positively charged head groups of gemini surfactant and the negatively charged amino acid residues of Hb. Meanwhile, the hydrophobic chains of GEM16-3-16 can penetrate into Hb molecules to unfold the electroactive centers of Hb due to the hydrophobic interactions between the surfactant hydrophobic chains and hydrophobic regions of Hb molecule. Thus, it is possible to achieve direct electron transfer of Hb on gemini surfactant film modified electrode. In conclusion, GEM16-3-16-Au film provided a favorable microenvironment both for immobilization of Hb and for electron transfer between Hb and the proposed electrode [159].

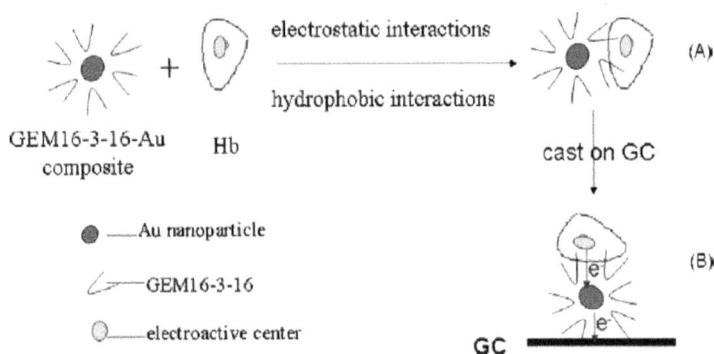

Fig. 4.28. The schemes of (A) the interactions of GEM16-3-16-Au nanocomposite and Hb molecule, (B) The electron transfer on Hb/GEM16-3-16-Au/GC electrode [159].

4.5. Conclusions

Surface-active agents are very effective to be used in the electroanalysis of organic compounds, biologically and pharmaceutically important compounds, drugs, some important inorganic ions and metals. In addition, micelles have been employed as selective masking agents to improve selectivity and sensitivity of electrochemical analysis. The effect of surfactants on the electrochemistry of different analytes can be summarized as following; in some cases synergistic preconcentration may occur for the studied analytes on the electrode surface, in other cases surfactant may act as an antifouling and homogenizing agent. The improved response exhibited by surfactants was found to be a function of the type of surfactant and its concentration, pH of solution and accumulation time.

On the other hand, the micro-heterogeneous-assemblies of surfactant aggregates are potential candidates for stabilizing inorganic complexes both in solution phase and on electrode surfaces for various applications especially sensing applications. Moreover, amphiphilic molecules can be used as solubilizing agents for CNTs and prevent them from the aggregation into bundles and ropes. Furthermore, the use of surfactant would offer a quick and effective method to disperse graphene and prevent its tendency towards aggregation on the electrode surface.

The use of surfactant modified electrodes as nanosensors with excellent reproducibility, high sensitivity, unique selectivity and exceptional stability was enhanced due to the synergistic effect between the conducting substrate and surfactants. The main advantages of the proposed methods utilizing surfactant modified electrodes are simple, cheap and fast compared to other determination methods of different studied analytes. Furthermore, the methods are sensitive enough for the determination of the studied compounds in clinical preparations (human urine) and in commercial tablets under physiological conditions with good precision, accuracy, selectivity and very low detection limit (sub-nanomolar concentrations).

Acknowledgments

The authors would like to acknowledge the financial support from Cairo University through the President Office for Research Funds.

References

[1].   A. Kaifer, M. G. Kaifer, Supramolecular Electrochemistry, *Wiley-VCH*, 1999.

[2].   N. F. Atta, A. Galal, F. M. Abu-Attia, S. M. Azab, Carbon Paste Gold Nanoparticles Sensor for the Selective Determination of Dopamine in Buffered Solutions, *J. Electrochem. Soc.*, Vol. 157, Issue 9, 2010, pp. F116-F123.

[3].   N. F. Atta, M. F. El-Kady, A. Galal, Simultaneous determination of catecholamines, uric acid and ascorbic acid at physiological levels using poly(N-methylpyrrole)/Pd-nanoclusters sensor, *Anal. Biochem.*, Vol. 400, 2010, pp. 78–88.

[4].   N. F. Atta, M. F. El-Kady, Novel poly(3-methylthiophene)/Pd, Pt nanoparticle sensor: Synthesis, characterization and its application to the simultaneous analysis of dopamine and ascorbic acid in biological fluids, *Sens. Actuators, B,* Vol. 145, 2010, pp. 299–310.

[5].   N. F. Atta, M. F. El-Kady, A. Galal, Palladium nanoclusters-coated polyfuran as a novel sensor for catecholamine neurotransmitters and paracetamol, *Sens. Actuators, B,* Vol. 141, 2009, pp. 566–574.

[6].   J. Mathiyarasu, S. Senthilkumar, K. L. N. Phani, V. Yegnaraman, PEDOT-Au nanocomposite film for electrochemical sensing, *Mater. Lett.,* Vol. 62, 2008, pp. 571–573.

[7].   A. Galal, N. F. Atta, H. K. Hassan, Graphene Supported-Pt-M (M = Ru or Pd) for Electrocatalytic Methanol Oxidation, *Int. J. Electrochem. Sci.,* Vol. 7, 2012, pp. 768–784.

[8]. N. F. Atta, A. Galal, S. M. Azab, Determination of morphine at gold nanoparticles/Nafion® carbon paste modified sensor electrode, *Analyst*, Vol. 136, 2011, pp. 4682–4691.

[9]. N. F. Atta, A. M. Abdel-Mageed, Smart electrochemical sensor for some neurotransmitters using imprinted sol–gel films, *Talanta*, Vol. 80, 2009, pp. 511–518.

[10]. M. F. El-Kady, V. Strong, S. Dubin, R. B. Kaner, Laser Scribing of High-Performance and Flexible Graphene-Based Electrochemical Capacitors, *Science*, Vol. 335, 2012, pp. 1326-1330.

[11]. A. Galal, N. F. Atta, S. A. Darwish, A. Abdel Fatah, S. M. Ali, Electrocatalytic evolution of hydrogen on a novel SrPdO$_3$ perovskite electrode, *J. Power Sources*, Vol. 195, 2010, pp. 3806–3809.

[12]. A. Galal, N. F. Atta, S. A. Darwish, S. M. Ali, Electrodeposited Metals at Conducting Polymer Electrodes. II:Study of the Oxidation of Methanol at Poly(3-methylthiophene) Modified with Pt–Pd Co-catalyst, *Top. Catal.*, Vol. 47, 2008, pp. 73–83.

[13]. J. Wang, Analytical electrochemistry, 2nd Edition, *Wiley-VCH*, 2000.

[14]. R. W. Murray, A. G. Ewing, R. A. Durst, Chemically Modified Electrodes: Molecular Design for Chemical Analysis, *Anal. Chem.*, Vol. 59, 1987, pp. 379A-390A.

[15]. J. Wang, Modified electrodes for electrochemical sensors, *Electroanalysis*, Vol. 3, 1991, pp. 255-259.

[16]. R. P. Baldwin, K. N. Thomsen, Chemically modified electrodes in liquid chromatography detection: a review, *Talanta*, Vol. 38, 1991, pp. 1-16.

[17]. C. M. A. Brett, A. M. O. Brett, Electroanalysis, *Oxford University Press*, Oxford, 1998.

[18]. N. F. Atta, A. Galal, E. H. El-Ads, A novel sensor of cysteine self-assembled monolayers over gold nanoparticles for the selective determination of epinephrine in presence of sodium dodecyl sulfate, *Analyst*, Vol. 137, 2012, pp. 2658–2668.

[19]. N. F. Atta, A. Galal, R. A. Ahmed, Poly(3, 4-ethylene-dioxythiophene) electrode for the selective determination of dopamine in presence of sodium dodecyl sulfate, *Bioelectrochemistry*, Vol. 80, 2011, pp. 132–141.

[20]. R. Nagarajan, E. Ruckenstein, Theory of Surfactant Self -Assembly: A Predictive Molecular Thermodynamic Approach, *Langmuir*, Vol. 7, 1991, pp. 2934-2969.

[21]. M. Malmsten, Surfactant and polymer in drug delivery, *Marcel Dekker*, New York, 2002.

[22]. D. A. Fridrikhsberg, A course in colloid Chemistry, *Mir Publishers*, Moscow, 1986.

[23]. F. Wang, S. Hu, Direct electron-transfer of myoglobin within a new zwitterionic Gemini surfactant film and its analytical application for H$_2$O$_2$ detection, *Colloids Surf.*, B, Vol. 63, 2008, pp. 262–268.

[24]. R. Vittal, H. Gomathi, K. Kim, Beneficial role of surfactants in electrochemistry and in the modification of electrodes, *Adv. Colloid Interface Sci.,* Vol. 119, 2006, pp. 55–68.

[25]. M. J. Rosen, Surfactants and Interfacial Phenomena, 2nd ed, *Wiley-Inter-Science Publication,* New York, 1989.

[26]. J. F. Rusling, Controlling electrochemical catalysis with surfactant microstructures, *Acc. Chem. Res.,* Vol. 24, 1991, pp. 75-81.

[27]. J. F. Rusling, Reactions and synthesis in surfactant systems, Electroanal. Chem., in Bard A. J. (Ed.), *Marcel Dekker,* New York, Vol. 18, 1994, pp. 267.

[28]. T. C. Franklin, S. Mathew, Surfactants in solution, In: Mittall K. L. (ed.), *Plenum,* New York, Vol. 10, 1989.

[29]. N. Shinozuka, S. Hayano, Solution chemistry of surfactants, in Mitall K. L. (ed.), *Plenum,* New York, Vol. 2, 1979.

[30]. A. Diaz, A. Z. Kaifer, Self-Assembled Surfactant Monolayers on Electrode Surfaces: The Formation of Surfactant Viologen Monolayers on Au and Pt, *J. Electroanal. Chem.,* Vol. 249, 1988, pp. 333-338.

[31]. C. A. Widrig, M. Majda, Self-assembly of ordered monolayers and bilayers of N-methyl-N'-octadecylviologen amphiphile on gold surfaces in aqueous solutions, *Langmuir,* Vol. 5, 1989, pp. 689-695.

[32]. H. C. D. Long, J. J. Donohue, D. A. Buttry, Ionic interactions in electroactive self-assembled monolayers of ferrocene species, *Langmuir,* Vol. 7, 1991, pp. 2196-2202.

[33]. L. M. Grant, W. A. Ducker, Effect of Substrate Hydrophobicity on Surface–Aggregate Geometry: Zwitterionic and Nonionic Surfactants, *Phys. Chem., B,* Vol. 101, 1997, pp. 5337-5345.

[34]. L. M. Grant, F. Tiberg, W. A. Duker, Nanometer-Scale Organization of Ethylene Oxide Surfactants on Graphite, Hydrophilic Silica, and Hydrophobic Silica, *J. Phys. Chem.,* B, Vol. 102, 1998, pp. 4288-4294.

[35]. A. R. Rennie, E. M. Lee, E. A. Simister, R. K. Thomas, Structure of A Cationic Surfactant Layer at the Silica-Water Interface, *Langmuir,* Vol. 6, 1990, pp. 1031-1034.

[36]. S. Manne, H. E. Gaub, Molecular organization of surfactants at solid-liquid interfaces, *Science,* Vol. 270, 1995, pp. 1480-1482.

[37]. J. C. Schulz, G. G. Warr, P. D. Bulter, W. A. Hamilton, Adsorbed layer structure of cationic surfactants on quartz, *Phys. Rev. E,* Vol. 63, 2001, pp. 041604-041608.

[38]. J. S. Facci, *Langmuir,* Vol. 3, 1987, pp. 525-529.

[39]. J. F. Rusling, C. N. Shi, D. K. Gosser, S. S. Shukla, Electrocatalytic Reactions in Organized Assemblies I. Reduction of 4-Bromobiphenyl in Cationic and Nonionic Micelles, *J. Electroanal. Chem.,* Vol. 240, 1988, pp. 201-216.

[40]. J. F. Rusling, Electrocatalytic Systems Organized by Micelles, *Trends Anal. Chem.,* Vol. 7, 1988, pp. 266-269.

[41]. S. Boussaad, N. J. Tao, Electron Transfer and Adsorption of Myoglobin on Self-assembled Surfactant Films: An Electrochemical Tapping-

Mode AFM Study, *J. Am. Chem. Soc.*, Vol. 121, 1999, pp. 4510-4515.

[42]. J. F. Liu, G. Min, W. A. Duker, AFM Study of Adsorption of Cationic Surfactants and Cationic Polyelectrolytes at the Silica–Water Interface, *Langmuir*, Vol. 17, 2001, pp. 4895-4903.

[43]. U. Retter, A. Avranas, On Anion-Induced Formation of Hemicylindrical and Hemispherical Surface Micelles of Amphiphiles at the Metal/Electrolyte Interface, *Langmuir*, Vol. 17, 2001, pp. 5039-5044.

[44]. J. C. Schulz, G. G. Warr, Adsorbed Layer Structure of Cationic and Anionic Surfactants on Mineral Oxide Surfaces, *Langmuir*, Vol. 18, 2002, pp. 3191-3197.

[45]. M. Petri, D. M. Kolb, Nanostructuring of a sodium dodecyl sulfate-covered Au(111) electrode, *Phys. Chem.*, Vol. 4, 2002, pp. 1211-1216.

[46]. E. J. Wanless, W. A. Duker, Organization of sodium dodecyl sulfate at the graphite-solution interface, *J. Phys. Chem.*, Vol. 100, 1996, pp. 3207.

[47]. W. A. Duker, L. M. Grant, Effect of Substrate Hydrophobicity on Surfactant Surface–Aggregate Geometry, *J. Phys. Chem.*, Vol. 100, 1996, pp. 11507-11511.

[48]. J. L. Wolgemuth, R. K. Workman, S. Manne, Surfactant Aggregates at a Flat, Isotropic Hydrophobic Surface, *Langmuir*, Vol. 16, 2000, pp. 3077-3081.

[49]. E. J. Wanless, W. A. Duker, Weak Influence of Divalent Ions on Anionic Surfactant Surface-Aggregation, *Langmuir*, Vol. 13, 1997, pp. 1463-1474.

[50]. V. Subramanian, W. A. Duker, Counterion Effects on Adsorbed Micellar Shape: Experimental Study of the Role of Polarizability and Charge, *Langmuir*, Vol. 16, 2000, pp. 4447-4454.

[51]. I. Burgess, C. A. Jeffrey, X. Cai, G. Szymanski, J. Lipkowski, Direct visualization of the potential controlled transformation of hemimicellar aggregates of dodecyl sulfate into a condenced monolayer at the electrode surface, *Langmuir*, Vol. 15, 1999, pp. 2607-2616.

[52]. I. Burgess, V. Zamlynny, G. Szymanski, J. Lipkowski, Electrochemical and neutron reflectivity characterization of dodecyl sulfate adsorption and aggregation at the gold-water interface, *Langmuir*, Vol. 17, 2001, pp. 3355-3367.

[53]. E. Cholewa, I. Burgess, J. Kunze, Adsorption of *N*-dodecyl-*N*, *N*-dimethyl-3-ammonio-1-propanesulfonate (DDAPS), a model zwitterionic surfactant, on the Au(111) electrode surface, *J. Solid State Electrochem.*, Vol. 8, 2004, pp. 693-705.

[54]. P. Chandar, P Somasundaram, N. J. Turro, Fluorescence probe studies on the structure of the adsorbed layer of dodecyl sulfate at the alumina—water interface, *Colloid Interface Sci.*, Vol. 117, 1987, pp. 31-46.

[55]. S. Manne, Visualizing Self-Assembly: Force Microscopy of Ionic Surfactant Aggregates at Solid-Liquid Interfaces, *Progr. Colloid*

Polym. Sci., Vol. 103, 1997, pp. 226-233.

[56].   Y. Gao, J. Du, T. Gu, Hemimicelle formation of cationic surfactants at silica gel–water interface, *Chem. Soc. Faraday Trans.,* Vol. 1, 1987, pp. 2671-2679.

[57].   A. Fan, P. Somasundaram, N. Turro, Adsorption of alkyltrimethylammonium bromides on negatively charged alumina, *Langmuir,* Vol. 13, 1997, pp. 506-510.

[58].   B. G. Sharma, S. Basu, M. M. Sharma, Characterization of Adsorbed Ionic Surfactants on a Mica Substrate, *Langmuir,* Vol. 12, 1996, pp. 6506-6512.

[59].   P. K. Singh, J. J. Adler, Y. I. Rabinovich, B. M. Moudgil, Investigation of Self-Assembled Surfactant Structures at the Solid-Liquid Interface Using FT-IR/ATR, *Langmuir,* Vol. 17, 2, 2001, pp. 468-473.

[60].   H. Li, C. P. Tripp, Spectroscopic identification and dynamics of adsorbed cetyltrimethylammonium bromide structures on TiO_2 surfaces, *Langmuir,* Vol. 18, 2002, pp. 9441-9446.

[61].   R. Atkin, V. S. J. Craig, E. J. Wanless, S. Biggs, Mechanism of cationic surfactant adsorption at the solid-aqueous interface, review article, *Adv. Colloid Interface Sci.,* Vol. 103, 2003, pp. 219-304.

[62].   L. Gonzalez-Macia, M. R. Smyth, A. Morrinb, A. J. Killard, Enhanced electrochemical reduction of hydrogen peroxide at metallic electrodes modified with surfactant and salt, *Electrochim. Acta,* Vol. 58, 2011, pp. 562–570.

[63].   R. W. Muray, ed. Molecular Design of Electrode Surfaces, *John Wiley and Sons,* New York, Vol. 22, 1992, pp. 18.

[64].   M. Goldenberg, Use of electrochemical techniques to study the Langmuir–Blodgett films of redox active materials, *Russ. Chem. Rev,* Vol. 66, 1997, pp. 1033-1052.

[65].   M. Gomez, J. Li, A. E. Kaifer, Surfactant Monolayers on Electrode Surfaces: Self-Assembly of a Series of Amphiphilic Viologens on Gold and Tin Oxide, *Langmuir,* Vol. 7, 1991, pp. 1797-1806.

[66].   A. Ulman, An introduction to Ultra Thin Organic Films from Langmuir-Boldgett to Self-Assembly, *Academic,* San Diego, CA, 1991.

[67].   J. F. Rusling, Molecular aspects of electron transfer at electrodes in micellar solutions, *Colloids Surf.,* Vol. 123, 1997, pp. 81-88.

[68].   R. A. Mackay, Electrochemistry in association colloids, *Colloids Surf.,* A, Vol. 82, 1994, pp. 1-28.

[69].   J. F. Rusling, H. Zhang, W. S. Willis, Properties of Octadecylsilyl-coated Electrodes In Ionic Micellar Media, *Anal. Chim. Acta,* Vol. 235, 1990, pp. 307-315.

[70].   R. Guidelli, M. L. Foresti, The inhibitory effect of neutral organic surfactants upon simple electrode reactions, *Electroanal. Chem.,* Vol. 77, 1977, pp. 73-96.

[71].   A. Marino, A. Brajter-Toth, Ionic surfactants as molecular spacers at graphite electrodes, *Anal. Chem.,* Vol. 65, 1993, pp. 370-374.

[72]. K. A Bunding. Lee, Electron transfer into self-assembling monolayers on gold electrodes, *Langmuir*, Vol. 6, 1990, pp. 709-712.

[73]. C. E. D. Chidsey, Free energy and temperature dependence of electron transfer at the metal-electrolyte interface, *Science*, Vol. 251, 1991, pp. 919-922.

[74]. A. P. Abbott, G. Gounili, J. M. Bobbitt, J. F. Rusling, T. F. Kumosinski, Electron transfer between amphiphilic ferrocenes and electrodes in cationic micellar solution, *J. Phys. Chem.*, Vol. 96, 1992, pp. 11091–11095.

[75]. R. A. Marcus, Chemical and Electrochemical Electron-Transfer Theory, *Annu. Rev. Phys. Chem.*, Vol. 15, 1964, pp. 155-196.

[76]. J. Georges, S. Desmettre, Electrochemistry of ferrocene in anionic, cationic and nonionic micellar solutions. Effet of the micelle solubilization of the half-wave potentials, *Electrochim. Acta*, Vol. 29, 1984, pp. 521-525.

[77]. J. H. Fendler, Membrane Mimetic Chemistry, *Wiley*, New York, 1982.

[78]. R. Hosseinzadeh, R. E. Sabzi, K. Ghasemlu, Effect of cetyltrimethyl ammonium bromide (CTAB) in determination of dopamine and ascorbic acid using carbon paste electrode modified with tin hexacyanoferrate, *Colloids Surf.*, B, Vol. 68, 2009, pp. 213–217.

[79]. S. Paria, K. C. Khilar, A review on experimental studies of surfactant adsorption at the hydrophilic solid–water interface, *Adv. Colloid Interface Sci.*, Vol. 110, 2004, pp. 75–95.

[80]. L. J. C. Love, J. G. Habarta, J. G. Dorsey, The Micelle-Analytical Chemistry Interface, *Anal. Chem.*, Vol. 56, 1984, pp. 1132A-1148A.

[81]. C. Yang, Q. Sang, S. Zhang, W. Huang, Voltammetric determination of estrone based on the enhancement effect of surfactant and a MWNT film electrode, *Mater. Sci. Eng.*, C, Vol. 29, 2009, pp. 1741–1745.

[82]. C. Hu, C. Yang, S. Hu, Hydrophobic adsorption of surfactants on water-soluble carbon nanotubes: A simple approach to improve sensitivity and antifouling capacity of carbon nanotubes-based electrochemical sensors, *Electrochem. Commun.*, Vol. 9, 2007, pp. 128–134.

[83]. R. Jain, R. Mishra, A. Dwivedi, Effect of surfactant on voltammetric behaviour of ornidazole, *Colloids Surf.*, A, Vol. 337, 2009, pp. 74–79.

[84]. R. Vittal, H. Gomathi, K. J. Kim, Beneficial role of surfactants in electrochemistry and in the modification of electrodes, *Adv. Colloid Interface Sci.*, Vol. 119, 1, 2006, pp. 55-68.

[85]. S. Liu, J. Li, S. Zhang, J. Zhao, Study on the adsorptive stripping voltammetric determination of trace cerium at a carbon paste electrode modified in situ with cetyltrimethylammonium bromide, *Appl. Surf. Sci.*, Vol. 252, 2005, pp. 2078-2084.

[86]. I. Svancara, P. Foret, K. Vytras, A study on the determination of chromium as chromate at a carbon paste electrode modified with surfactants, *Talanta*, Vol. 64, 2004, pp. 844-852.

[87]. B. Hoyer, N. Jensen, Use of sodium dodecyl sulfate as an antifouling and homogenizing agent in the direct determination of heavy metals by anodic stripping voltammetry, *Analyst*, Vol. 129, 2004, pp. 751-754.

[88]. R. Ahmad Dar, P. Kumar Brahman, S. Tiwari, K. S. Pitre, Electrochemical studies of quinine in surfactant media using hanging mercury drop electrode: A cyclic voltammetric study, *Colloids Surf.*, B, Vol. 98, 2012, pp. 72–79.

[89]. M. Stadlober, K. Kalcher, G. Raber, A new method for the voltammetric determination of molybdenum(V1) using carbon paste electrodes modified in situ with cetyltrimethylammonium bromide, *Anal. Chim. Acta*, Vol. 350, 1997, pp. 319-328.

[90]. P. Deng, Y. Feng, J. Fei, A new electrochemical method for the determination of trace molybdenum(VI) using carbon paste electrode modified with sodium dodecyl sulfate, *J. Electroanal. Chem.*, Vol. 661, 2011, pp. 367–373.

[91]. I. Švancara, M. Galik, K. Vytřas, Stripping voltammetric determination of platinum metals at a carbon paste electrode modified with cationic surfactants, *Talanta*, Vol. 72, 2007, pp. 512–518.

[92]. I. Svancara, P. Foret, K. Vytras, A study on the determination of chromium as chromate at a carbon paste electrode modified with surfactants, *Talanta*, Vol. 64, 2004, pp. 844–852.

[93]. M. Stadlober, K. Kalcher, G. Raber, C. Neuhold, Anodic stripping voltammetric determination of titanium(IV) using a carbon paste electrode modified with cetyltrimethylammonium bromide, *Talanta*, Vol. 43, 1996, pp. 1915-1924.

[94]. P. K. Brahman, R. A. Dar, S. Tiwari, K. S. Pitre, Voltammetric determination of anticancer drug flutamide in surfactant media at polymer film modified carbon paste electrode, *Colloids Surf.*, A, Vol. 396, 2012, pp. 8–15.

[95]. A. Galal, N. F. Atta, E. H. El-Ads, Probing cysteine self-assembled monolayers over gold nanoparticles – Towards selective electrochemical sensors, *Talanta*, Vol. 93, 2012, pp. 264–273.

[96]. N. F. Atta, A. Galal, R. A. Ahmed, Simultaneous Determination of Catecholamines and Serotonin on Poly(3, 4-ethylene dioxythiophene) Modified Pt Electrode in Presence of Sodium Dodecyl Sulfate, *J. Electrochem. Soc.*, Vol. 158, Issue 4, 2011, pp. F52-F60.

[97]. N. F. Atta, A. Galal, F. M. Abu-Attia, S. M. Azab, Characterization and electrochemical investigations of micellar/drug interactions, *Electrochim. Acta*, Vol. 56, 2011, pp. 2510–2517.

[98]. M. Castilho, L. E. Almeida, M. Tabak, L. H. Mazo, The electrochemical oxidation of the antioxidant drug dipyridamole at glassy carbon and graphite electrodes in micellar solutions, *Electrochim. Acta*, Vol. 46, 2000, pp. 67-75.

[99]. R. N. Goyal, N. Jain, V. Gurnani, Electrooxidation of chlorpromazine in aqueous and micellar media and spectroscopic studies of the derived cationic free radical and dication species, *Monatsh. Chem.*, Vol. 132, 2001, pp. 575-585.

[100]. L. H. Wang, S. W. Tseng, Direct determination of D-panthenol and salt of pantothenic acid in cosmetic and pharmaceutical preparations by differential pulse voltammetry, *Anal. Chim. Acta,* Vol. 432, 2001, pp. 39-48.

[101]. S. Zhang, K. Wu, S. Hu, Voltammetric determination of diethylstilbestrol at carbon paste electrode using cetylpyridine bromide as medium, *Talanta,* Vol. 58, 2002, pp. 747-754.

[102]. S. Zhang, K. Wu, S. Hu, Carbon Paste Electrode Based on Surface Activation for Trace Adriamycin Determination by a Preconcentration and Voltammetric Method, *Anal. Sci.,* Vol. 18, 2002, pp. 1089-1092.

[103]. S. G. Fernandez, M. C. B. lopez, M. J. L. Castanon, A. J. M. Ordieres, P. T. Blanco, Adsorptive Stripping Voltammetry of Rifamycins at Unmodified and Surfactant-Modified Carbon Paste Electrodes, *Electroanalysis,* Vol. 16, 2004, pp. 1660-1666.

[104]. N. F. Atta, A. Galal, R. A. Ahmed, Direct and Simple Electrochemical Determination of Morphine at PEDOT Modified Pt Electrode, *Electroanalysis,* Vol. 23, Issue 3, 2011, pp. 737–746.

[105]. N. F. Atta, A. Galal, R. A. Ahmed, Voltammetric Behavior and Determination of Isoniazid Using PEDOT Electrode in Presence Of Surface Active Agents, *Int. J. Electrochem. Sci.,* Vol. 6, 2011, pp. 5097–5113.

[106]. C. Li, Y. Ya, G. Zhan, Electrochemical investigation of tryptophan at gold nanoparticles modified electrode in the presence of sodium dodecylbenzene sulfonate, *Colloids Surf.,* B, Vol. 76, 2010, pp. 340–345.

[107]. G. Yang, X. Qu, M. Shen, C. Wang, Q. Qu, X. Hu, Preparation of glassy carbon electrode modified by hydrophobic gold nanoparticles and its application for the determination of ethamsylate in the presence of cetyltrimethylammonium bromide, *Sens. Actuators, B,* 128, 2007, pp. 258–265.

[108]. N. F. Atta, A. Galal, E. H. El-Ads, Gold nanoparticles-coated poly(3, 4-ethylene-dioxythiophene) for the selective determination of sub-nano concentrations of dopamine in presence of sodium dodecyl sulfate, *Electrochim. Acta,* Vol. 69, 2012, pp. 102–111.

[109]. N. F. Atta, S. A. Darwish, S. E. Khalil, A. Galal, Effect of surfactants on the voltammetric response and determination of an antihypertensive drug, *Talanta,* Vol. 72, 2007, pp. 1438–1445.

[110]. G. Alarcón-Angeles, S. Corona-Avendaño, M. Palomar-Pardavé, A. Rojas-Hernàndez, M. Romero-Romob, M. T. Ramírez-Silva, Selective electrochemical determination of dopamine in the presence of ascorbic acid using sodium dodecyl sulfate micelles as masking agent, *Electrochim. Acta,* Vol. 53, 2008, pp. 3013–3020.

[111]. S. Corona-Avendaño, G. Alarcón-Angeles, M. T. Ramírez-Silva, G. Rosquete-Pina, M. Romero-Romo, M. Palomar-Pardavé, On the electrochemistry of dopamine in aqueous solution. Part I: The role of

[SDS] on the voltammetric behavior of dopamine on a carbon paste electrode, *J. Electroanal. Chem.,* Vol. 609, 2007, pp. 17–26.

[112]. T. V. Sathisha, B. E. K. Swamy, B. N. Chandrashekar, N. Thomas, B. Eswarappa, Selective determination of dopamine in presence of ascorbic acid and uric acid at hydroxy double salt/surfactant film modified carbon paste electrode, *J. Electroanal. Chem.,* Vol. 674, 2012, pp. 57–64.

[113]. S. S. Shankar, B. E. K. Swamy, B. N. Chandrashekar, Electrochemical selective determination of dopamine at TX-100 modified carbon paste electrode: A voltammetric study, *J. Mol. Liq.,* Vol. 168, 2012, pp. 80–86.

[114]. K. R. Mahanthesha, B. E. K. Swamy, U. Chandra, S. S. Shankar, K. V. Pai, Electrocatalytic oxidation of dopamine at murexide and TX-100 modified carbon paste electrode: A cyclic voltammetric study, *J. Mol. Liq.,* Vol. 172, 2012, pp. 119–124.

[115]. J. Zheng, X. Zhou, Sodium dodecyl sulfate-modified carbon paste electrodes for selective determination of dopamine in the presence of ascorbic acid, *Bioelectrochemistry,* Vol. 70, 2007, pp. 408–415.

[116]. S. Reddy, B. E. K. Swamy, H. Jayadevappa, CuO nanoparticle sensor for the electrochemical determination of dopamine, *Electrochim. Acta,* Vol. 61, 2012, pp. 78–86.

[117]. R. A. Dar, P. K. Brahman, S. Tiwari, K. S. Pitre, Electrochemical determination of atropine at multi-wall carbon nanotube electrode based on the enhancement effect of sodium dodecyl benzene sulfonate, *Colloids Surf.,* B, Vol. 91, 2012, pp. 10–17.

[118]. B. J. Sanghavi, A. K. Srivastava, Simultaneous voltammetric determination of acetaminophen, aspirin and caffeine using an in situ surfactant-modified multiwalled carbon nanotube paste electrode, *Electrochim. Acta,* Vol. 55, 2010, pp. 8638–8648.

[119]. C. Yuan, Y. Wang, O. Reiko, Improving the detection of hydrogen peroxide of screen-printed carbon paste electrodes by modifying with nonionic surfactants, *Anal. Chim. Acta,* Vol. 653, 2009, pp. 71–76.

[120]. Y. Xu, C. Hu, S. Hu, Single-chain surfactant monolayer on carbon paste electrode and its application for the studies on the direct electron transfer of hemoglobin, *Bioelectrochemistry,* Vol. 74, 2009, pp. 254–259.

[121]. L. D. Mello, A. P. d. Reis, L. T. Kubota, Improvement of the electrochemical determination of antioxidant using cationic micellar environment, *Acta Sci. Technol.,* Vol. 32, Issue 4, 2010, pp. 421-425.

[122]. X. Wen, Y. Jia, Z. Liu, Micellar effects on the electrochemistry of dopamine and its selective detection in the presence of ascorbic acid, *Talanta,* Vol. 50, 1999, pp. 1027–1033.

[123]. S. Chen, W. Chzo, Simultaneous voltammetric detection of dopamine and ascorbic acid using didodecyldimethylammonium bromide (DDAB) film-modified electrodes, *J. Electroanal. Chem.,* Vol. 587, 2006, pp. 226–234.

[124]. R. Jain, R. K. Yadav, J. A. Rather, Voltammetric assay of anti-vertigo drug betahistine hydrochloride in sodium lauryl sulphate, *Colloids Surf., A*, Vol. 366, 2010, pp. 63–67.

[125]. F. Wang, J. Fei, S. Hu, The influence of cetyltrimethyl ammonium bromide on electrochemical properties of thyroxine reduction at carbon nanotubes modified electrode, *Colloids Surf.*, B, Vol. 39, 2004, pp. 95–101.

[126]. B. Hoyer, N. Jensen, Stabilization of the voltammetric serotonin signal by surfactants, *Electrochem. Commun.*, Vol. 8, 2006, pp. 323–328.

[127]. J. Rajbongshi, D. K. Das, S. Mazumdar, Direct electrochemistry of dinuclear CuA fragment from cytochrome c oxidase of Thermus thermophilus at surfactant modified glassy carbon electrode, *Electrochim. Acta*, Vol. 55, 2010, pp. 4174–4179.

[128]. K. Chattopadhyay, S. Mazumdar, Direct electrochemistry of heme proteins: effect of electrode surface modification by neutral surfactants, *Bioelectrochemistry*, Vol. 53, 2000, pp. 17–24.

[129]. V. K. Gupta, R. Jain, K. Radhapyari, N. Jadon, S. Agarwal, Voltammetric techniques for the assay of pharmaceuticals—A review, *Anal. Biochemistry*, Vol. 408, 2011, pp. 179–196.

[130]. R. Jain, R. K. Yadav, A. Dwivedi, Square-wave adsorptive stripping voltammetric behaviour of entacapone at HMDE and its determination in the presence of surfactants, *Colloids Surf.*, A, Vol. 359, 2010, pp. 25–30.

[131]. J. C. Cardoso, B. M. L. Armondes, T. A. d. Araújo, J. L. R. Jr, N. R. Poppi, V. S. Ferreira, Determination of 4-methylbenzilidene camphor in sunscreen by square wave voltammetry in media of cationic surfactant, *Microchem. J.*, Vol. 85, 2007, pp. 301–307.

[132]. S. Dong, J. Zheng, H. Gao, Voltammetric behavior of 40, 7-dimethoxy-30-isoflavone sulfonic sodium and its enhancement determination in the presence of surfactant, *Anal. Biochem.*, Vol. 323, 2003, pp. 151–155.

[133]. A. Levent, Y. Yardim, Z. Senturk, Voltammetric behavior of nicotine at pencil graphite electrode and its enhancement determination in the presence of anionic surfactant, *Electrochim. Acta*, Vol. 55, 2009, pp. 190–195.

[134]. P. Xie, X. Chen, F. Wang, C. Hu, S. Hu, Electrochemical behaviors of adrenaline at acetylene black electrode in the presence of sodium dodecyl sulfate, *Colloids Surf.*, B, Vol. 48, 2006, pp. 17–23.

[135]. S. Liu, W. Sun, F. Hu, Graphene nano sheet-fabricated electrochemical sensor for the determination of dopamine in the presence of ascorbic acid using cetyltrimethylammonium bromide as the discriminating agent, *Sens. Actuators, B*, Vol. 173, 2012, pp. 497–504.

[136]. P. Rattanarat, W. Dungchai, W. Siangproh, O. Chailapakul, C. S. Henry, Sodium Dodecyl Sulfate Modified Electrochemical Paper-Based Analytical Device for Determination of Dopamine Levels in Biological Samples, *Anal. Chim. Acta*, Vol. 744, 2012, pp. 1–7.

[137]. S. Shahrokhian, H. R. Zare-Mehrjardi, Cobalt salophen-modified carbon-paste electrode incorporating a cationic surfactant for simultaneous voltammetric detection of ascorbic acid and dopamine, *Sens. Actuators, B*, Vol. 121, 2007, pp. 530–537.

[138]. E. Niranjana, B. E. K. Swamy, R. R. Naik, B. S. Sherigara, H. Jayadevappa, Electrochemical investigations of potassium ferricyanide and dopamine by sodium dodecyl sulphate modified carbon paste electrode: A cyclic voltammetric study, *J. Electroanal. Chem.*, Vol. 631, 2009, pp. 1–9.

[139]. X. Cao, L. Luo, Y. Ding, X. Zou, R. Bian, Electrochemical methods for simultaneous determination of dopamine and ascorbic acid using cetylpyridine bromide/chitosan composite film-modified glassy carbon electrode, *Sens. Actuators, B*, Vol. 129, 2008, pp. 941–946.

[140]. J. R. Posac, M. D. Vàzquez, M. L. Tascón, J. A. Acuña, C. D. Fuente, E. Velasco, P. Sànchez-Batanero, Determination of Aceclofenac using adsorptive stripping voltammetric techniques on conventional and surfactant chemically modified carbon paste electrodes, *Talanta*, Vol. 42, Issue 2, 1995, pp. 293-304.

[141]. M. Kawakami, K. Tanaka, N. Uriuda, S. Gondo, Effects of nonionic surfactants on electrochemical behavior of ubiquinone and menaquinone incorporated in a carbon paste electrode, *Bioelectrochemistry*, Vol. 52, 2000, pp. 51–56.

[142]. A. Nezamzadeh-Ejhieh, E. Afshari, Modification of a PVC-membrane electrode by surfactant modified clinoptilolite zeolite towards potentiometric determination of sulfide, *Microporous Mesoporous Mater.*, Vol. 153, 2012, pp. 267–274.

[143]. A. Nezamzadeh-Ejhieh, Z. Nematollahi, Surfactant modified zeolite carbon paste electrode (SMZ-CPE) as a nitrate selective electrode, *Electrochim. Acta*, Vol. 56, 2011, pp. 8334–8341.

[144]. A. Nezamzadeh-Ejhieh, E. Mirzaeyan, Oxalate membrane-selective electrode based on surfactant-modified zeolite, *Electrochim. Acta*, Vol. 56, 2011, pp. 7749–7757.

[145]. Z. Mo, Y. Zhang, F. Zhao, F. Xiao, G. Guo, B. Zeng, Sensitive voltammetric determination of Sudan I in food samples by using gemini surfactant–ionic liquid–multiwalled carbon nanotube composite film modified glassy carbon electrodes, *Food Chem.*, Vol. 121, 2010, pp. 233–237.

[146]. C. R. Molina, M. Boujtita, N. El Murr, *Electroanalysis*, Vol. 15, 2003, pp. 1059.

[147]. R. Wang, J. Zhang, Y. Hu, Liquid phase deposition of hemoglobin/SDS/TiO$_2$ hybrid film preserving photoelectrochemical activity, *Bioelectrochemistry*, Vol. 81, 2011, pp. 34–38.

[148]. J. Zhang, L. Gao, Dispersion of multiwall carbon nanotubes by sodium dodecyl sulfate for preparation of modified electrodes toward detecting hydrogen peroxide, *Mater. Lett.*, Vol. 61, 2007, pp. 3571–3574.

[149]. Y. Li, J. Du, J. Yang, D. Liu, X. Lu, Electrocatalytic detection of dopamine in the presence of ascorbic acid and uric acid using single-

walled carbon nanotubes modified electrode, *Colloids Surf.*, B, Vol. 97, 2012, pp. 32–36.

[150]. A. L. Ndiaye, C. Varenne, P. Bonnet, E. Petit, L. Spinelle, J. Brunet, A. Pauly, B. Lauron, Elaboration of SWNTs-based gas sensors using dispersion techniques: Evaluating the role of the surfactant and its influence on the sensor response, *Sens. Actuators, B*, Vol. 162, 2012, pp. 95–101.

[151]. W. Wen, W. Chen, Q. Rena, X. Hub, H. Xiongb, X. Zhangb, S. Wangb, Y. Zhao, A highly sensitive nitric oxide biosensor based on hemoglobin–chitosan/graphene–hexadecyltrimethylammonium bromide nanomatrix, *Sens. Actuators, B*, Vol. 166, 2012, pp. 444–450.

[152]. G. Fan, J. Huang, X. Fan, S. Xie, Z. Zheng, Q. Cheng, P. Wang, Enhanced oxidation and detection of toxic clenbuterol on the surface of acetylene black nanoparticle-modified electrode, *J. Mol. Liq.*, Vol. 169, 2012, pp. 102–105.

[153]. P. Zhou, L. He, G. Gan, S. Ni, H. Li, W. Li, Fabrication and evaluation of $[Co(phen)2L]^{3+}$-modified DNA-MWCNT and SDS-MWCNT electrodes for electrochemical detection of 6-mercaptopurine, *J. Electroanal. Chem.*, Vol. 665, 2012, pp. 63–69.

[154]. R. Ojani, J. Raoof, S. Zamani, A novel sensor for cephalosporins based on electrocatalytic oxidation by poly(o-anisidine)/SDS/Ni modified carbon paste electrode, *Talanta*, Vol. 81, 2010, pp. 1522–1528.

[155]. Q. Xu, C. Xu, Q. Wang, K. Tanaka, H. Toada, W. Zhang, L. Jin, A pplication of a single electrode, modified with polydiphenylamine and dodecyl sulfate, for the simultaneous amperometric determination of electro-inactive anions and cations in ion chromatography, *J. Chromatogr.*, A, Vol. 997, 2003, pp. 65–71.

[156]. S. M. S. Kumar, K. C. Pillai, Compositional changes in unusually stabilized Prussian blue by CTAB surfactant: Application to electrocatalytic reduction of H_2O_2, *Electrochem. Commun.*, Vol. 8, 2006, pp. 621–626.

[157]. P. Salazar, M. Martìn, R. D. O'Neill, R. Roche, J. L. Gonzàlez-Mora, Improvement and characterization of surfactant-modified Prussian blue screen-printed carbon electrodes for selective H_2O_2 detection at low applied potentials, *J. Electroanal. Chem.*, Vol. 674, 2012, pp. 48–56.

[158]. P. Salazar, M. Martìn, R. D. O'Neill, R. Roche, J. L. Gonzàlez-Mora, Surfactant-promoted Prussian Blue-modified carbon electrodes: Enhancement of electro-deposition step, stabilization, electrochemical properties and application to lactate microbiosensors for the neurosciences, *Colloids Surf.*, B, Vol. 92, 2012, pp. 180–189.

[159]. J. Li, L. Zhou, X. Han, H. Liu, Direct electrochemistry of hemoglobin based on Gemini surfactant protected gold nanoparticles modified glassy carbon electrode, *Sens. Actuators, B*, Vol. 135, 2008, pp. 322–326.

Chapter 5

Synthesis and Sensing Applications of Nano-structured Conducting Polymers and Conducting Polymers-based Nanocomposites

Ahmed Galal, Nada F. Atta and Shimaa M. Ali

5.1. Introduction

Intrinsically conducting polymers, also known as "synthetic metals", are polymers with a highly π-conjugated polymeric chain. For the discovery of conducting polymers, Alan J. Heeger, Alan G. MacDiarmid and Hideki Shirakawa were awarded the Nobel Prize in Chemistry in 2000. The conjugated polymers can be electrical insulators, semiconductors or conductors, depending on the level of doping and nature of the dopants. Upon treating with dopants and/or subjecting to chemical or electrochemical redox reactions, the electrical conductivity of these conjugated polymers can increase by several orders of magnitude. The high conductivity upon doping makes these polymers promising materials for applications ranging from electro-optic, molecular and nanoelectronic devices to microwave absorbing and corrosion protection coatings. In addition, the ability of the reversible doping–dedoping properties provides them for the applications in sensors, actuators and separation membranes [1–6]. Recently, nanostructured conducting polymers have been subjected to intensive investigation owing to the unique combination of electronic properties of conductive polymers and large surface area of nanomaterials. There are mainly four types of nanostructures: zero, one, two and three dimension structures. Among them, one-dimensional (1D) nanostructures have been the focus of quite extensive studies worldwide, partially because of their unique physical and chemical properties. Compared to the other three dimensions, the first characteristic of 1D nanostructure is its smaller dimension structure and high aspect ratio, which could efficiently transport electrical carriers along one controllable direction, thus are highly suitable for moving charges in intergrated nanoscale systems. The second charateristic of 1D nanostructure is its device function, which can be exploited as device elements in many kinds of nanodevices. In particular, 1D conducting polymeric nanomaterials are exploited as nanowires in electronic devices [7–13]. Over the last few years, many synthetic strategies have been derived for the fabrication of 1D conducting polymer nanomaterials, including template-directed approach, self-assembly, interfacial polymerization, seeded polymerization, rapid mixing polymerization, radialysis or ultrasonic assisted method, etc. [14–20]. In addition to the formation of 1D nanostructures, incorporation of at least one secondary component into conducting polymers to form nanocomposite is another useful approach to improve or extend the functionality of conducting polymers [21]. It is anticipated that distinct properties from the synergistic effect of each component will be observed in the conducting polymer nanocomposites. This may include improved chemical properties or combined multi-functionalized chemical/physical/biological properties.

The polymer-based nanocomposite retains the inherent properties of nanoparticles, while the polymer support materials provide higher stability, processability and some interesting improvements caused by the nanoparticle–matrix interaction. The generally used nanoparticles include zero-valent metals [22–29], metallic oxides [30–40], biopolymers [41–47], and single-enzyme nanoparticles [48–54]. The choice of the polymeric supports is usually guided by their mechanical and thermal behavior. Other properties such as hydrophobic/hydrophilic balance, chemical stability, bio-compatibility, optical and/or electronic properties and chemical functionalities (i.e. solvation, wettability, templating effect, etc.) have to be considered to select the organic hosts [55]. This chapter presents an updated review on synthetic methodologies of nano-structured CPs and polymer nanocomposites and their potential applications in the field of nanosensors/biosensors.

222

5.2. Synthesis Methods of Nanostructured CPs

5.2.1. Hard Physical Template Method

The template method of polymerization proposed by Martin et al.
[56-64] is an effective technique to synthesize arrays of aligned
polymer micro-/nanotubes and wires with controllable length and
diameter. The disadvantage of this method is that a post-synthesis
process is needed in order to remove the template. Many porous
materials have been and are currently being used as templates for the
fabrication of nanofibers and tubes, but anodic aluminum oxide
templates and particle track-etched membranes are the most commonly
used nanoporous materials. By now, nanotubes/wires of a variety of
CPs such as polyaniline [56, 57], polypyrrole [56-60],
poly(3-methylthiophene) (P3MT) [59], PEDOT [58, 61], PPV [64]
have been chemically or electrochemically synthesized inside the pores
of these membranes. Furthermore, CdS-PPY heterojunction nanowires
[65, 66], multi-segmented Au-PEDOT-Au and Au-PEDOT-PPY-Au
nanowires [67, 68], MnO_2/PEDOT [69] and Ni/PPV [70] coaxial
nanowires have also been prepared by the hard template method. For
example, core-shell Ni/PPV nanowires were fabricated by a two-step
process: tubular PPV shells were prepared first within nanoporous
membranes by a wet-chemical technique; the PPV tubules then were
operated as a secondary template for the electrodeposition of the Ni
cores [70]. The coaxial MnO_2/PEDOT nanowires were obtained by a
simple one-step method of coelectrodeposition in a porous alumina
template, the phase segregation of these two materials may help
formation of the core-shell structures [69]. Besides these hard templates
with channels inside pores, many kinds of pre-existing nanostructures
can serve as seeds or templates to synthesize CP nanostructures. For
example, polyaniline, polypyrrole, and PEDOT nanofibers/tubes have
been prepared by using V_2O_5 nanofibers [71–73] or MnO_2 nanowires
[74] as seeds. In addition, poly(styrene-block-2-vinylpyridine) diblock
copolymers [75] and a variety of biological templates such as DNA
[76–78] and tobacco mosaic virus [79] have also been used to fabricate
conducting polymer nanofibers. Particularly, Abidian et al. [80-83]
recently developed a method for the fabrication of PEDOT and
polypyrrole nanotubes from biodegradable electrospun nanofibers as
templates for drug delivery and neural interface applications.

5.2.2. Soft Chemical Template Method

The soft-template method is another powerful and popular technique to produce CP nanomaterials. By now, surface micelles, surfactants, colloidal particles, liquidcrystalline phases, structure-directing molecules, and aniline oligomers have served as soft templates, and various soft-template methods have been developed: interfacial polymerization [84, 85], dilute polymerization [86], template-free method [87, 88], rapidly mixed reactions [89], reverse emulsion polymerization [90], ultrasonic irradiation [91], and radiolytic synthesis [92–94]. The interfacial polymerization method proposed by Kaner et al. [84, 85] involves step polymerization of two monomers or agents, which are dissolved respectively in two immiscible phases so that the reaction takes place at the interface between the two liquids. These methods are usually based on self-assembly mechanisms due to hydrogen bonding, π–π stacking, van der Waals forces, and electrostatic interactions as driving forces. The disadvantage of the soft-template method is poor control of the morphology, orientation, and diameter of the 1D CP nanostructures.

The template-free method developed by Wan et al. [95–102] is a simple self-assembly (soft-template) method without an external template. By controlling synthesis conditions, such as temperature and molar ratio of monomer to dopant, polyaniline and polypyrrole nanostructures can be prepared by in situ doping polymerization in the presence of protonic acids as dopants. In the self-assembled formation mechanism in this approach the micelles formed by dopant and/or monomerdopant act as soft templates in the process of forming tubes/wires [87]. Up to now, a variety of polyaniline micro/nanostructures such as micro/nanotubes [88, 95, 96], nanowires/fibers [97-99], hollow microspheres [100, 101], nanotube junctions and dendrites [95, 102] have been prepared by the template-free method. Furthermore, it was reported that the magnetic and optical properties of polymer micro/nanostructures synthesized by this method can be significantly improved by using a functional dopant (e.g., Fe_3O_4, γ-Fe_2O_3, carbonyl iron, $CoFe_2O_4$ and TiO_2 nanoparticles, azobenzene sulfuric acid, and Rhodamine B, etc.). Namely, (electrical, magnetic, optical, etc.) multifunctionalized micro/nanostructures of polyaniline can be fabricated by this approach. In addition, self-assembled polyaniline microstructures with high orientation were also reported [103].

5.2.3. Electrospinning

Electrospinning is an effective approach to fabricate long polymer fibers with diameter from micrometers down to 100 nm or even a few nanometers by using strong electrostatic forces [104, 105]. As shown in Fig. 5.1, in the usual electrospinning process, polymer solution is extruded from an orifice to form a small droplet in the presence of an electric field, and then the charged solution jets are extruded from the cone.

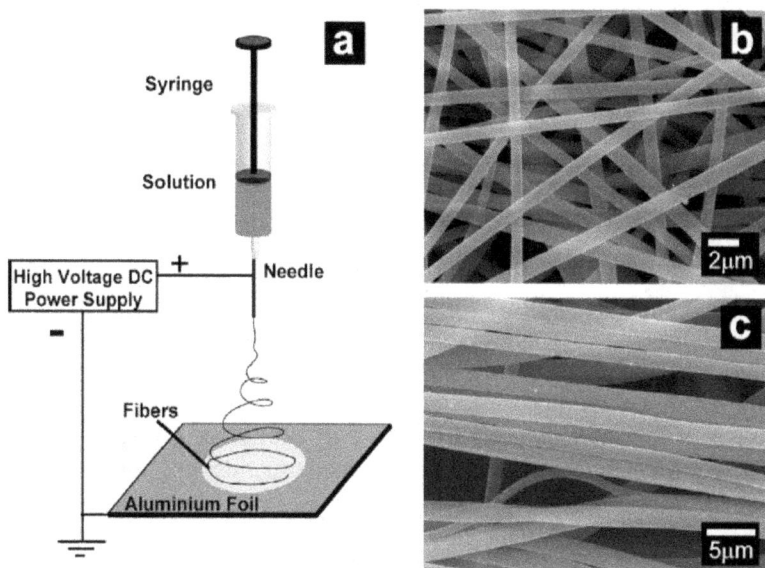

Fig. 5.1. (a) Schematic diagram of electrospinning method and SEM images of electrospun polymer nanofibers (b) without orientation, and (c) with preferential orientation [106].

Generally the fluid extension occurs first in uniform, and then the straight flow lines undergo vigorous whipping and/or splitting motion due to fluid instability and electrically driven bending instability. Finally, the spun fibers are deposited commonly as a nonwoven web on a collector. Through an improved or modified electrospinning device, nanofibers with part or even good orientation could be fabricated [104, 105]. By now, micro- and nano-scale fibers of polyaniline/polyethylene oxide (PEO) [107, 108], polypyrrole/PEO [108], pure polyaniline [109] and polypyrrole [110, 111], poly(3-hexyl-thiophene)/PEO [112], and polyaniline/PEO/carbon nanotubes [113] have been prepared by this technique. In addition, field-effect transistor [114, 115] and chemical sensor [116] based on individual electrospun nanofibers have also been reported. Compared with other synthetic approaches, the electrospinning process seems to be the only method that can mass-produce continuous long nanofibers. However, in order to assist in the fiber formation, some non-conducting polymers or chemicals (e.g. PEO) are usually added intospinning solution, which may result in a decrease of the conductivity (10^{-1}–10^{-4} S/cm [107, 109–111]) of the electrospun composite fibers. It is found that through reducing or eliminating PEO content [109–111, 114] or embedding carbon nanotubes in the fibers [113], their conductivity could be increased by one or several orders of magnitude.

5.2.4. Nanoimprint Lithography or Embossing

Soft lithography or embossing is a rapid and low-cost approach to shape an initially flat polymer film by using a micro-mold with the assistance of temperature or solvent vapors. Recently, it was reported that conducting polymer nanowires can easily be fabricated by this technique [117–120]. For example, conducting PEDOT doped with poly(4-styrenesulfonate) was patterned in the form of nanowires on a glass or a Si wafer by micro-molding in capillaries [117]. Nanowires and two-dimensional nanodots of semiconducting polymer were also achieved by a liquid embossing technique [117]. In addition, Hu et al. [118] demonstrated that arrays of CP nanowires with internal preferential alignment can be produced by a simple embossing protocol. Recently, Huang et al. [119] proposed a technique based on nanoimprint lithography and a lift-off process for patterning CPs.

5.2.5. Directed Electrochemical Nanowire Assembly

Directed electrochemical nanowire assembly technique has been employed to grow metal nanowires [121] as well as CP micro/nanowires [122–126]. In this method, CP wire is electrochemically polymerized and assembled onto two biased electrodes (anode and cathode) immersed inaqueous monomer solutions. The essence of this method is an electrode-wire-electrode or electrode-wire-target assembly. Directional growth of polypyrrole, polyaniline, and PEDOT micro/nanowires with knobby structures (e.g., varying from 90 to 700 nm in thickness with a lengthwise averaged diameter of 340 nm for PEDOT wire [126]) between electrodes by this technique was reported recently [122–126]. It was found that these assembled polymer wires can be used as pH sensors [122] and in cell stimulation studies [126].

5.2.6. Other Methods

Some other methods were also reported to produce CP micro/nanostructures. For example, Noy et al. [127] reported the fabrication of luminescent nanostructures and conductive polymer nanowires of controlled size using dippen nanolithography. Conducting polymer nanowires of poly(styrenesulfonate) doped PEDOT with diameters under 10 nm were prepared by a molecular combing method [128], which is a well-known technique to stretch a single DNA molecule on various substrates. Recently, Samitsu et al. reported the production of self-assembled conducting polymer nanofibers by a whisker method using anisotropic crystallization in a nematic liquid crystal, and obtained large-scale alignment of the nanofibers [129]. In addition, nanowires consisting of poly(3-hexylthiophene) with part magnetic orientation were prepared by using strong magnetic field [130].

5.3. Synthesis Methods of Polymer Nanocomposites

5.3.1. 1D Conducting Polymer Nanocomposites
with Metal Nanomaterials

Metal nanoparticles or 1D structured nanomaterials are of great importance for their numerous electrical, optical and catalytical properties and a wide range of applications including nanoelectronics and sensing devices [131]. In the past decade or so, the composition and shape control of metal nanoparticles have been extensively studied. Combining the metal nanomaterials with 1D conducting polymers, in addition to the anisotropic electrical properties of 1D conducting polymers and optical properties of metal nanostructures, new and yet-to-be imagined properties can be attained because of the electron transfer between the two components. One of the most important properties for conducting polymers is their reversible oxidation/reduction (i.e., redox) chemistry. It means that conducting polymers can be oxidized by strong oxidants such as $HAuCl_4$, H_2PdCl_4,

H_2PtCl_6, $AgNO_3$, etc., to an over-oxidized state. This property can be applied for the fabrication of 1D conducting polymer nanocomposites with metal component. Kaner and co-workers have demonstrated that PANI/Ag composite nanofibers can be prepared by the redox process between PANI nanofibers and the corresponding $AgNO_3$ [132]. Similarly, gold and Pd nanoparticles were also easily produced inside or on the surface of the PANI nanofibers via such an in situ redox reaction [133, 134]. In addition to PANI nanofibers, PPy nanostructures were also proven a good matrix for loading metal nanoparticles to form 1D PPy/metal nanocomposite. Ag and Au nanoparticles 3–5 nm in size can be deposited into PPy nanotubes with an inner diameter of less than 10 nm [135]. The results demonstrated Ag nanoparticles synthesized via a rapid reduction by PPy were on the surface and in the interior of the PPy sheath. Cable-like Ag/PPy nanocomposites could also be obtained by the agglomeration of Ag nanoparticles in the pore of PPy nanotubes. In contrast, Au nanoparticles were mainly formed in the pore region of PPy nanotubes. Wei and co-workers have reported the fabrication of PPy nanotubes with the inner diameter of about 20 nm or more by a self-degraded template method. These were used matrices for loading Au nanoparticles to form 1D PPy/Au nanocomposites [136]. The results represented a somewhat different loading pathway compared to the above mentioned PPy nanotubes with an inner diameter of less than 10 nm, in which the as-synthesized Au nanoparticles were formed on every region of PPy nanotubes (i.e., in the pore and sheath, and on the surface). Furthermore, the size of the as-synthesized Au nanoparticles was not uniform, the average diameter for small Au nanoparticles was about 13 nm, while that for the large nanoparticles could reach about 80 nm. However, adding surfactant molecules such as Tween-80 into the reaction system could make monodispersed Au nanoparticles with a diameter of about 13 nm on PPy nanotubes.

In addition to in situ reduction by the conducting polymers, metal nanoparticles can also be synthesized on the surface of the 1D conducting polymers via a reduction process in the presence of other reducing agents. Yan and co-workers prepared PANI/Pt composite nanofibers by reducing a Pt salt with ethylene glycol on the surface of PANI nanofibers [137]. The diameter of PANI nanofibers was about 60 nm, and that of as-synthesized Pt nanoparticles was only about 1.8 nm as calculated from XRD data. The small size of Pt nanoparticles on the 1D conducting polymer matrix enhanced the electrocatalytic activity for the methanol oxidation reaction. Through the polyol reduction approach with assistance by microwave technique, Hu and co-workers have also fabricated PPy/Pt composite nanofibers [138]. They found that the average size of the Pt nanoparticles was about 2.5 nm. Besides ethylene glycol, Wang and co-workers reported a HCOOH reduction method for the fabrication of PANI nanofibers supported Pt nanoparticles with a diameter of about 3 nm [139]. Importantly, the amount of Pt nanoparticles on the surface of PANI nanofibers can be well controlled by adjusting the molar ratio of PANI to Pt precursors. Furthermore, Pt/Pd hybrid nanoparticles could also be readily loaded on the surface of PANI nanofibers to form 1D nanocomposites via simultaneously reducing Pt and Pd precursors by HCOOH reducing agent. The ease in the control of density and composition rendered such 1D conducting polymers/metal nanocomposites of interest for potential applications in sensors, nanoelectronics and other electrochemical devices.

Template techniques are among the most versatile and simple approaches for the fabrication of 1D nanostructures from many kinds of materials, including conducting polymers. The processes based on these techniques include deposition of the desired materials within the pores of the self-assembled template molecules, followed by the dissolution of the template. Metal/conducting polymers with core-sheath structure have been fabricated by template techniques. For example, Xuand co-workers prepared metal/PANI nanotubules using the template approach via a synthetic strategy consisting of three steps: (1) polymerization of aniline using ammonium persulphate (APS) as an oxidant in the pores of anodic aluminum oxide (AAO) template to form PANI nanotubules; (2) preparation of metal nanowires within PANI nanotubules by electrodeposition; (3) dissolution of AAO template with an alkali solution [140–142]. The PANI might also protect the metal nanowires from oxidation and corrosion, a deirable property for a candidate for use in electronic devices. The Au/PTh composite nanowires with core sheath structures can also be fabricated by electrochemical preparation of PTh, followed by electrochemical depositon of Au into the pores of AAO [143]. To avoid amulti-step synthesis, metal/conducting polymers composite nanowires can be prepared by simultaneous polymerization of the monomers of conducting polymers such as PANI and the reduction of metal within the pores of the AAO template. High-resolution transmission electron microscopy (HRTEM) images showed that cable-like nanocomposites were formed with a metal core surrounded by PANI coating [144]. Besides the core-sheath structures, segmented architectures of metal/conducting polymers can be fabricated electrochemically within AAO template [145–148]. In addition to porous templates such as AAO, polymer or organic fibers can also be employed as templates to fabricate 1D PANI/metal composites [149,150]. To prevent leaching of metal nanoparticles from conducting polymers matrices, Zhang and co-workers have developed a template method to synthesize PANI/Pd nanotubes with Pd nanoparticles attached to the inner walls of PANI nanotubes [149]. The first step was to coat Pd nanoparticles on the surface of sulfonated polystyrene (PS) nanofibers. Then the PANI layer was coated on the surface of PS/Pd composite nanofibers by a self-assembling polymerization method. Finally, the PS nanofibers as the templates were removed by dissolution in tetrahydrofuran. TEM images revealed that Pd nanoparticles with average size of about 3.4 nm were attached onto the inner walls of PANI nanotubes. This approach can be extended to the fabrication of other 1D conducting polymers/metal nanocomposites with metal nanoparticles on the inner walls of conducting polymer nanotubes.

As mentioned above, metal/conducting polymers with core-sheath structure can be prepared by the template method. However, the approach based on the template technique is complicated and non-economical because of the need to remove the templates. In fact, metal/conducting polymers with core-sheath structure can be fabricated via a one-step chemical polymerization [151–155]. It is well known that the standard reduction potential of most common noble metal salts, such as $HAuCl_4$, H_2PtCl_6, H_2PdCl_4, and $AgNO_3$, is higher than that of the monomers of conducting polymers, including aniline, pyrrole and ethylenedioxythiophene (EDOT), so that the noble metal salts have the ability to oxidatively polymerize such monomers. Cable-like structures of metal/conducting polymers can be synthesized by controlling the reaction conditions. Li and co-workers were the first to report that Ag/PPy nanocables can be prepared through the redox reaction between pyrrole and silver nitrite in aqueous solution in the presence of poly(vinyl pyrrolidone) (PVP) [151]. During the reaction process, the formation of Ag nanowires and the polymerization of a pyrrole sheath proceeded simultaneously. Fig. 5.2 shows the TEM images of the as-synthesized Ag/PPy nanocables, displaying the coaxial nanostructures with a dark core and a lighter sheath. The diameter of the Ag nanowires core was about 20 nm, while the outer diameter of PPy sheath reached about 50 nm.

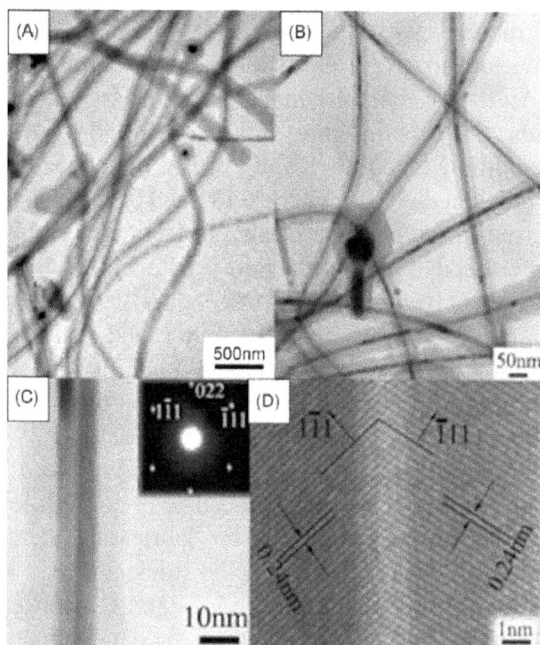

Fig. 5.2. (A, B) TEM images of Ag/PPy nanocables with low-magnification,
(C) High-magnification TEM image of a single Ag/PPy nanocable; inset: ED
patten of silver core, showing the diffracton dot of silver. (D) HRTEM image
of Ag core, showing the silver grows as a crystal twinned at the (111) planes
[151]. Copyright 2005, the Royal Society of Chemistry.

In such a system for the preparation of Ag/PPy nanocables, PVP played
an important role in kinetic control of the growth rates of various faces
of silver. In the absence of PVP, the growth rate of the redox reaction
can be controlled by lowering reduction potential of the oxidants.
Replacing $AgNO_3$ with Ag_2O as the oxidant, snake-like Ag/PPy core-
sheath structures have been prepared by one-pot procedure involving
hydrothermal reactions in the absence of any surfactant [152]. The
mechanism of the formation of Ag/PPy core-sheath nanostructures
might be related to the self-assembly of the silver nanoparticles reduced
from Ag_2O particles inside the PPy matrix oxidatively polymerized
from pyrrole monomer. Niu and co-workers demonstrated that
Au/PANI coaxial nanocables could also be fabricated by the redox
reaction between chlorauric acid and aniline in the presence of d-CSA
[153]. In that case, CSA acted not only as a dopant, but also as a
surfactantsor a soft template, just as the role of PVP in the system to the
fabrication of Ag/PPy nanocables. TEM images exhibited coaxial

nanocables structures with an outer (PANI–CSA sheath) diameter of 50–60 nm and an inner (Au core) diameter of about 20 nm. The as-synthesized Au/PANI coaxial nanocables can be converted to hollow PANI nanotubes by dissolution of Au nanowire core by a saturated I_2 solution. In addition to Ag/PPy and Au/PANI nanocables, cable-like Au/poly(3,4-ethylenedioxythiophene) (PEDOT) nanostructures have been synthesized in the absence of any surfactant or stabilizer through one-step interfacial polymerization of EDOT dissolved in dichloromethane solvent and $HAuCl_4$ dissolved in water [154]. Microscopy studies showed that the outer and inner diameters of Au/PEDOT nanocables were around 50 and 30 nm, respectively. Similar to the Au/PANI nanocables, PEDOT nanotubes can be obtained by dissolution of the Au component in saturated I_2 solution. Furthermore, pure gold nanowires could also be produced by removing PEDOT sheath via oxygen plasma decomposition. While the greatest advantage of the one-step approach for fabricating metal/conducting polymer nanocables is its simplicity, the thickness of the conducting polymer sheaths is not easily controlled. Li and co-workers developed an interesting post-polymerization method to coat conducting polymers on the surface of metal nanowires [156]. To ensure the polymerization occurs on the surface of silver nanowires, Ag ions were firstly adsorbed onto the closest surface of silver nanowires through the common ions adsorption effect. Pyrrole monomer was then added into the system of Ag^+/Ag nanowires and the polymerization would begin between the pyrrole monomer and Ag^+ to produce Ag/PPy nanocables. TEM images proved the coresheath structures of the as-prepared Ag/PPy nanocables. More importantly, the thickness of PPy sheath layer can be well controlled by changing the concentration of the $AgNO_3$ aqueous solution used to adsorb Ag^+ onto the surface of the Ag nanowires. The thickness of the PPy layer increased from 3 nm to more than 20 nm as the concentration of $AgNO_3$ solution was increased from 1 to 10 %. In addition to the core-sheath structures, 1D conducting PANI decorated with metal nanoparticles can be obtained by γ radiolysis. The size of metal nanoparticles can be varied by adjusting the ratio of aniline to the metal precursors [157].

5.3.2. 1D Conducting Polymer Nanocomposites with Metal Oxides

The fabrication of 1D conducting polymer nanocomposites with metal oxides has become an alluring aspect for their potential applications in sensors and energy devices. The synergy between 1D conducting polymers and metal oxides provides the resultant nanocomposites with added functionalities compared to their individual parent materials. The self-assembly method, also called "template-free" method, is a facile and versatile technique for the preparation of 1D conducting polymers and their nanocomposites. Wan and co-workers demonstrated that 1D PANI/TiO_2, PANI/Fe_3O_4 and PANI/γ-Fe_2O_3 nanocomposites can be readily fabricated by a selfassembling process [158, 159]. The mechanism of the self-assembly can be related to the formation of micelles by β-naphthalene sulfonic acid (β-NSA) anions and anilinium cations as soft templates. As an example for the preparation of 1D PANI/TiO_2 nanocomposites, TiO_2 nanoparticles can be well dispersed in a β-NSA solution prior to the polymerization of aniline monomer. The micelles contained TiO_2 nanoparticles act as the "core", with the aniline/ β-NSA salt the "shell", to function as a soft template in the formation of PANI/TiO_2 composite nanofibers or nanotubes. After adding APS to the system to induce the polymerization, the micelles containing TiO_2 become larger spheres or tubes/fibers by elongation, resulting in the formation of a 1D PANI/TiO_2 nanocomposite. Microscopic studies indicated that PANI/TiO_2 nanotubes or nanofibers were formed at low concentrations of TiO_2. However, granular PANI/TiO_2 composites would form when the concentration of TiO_2 wasincreased to 0.12 M. Introducing ultrasonic irradiation into the self-assembly process for the fabrication of 1D conducting polymers/metal oxide nanocomposites dispersed the inorganic nanocomponents in the conducting polymer matrix. Our group reported the synthesis of PANI/Fe_3O_4 composite nanotubes by chemical polymerization assisted with ultrasonic irradiation [160]. To enhance the compatibility between PANI and Fe_3O_4, Fe_3O_4 nanoparticles were capped with an aniline dimmer, which can be covalently bonded with PANI. TEM images showed a morphology with the nanotubes and Fe_3O_4 nanoparticles well dispersed in the PANI matrix. During the preparation of conducting polymers, an inorganic or organic acid is often added to the reaction system as the doping agent to increase the conductivity of the conducting polymers. Recently, scientists have found that novel conducting polymer nanostructures can be synthesized in the absence of inorganic or organic acid, or even in a basic solution [161–163]. 1D conducting polymers/metal oxide nanocomposites have also been fabricated in the absence of acids [164–167]. Guo and co-workers demonstrated that rectangular tubes of PANI/NiO composites can be prepared in the presence of SDBS without adding any acid through a self-assembly process [165].TEM images showed that the wall thickness of rectangular PANI/NiO composite tubes was approximately

150 nm, with a length of several micrometers. By changing the molar ratio of aniline to NiO, nanobelts of PANI/NiO composite could also be obtained [166]. Ciric-Marjanovic et al. reported the similar self-assembly technique for the fabrication PANI/silica composite nanotubes without any added acid, also in the absence of any surfactants [167]. The weight ratio of silica to aniline was found to influence the morphology of the as-prepared PANI/silica nanocomposites. With an initial weight ratio of silica to aniline lower than 0.2, nanotubes of PANI/silica with an outer diameter of 100–250 nm and an inner diameter of 10–80 nm can be observed, while only nanogranules have been obtained as the weight ratio of silica to aniline reaches about 2. Similar to the fabrication of conducting polymer/metal nanocomposites, 1D conducting polymer/metal oxide nanocomposites could also be prepared via a one-step approach. Zhang et al. demonstrated that PEDOT/ β-Fe^{3+}O(OH,Cl) nanospindles can be synthesized through chemical oxidation polymerization using FeCl$_3 \cdot$6H$_2$O as an oxidant in the presence of CTAB and poly(acrylic acid) (PAA) [168]. At the same time of the polymerization of EDOT, FeCl$_3$ would hydrolyze to form β-Fe^{3+}O(OH,Cl), resulting in the formation of PEDOT/ β-Fe^{3+}O(OH,Cl) nanospindles. Microscopy revealed that the nanospindles had a length in the range of 350–370 nm and a width of about 80–90 nm. The reaction conditions, such as the concentration of EDOT, the molar ratio of EDOT to FeCl$_3 \cdot$6H$_2$O, the temperature, the type of oxidant, surfactants and even solvents all influenced the morphology of the as-synthesized products. The core-sheath structure of conducting polymers/metal oxides is one of the most interesting features of nanocomposites for the potential synergic behavior between the good electrical conducting polymers and a functional metal oxide. In a one-step process, Li et al. presented an in situ polymerization method for the fabrication of V$_2$O$_5$/PANI nanobelts with core-sheath structure, without adding any other oxidant or initiator [169]. In a typical polymerization, V$_2$O$_5$ acted not only as a template, but also the oxidant for the preparation of PANI. In an appropriate acidic solution, V$_2$O$_5$ can be partly dissolved to form vanadic acid, which could function as an oxidant for the polymerization of aniline. The results showed that V$_2$O$_5$/PANI core-sheath nanobelts with a width and thickness of 400 and 50 nm had been formed. Furthermore, the morphologies of the V$_2$O$_5$/PANI core-sheath nanobelts are influenced by the pH of the solution and the additional initiator. Vapor-phase polymerization gives another versatile approach for the fabrication of metal oxide/conducting polymers core-sheath structures. Our group demonstrated that core-sheath nanocables of TiO$_2$/PPy can be obtained

by the polymerization of pyrrole on the surface of electrospun TiO_2 nanofibers, with $FeCl_3$ as the oxidant [170]. The electrospun TiO_2 nanofibers have a rough surface and some voids, which favor the existence of $FeCl_3$ oxidant on their surface; thus pyrrole can be polymerized on the surface of TiO_2 nanofibers to form core-sheath structures. TEM images showed a typical coaxial nanocable structure of the TiO_2/PPy nanocomposites obtained. The thickness of the sheath of as-synthesized TiO_2/PPy composite nanocables is about 20 nm under such an experimental procedure. However, the thickness of the conducting polymer layers is not well controlled in the vapor-phase polymerization. Surfactant-directed in situ polymerization approach can be used to prepare TiO_2/PPy coaxial nanocables with controlled thickness of the PPy layer [171]. Very uniform and smooth PPy layer was coated on the surface of electrospun TiO_2 nanofibers via the surfactant-directed polymerization with the assistance of the ultrasonic irradiation. In addition to the surfactants, block copolymers have also been used to direct the formation of metal oxide/conducting polymers nanocomposites with core-sheath structure [172]. A poly(ethylene oxide)–poly(propylene oxide)–poly(ethylene oxide) (PEO–PPO–PEO) triblock copolymer, adsorbed on the surface of titanate nanotubes during the process of the polymerization of aniline, acting as a soft template for the fabrication of titanate/PANI core-sheath nanotubes. TEM images indicated that the outer layer was amorphous PANI, while the inner layer was crystalline of titanate. Furthermore, PANI was almost uniformly grown on the surface of titanate nanotubes to form titanate/PANI core-sheath nanotubes with a smooth surface. Hydrothermal reaction offers another simple approach for the preparation of metal oxide/conducting polymer core-sheath nanostructures. Li and co-workers synthesized a series of metal oxide/conducting polymers, including MoO_3/PPy, MoO_3/PANI, VO_2/PPy, SnO_2/PPy and so on, via the hydrothermal reaction [173]. For example, TEM images of MoO_3/PPy proved that the core-sheath structure had formed. ED patterns and EELS spectra showed that the core phase was the single crystalline MoO_3 and the amorphous PPy shell with average thickness of around 70 nm was coated on the surface of MoO_3 nanobelts. The morphology and chemical structures of the core-sheath MoO_3/PPy nanostructures were dependent on the pH and the reaction temperature. This approach provides a general methodology for preparing many different kinds of metal oxide/conducting polymer nanocomposites with core-sheath structure.

5.4. Applications of Nano-structured CPs/Nanocomposites in Sensors/Biosensors

Nanomaterials can be used in a variety of electrochemical biosensing schemes thereby enhancing the performance of these devices and opening new horizons in their applications. Nanoparticles, nanowires and nanotubes have already made an impact on the field of electrochemical biosensors, ranging from glucose enzyme electrodes to genoelectronic sensors. As conducting polymer nanomaterials are light weight, have large surface area, adjustable transport properties, chemical specificities, low cost, easy processing and scalable productions, they are used for applications in nanoelectric devices, chemical and biological sensors [174]. Thin polypyrrole nanofilms doped with sulphate were prepared chemically by interfacial polymerization which makes insertion of various functional groups to pyrrole films possible and provided various applications in developing chemical and biological sensors [175]. Currently, nanoparticle based protocols are being exploited for detection of proteins. The property associated with nanowires and nanotubes, which enable us to modify them with biological recognition elements, imparts high selectivity to these devices. Conducting polymers particularly in the form of thin films or blends or composite as sensors for air-borne volatiles (alcohols, NH_3, NO_2, CO) has also been used widely. Polythiophene based sensor has shown the detection of ppb of hydrazine gases [176]. Also, polyaniline–SnO_2/TiO_2 nanocomposite ultra thin films have been fabricated for CO gas sensing [177]. Nanomaterials based electrochemical sensors are expected to create a major impact upon clinical diagnosis, environmental monitoring, security surveillance, or for ensuring our food safety. The use of biological elements in biosensor construction comes with a challenge of preserving their biological integrity outside their natural environment. For this reasons these biological components of biosensors are generally immobilized onto supports by physical, covalent or electrochemical methods, as will be discussed latter.

5.4.1. Gas Sensor

The emission of gaseous pollutants such as sulfur oxide, nitrogen oxide and toxic gases from related industries has become a serious environmental concern. Sensors are needed to detect and measure the concentration of such gaseous pollutants. In fact analytical gas sensors

236

offer a promising and inexpensive solution to problems related to hazardous gases in the environment. Conducting polymers showed promising applications for sensing gases having acid–base or oxidizing characteristics.

Polyacetylene (III) is known to be the first organic conducting polymer. Exposure of this normally resistive polymer to iodine vapor altered the conductivity by up to 11 orders of magnitude [178, 179]. Polyacetylene is doped with iodine on exposure to iodine vapor. Then, charge transfer occurs from polyacetylene chain (donor) to the iodine (acceptor) leads to the formation of charge carriers. Above approximately 2 % doping, the carriers are free to move along the polymer chains resulting in metallic behavior. Later heterocyclic polymers, which retain the p-system of polyacetylene but include heteroatom bonded to the chain in a five membered ring were developed [180]. Such heterocyclic CPs include polyfuran, polythiophene [181], and polypyrrole. The intrinsically conducting polymers are p-conjugated macromolecules that show electrical and optical property changes, when they are doped/dedoped by some chemical agent. These physical property changes can be observed at room temperature, when they are exposed to lower concentrations of the chemicals, which make them attractive candidates for gas sensing elements. Nylander et al. [182] investigated the gas sensing properties of polypyrrole by exposing polypyrrole impregnated filter paper to ammonia vapor. The performance of the sensor was linear at room temperature with higher concentrations (0.5–5 %), responding within a matter of minutes. Persaud and Pelosi reported conducting polymer sensor arrays for gas and odor sensing based on substituted polymers of pyrrole, thiophene, aniline, indole and others in 1984 at the European Chemoreception Congress (ECRO), Lyon, followed by a detailed paper in 1985 [183, 184]. It was observed that nucleophilic gases (ammonia and methanol, ethanol vapors) cause a decrease in conductivity, with electrophilic gases (NOx, PCl_3, SO_2) having the opposite effect [185]. Most of the widely studied conducting polymers in gas sensing applications are polythiophene and its derivatives [186, 187], polypyrroles [188, 189], polyaniline and their composites [186, 190–192]. These polymers have characteristics of low power consumption, optimum performance at low to ambient temperature, low poisoning effects, sensor response proportional to analyte concentration and rapid adsorption/desorption kinetics.

Electroactive nanocomposite ultrathin films of polyaniline and isopolymolybdic acid (PMA) for detection of NH_3 and NO_2 gases were fabricated by alternate deposition of PAN and PMA following Langmuir–Blodgett (LB) and self-assembly techniques [193]. The process was based on doping induced deposition effect of emeraldine base. The NH_3-sensing mechanism was based on dedoping of PAN by basic ammonia, since the conductivity is strongly dependent on the doping level. In NO_2 sensing, NO_2 played the role of an oxidative dopant, causing an increase in the conductivity when emeraldine base is exposed to NO_2. Nicho et al. [194] found that the optical and electrical properties of p-conjugated polyaniline change due to interaction of the emeraldine salt with NH_3 gas. Silicon substrates were surface modified using an amino-silane $((CH_3O)_3\text{-}Si\text{-}(CH_2)_3NH(CH_2)_2\text{-}NH_2)$ self-assembled monolayer (SAM) [195]. The amino groups (NH_2) of the amino-silane SAM were employed as artificial seeds for self-organization of polyaniline during chemical polymerization of polyaniline. Scanning electron microscopy showed formation of crystalline nanofibrous structure with nanofibers roughly perpendicular to the substrate surface. Chemiresistor sensors formed using these films as a sensitive layer showed sensitivity to very low concentration (0.5 ppm) of ammonia and better response and recovery times in comparison to earlier reports. Yadong et al. [196] reported that submicrometer polypyrrole film exhibits a useful sensitivity to NH_3. The NH_3 sensitivity was detected by the change in resistance of the polypyrrole film. They interpreted the resistance change of the film in terms of the formation of a positively charged electric barrier of NH_4^+-ion in the submicrometer film. The electrons of the NH3 gas act as the donor to the p-type semiconductor polypyrrole, with the consequence of reducing the number of holes in the polypyrrole and increasing the resistivity of the sbmicrometer film. A polypyrrole–poly(vinyl alcohol)(PVA) composite prepared by electropolymerizing pyrrole in a cross-linked matrix of pyrrole was found to posses significant NH_3 sensing capacity [197]. The ammonia sensing mechanism of the polypyrrole electrode has been addressed by La¨hdesma¨ki et al. [198], with evidence that a mobile counter ion may be required for proper sensor operation. Such evidence supports the idea that polypyrrole undergoes a reversible redox reaction when ammonia is detected at submillimolar concentrations. A self-patterning method of conducting polymers consisted of magnetic alignment, electropolymerization and selective etching process was developed by Yoo [199]. This method provides simpler, faster and more cost-effective process compare with other process using e-beam lithography and focused ion beam deposition techniques and enables to easily scale down to nanometer-size. Specifically, the Ni nanowires were grown into AAO template by electrodeposition. The Ni nanowire was magnetically aligned on Au electrodes as a sacrificial seed layer to grow the Ppy nanowire. Then, the Ppy nanowire was successfully synthesized by electropolymerization. Finally, we obtained the Ppy nanowire by selectively etching the Ni nanowire in acidic solution (Fig. 5.3). The electrical resistance after removal of the Ni nanowire was increased by two orders of magnitude, clearly indicating the complete dissolution of the Ni nanowires. The contact resistance between the Ppy nanowire and Au electrodes was measured by extrapolation of serial resistance of the Ppy nanowires by varying dimension of Ppy nanowires and found to be quite low, which effectively improved the sensitivity of the Ppy-based gas sensor for the detection of sub-ppm level concentration of NH_3 gas.

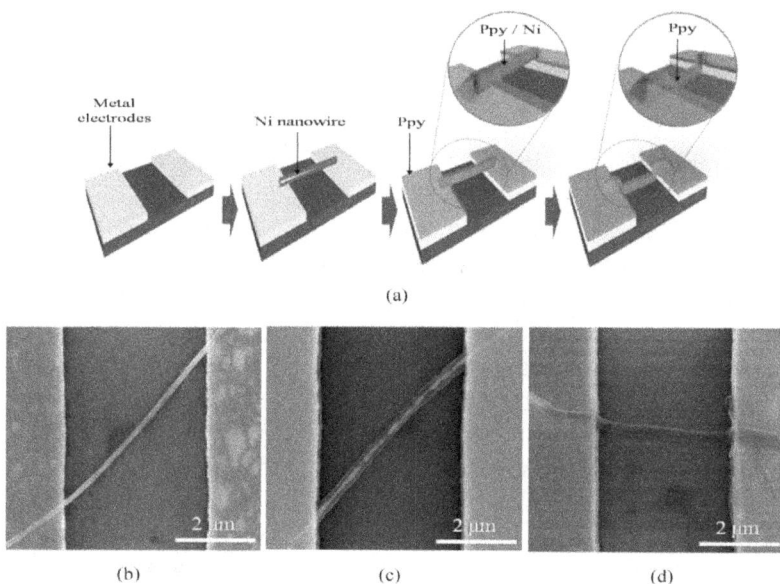

Fig. 5.3. (a) Schematic of Ppy nanowire synthesis by new electropolymerization method; (b) SEM images of magnetically aligned Ni nanowire on Au electrodes; (c) after electropolymerization of Ppy, and (d) Ppy nanowire after removing the Ni nanowire [199].

For the detection of sulfur dioxide in both gas and solution a novel electrochemical sensor has been described by Shi et al. [200]. They constructed the chemically modified electrode by polymerizing 4-vinyl pyridine (4-VP), palladium and iridium oxide (PVP(VI)/Pd/IrO$_2$) onto a platinum microelectrode, which exhibits excellent catalytic activity toward sulfite with an oxidation potential of 0.50 V. The effects of different internal electrolyte solutions of hydrochloric acid, sulfuric acid, phosphates buffer solution, mixed solution of dimethyl sulfoxide and sulfuric acid to the determination of SO_2 were also studied. The sensor was found to have a high current sensitivity, a short response time and a good reproducibility for the detection of SO_2, and showed good potential for use in the field of environmental monitoring and controlling.

Christensen et al. [201] developed a novel NO_2 sensing device using a polystyrene film. When the film was exposed to a 1:10 v/v mixture of NO_2/N_2, the conductivity of the film increased irreversibly and rapidly by several orders of magnitude. They believed that the increase in conductivity of the film might be due to self-ionization of N_2O_4, the form of NO_2 within the film, to NO^+NO3^-. Xie et al. [202] reported the fabrication and characterization of a polyaniline-based gas sensor by ultra-thin film technology. They prepared a pure polyaniline (PAN) film, PAN and acetic acid (AA) mixed films, as well as PAN and polystyrenesulfonic acid (PSSA) composite films, with various number of layers, by LB and self-assembly (SA) techniques. The authors studied the gas sensitivity of these ultra-thin films with various layers to NO_2 gas. They found that pure PAN films prepared by the LB technique had good sensitivity to NO_2, while SA films exhibited faster recovery. PAN is oxidized by contact with NO_2, a well-known oxidizing gas. Contact of NO_2 with the p-electron network of polyaniline is likely to result in the transfer of an electron from the polymer to the gas, making the polymer positively charged. The charge carriers give rise to increased conductivity of the films. They also found that PAN–AA mixed films showed reduced sensitivity, due to the fact that acetic acid molecules had occupied and chemically blocked sensitive sites responsive to NO_2. A highly sensitive NOx sensor was designed and developed by electrochemical incorporation of copper nanoparticles (CuNP) on single-walled carbon nanotubes (SWCNT)-polypyrrole (PPy) nanocomposite modified Pt electrode (Fig. 5.4) [203]. The electrochemical behavior of the CuNP-SWCNT-PPy-Pt electrode was investigated by cyclic voltammetry. It exhibited the characteristic CuNP reversible redox peaks at −0.15 V and −0.3 V vs. Ag/AgCl respectively. The electrocatalytic activity of the CuNP-SWCNT-PPy-Pt electrode towards NOx is four-fold than the CuNP-PPy-Pt electrode. These results clearly revealed that the SWCNT-PPy nanocomposite facilitated the electron transfer from

CuNP to Pt electrode and provided an electrochemical approach for the determination of NO_x. A linear dependence on the NO_x concentrations ranging from 0.7 to 2000 μM, with a sensitivity of 0.22±0.002 μA μM^{-1} cm^{-2} and detection limit of 0.7 μM was observed for the CuNP-SWCNT-PPy-Pt electrode. In addition, the sensor exhibited good reproducibility and retained stability over a period of one month.

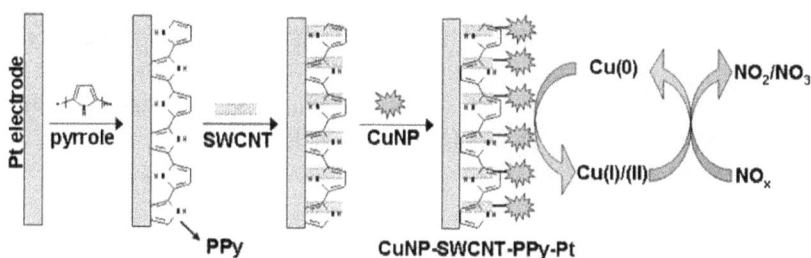

Fig. 5.4. Schematic stepwise preparation of the CuNP-SWCNT-PPy-Pt electrode and illustration of the reaction processes occurring at the surface of the sensor during the determination of NOx [203].

Torsi et al. [204] doped electrochemically synthesized conducting polymers, such as polypyrrole and poly-3-methylthiophene, with copper and palladium inclusions. These metals were deposited potentiostatically, either on pristine conducting films or on partially reduced samples. Exposure of PPy and Cudoped PPy sensors to H_2 and CO reducing gas produced an expected enhancement of the film resistance. On the other hand, the electrical response of the Pd–PPy sensor to H_2, and CO was a drastic drop in resistivity, while a resistivity enhancement is produced upon ammonia exposure. Moreover, the CO and H_2 responses of Pd–PPy sensor are highly reversible and reproducible. Roy et al. [205] has reported the hydrogen gas sensing characteristics of doped polyaniline and polypyrrole films. A thin film of 1,4-polybutadiene has been used to construct a small and very sensitive (10 ppb) ozone sensor [206]. A novel flexible H_2 gas sensor was fabricated by the layer-by-layer (LBL) self-assembly of a polypyrrole (PPy) thin film on a polyester (PET) substrate [207]. A Pt-based complex was self-assembled in situ on the as-prepared PPy thin film, which was reduced to form a Pt-PPy thin film. Microstructural observations revealed that Pt nanoparticles formed on the surface of the PPy film. The sensitivity of the PPy thin film was improved by the Pt nanoparticles, providing catalytically active sites for H_2 gas molecules. The interfering gas NH_3 affected the limit of detection of a targeted H_2 gas in a real-world binary gas mixture. A plausible H_2 gas sensing mechanism involves catalytic effects of Pt particles and the formation of charge carriers in the PPy thin film. The flexible H_2 gas sensor exhibited a strong sensitivity that was greater than that of sensors that were made of Pd-MWCNTs at room temperature.

Polyaniline, nanofibres can also be used for gas sensing application [208]. Thin films of conventional PANI and PANI nanofibres were compared by depositing on interdigitated gold electrode, where PANI nanofibre films showed an enormous increase in response and sensitivity towards HCl vapors (Fig. 5.5) [209].

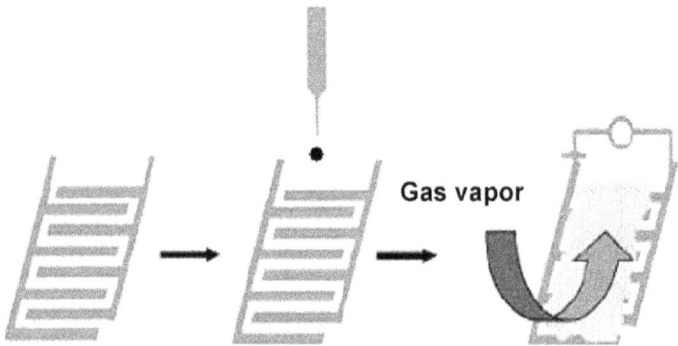

Fig. 5.5. Schematic diagram showing a typical sensor experiment: gold interdigitated electrodes (left) are coated with polyaniline film by drop casting (middle), and the resistance of the film is monitored as the sensor is exposed to vapor (right) [210].

5.4.2. pH Sensor

The pH indicates the amount of hydrogen ion in a solution. Since the solution pH has a significant effect on chemical reactions, the measurement and control of pH is very important in chemistry, biochemistry, clinical chemistry and environmental science. Amongst various organic materials, polyaniline has been found as most suitable for pH sensing in aqueous medium [211–214]. The use of conducting polymers in the preparation of optical pH sensor has eliminated the need for organic dyes. Demarcos and Wolfbeis [215] developed an optical pH sensor based on polypyrrole by oxidative polymerization.

Since the polymer film has suitable optical properties for optical pH sensor, the immobilization step for an organic dye during preparation of the sensor layer was not required. Others [216–218] have also developed optical pH sensors based on polyaniline for measurement of pH in the range 2–12. They reported that the polyaniline films synthesized within a time span of 30 min are very stable in water. Jin et al. [219] reported an optical pH sensor based on polyaniline. While they prepared polyaniline films by chemical oxidation at room temperature, they improved the stability of the polyaniline film significantly by increasing the reaction time up to 12 h. The film showed rapid reversible color change upon pH change. The solution pH could be determined by monitoring either absorption at a fixed wavelength or the maximum absorption wavelength of the film. The effect of pH on the change in electronic spectrum of polyaniline polymers was explained by the different degree of protonation of the imine nitrogen atoms in the polymer chain [220]. The optical pH sensors could be kept exposed in air for over 1 month without any deterioration in sensor performance.

5.4.3. Alcohol Sensor

The determination of alcohol is important in industrial and clinical analyses, as well as in biochemical applications. Since conducting polymers gained popularity as competent sensor material for organic vapors, few reports are available describing the use of polyaniline as a sensor for alcohol vapors, such as methanol, ethanol and propanol [221, 222]. Polyaniline doped with camphor sulphonic acid (CSA) also showed a good response for alcohol vapors [223-226]. These reports discussed the sensing mechanism on the basis of the crystallinity of polyaniline. Polyaniline and its substituted derivatives (XI) such as poly(o-toluidine), poly(o-anisidine), poly(N-methyl aniline), poly(N-ethyl aniline), poly(2,3 dimethyl aniline), poly(2,5 dimethyl aniline) and poly(diphenyl amine) were found by Athawale and Kulkarni [227] to be sensitive to various alcohols such as methanol, ethanol, propanol, butanol and heptanol vapors. All the polymers respond to the saturated alcohol vapors by undergoing a change in resistance. While the resistance decreased in presence of small chain alcohols, methanol, ethanol and propanol, an opposite trend in the change of resistance was observed with butanol and heptanol vapors. The change in resistance of the polymers on exposure to different alcohol vapors was attributed to their chemical structure, chain length and dielectric nature. All the polymers showed measurable responses (sensitivity, 60 %) for short chain alcohols, at concentrations up to 3000 ppm, but none of them are suitable for long chain alcohols. They explained the results based on the vapor-induced change in the crystallinity of the polymer. The polypyrrole was also studied as a sensing layer for alcohols. Polypyrrole [228] incorporated with dodecyl benzene sulfonic acid (DBSA) and ammonium persulfate (APS) showed a linear change in resistance when exposed to methanol vapor in the range 87–5000 ppm. Bartlett et al. [229] also detected methanol vapor by the change in resistance of a polypyrrole film. The response is rapid and reversible at room temperature. They investigated the effects of methanol concentration, operating temperature and film thickness on the response. Polyaniline–silver (polyaniline/Ag) nanocomposite was prepared by an ultrasound assisted in situ miniemulsion polymerization of aniline along with different loading of silver nanoparticles [230]. Colloidal silver nanoparticles were synthesized by reduction of silver nitrate with sodium borohydride using sodium dodecyl sulfate. Films of Polyaniline/Ag were casted using 1-methyl-2-pyrrolidone by spin coating, which were further tested for ethanol vapor sensing. In-house prototype batch set-up was fabricated for ethanol vapor sensing (Fig. 5.6). The response in terms of change in electrical conductivity on exposure to ethanol vapors was reported. Ethanol vapor sensing ability of the nanocomposite was found to increase on addition of silver nanoparticles in the polymer matrix. Response time of the sensor for different loadings of Ag was also studied, which show decrement in response time with increase in amount of Ag.

5.4.4. Neurotransmitters and Drugs Sensors

An ascorbic acid sensor was fabricated via the drop-casting of dodecylbenzene sulphonic acid (DBSA)-doped polyaniline nanoparticles onto a screen-printed carbon-paste electrode [231]. The modified electrode was characterised with respect to the numbers of drop cast layers, optimum potential and operating pH. The sensor was found to be optimal at neutral pH and at 0 V vs. Ag/AgCl. Under these conditions, the sensor showed good selectivity and sensitivity in that it did not respond to a range of common interferents such as dopamine, acetaminophen, uric acid and citric acid, but was capable of the detection of ascorbic acid at a sensitivity of 0.76 μA mM^{-1} or 10.75 μA mM^{-1} cm^{-2} across a range from 0.5 to 8 mM, and a limit of detection of 8.3 μM.

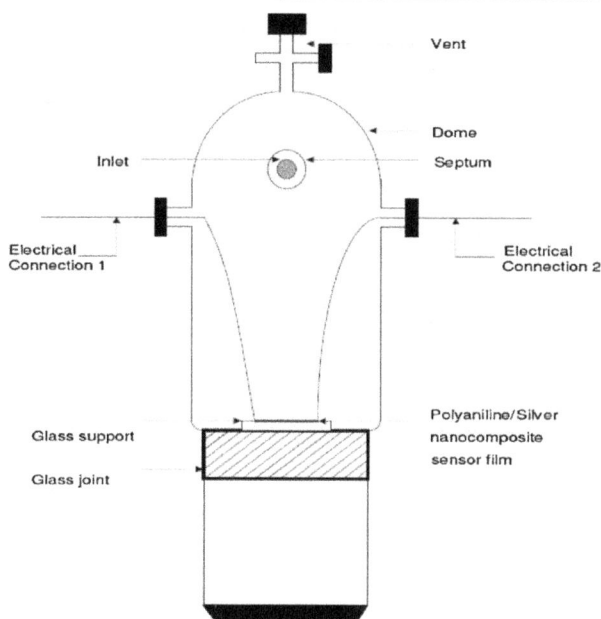

Fig. 5.6. Experimental set up for ethanol sensing [230].

Composite materials consisting of poly(3,4-ethylendioxythiophene) including Au nanoparticles have been synthesised in order to develop new amperometric sensors with improved performances with respect to bare and to pure organic modified electrodes [232]. The composite materials were prepared by including Au nanoparticles during the electropolymerisation process, by taking advantages to the anionic nature of the species surrounding the metal core. The resulting modified electrodes are found to preserve the repeatability and reproducibility of pure organic modified electrodes and to improve the selectivity toward anionic and cationic species chosen as the benchmark analytes, namely ascorbic acid and dopamine, respectively. A novel electrochemical sensor; gold nanoparticles-coated poly(3,4-ethylenedioxythiophene) polymer modified gold electrode in presence of SDS (Au/PEDOT-Au$_{nano}$...SDS) was developed by the electrodeposition of gold nanoparticles on poly(3,4-ethylene-dioxythiophene) (PEDOT) modified gold electrode for the selective determination of dopamine (DA) in presence of uric acid (UA) and ascorbic acid (AA) in presence of sodium dodecyl sulfate (SDS) [233]. Synergism between the composite of conducting polymer matrix and gold nanoparticles in presence of SDS for electron transfer enhancement of DA is explored. Electrochemical investigation and

245

characterization of the modified electrode are achieved using cyclic voltammetry, electrochemical impedance spectroscopy, scanning electron, and atomic force microscopies. The oxidation current signal of DA is remarkably stable via repeated cycles and has unique long term stability. Very small peak potential separation, almost zero or 15 mV is also obtained by repeated cycles indicating unusual high reversibility. The use of SDS in the electrochemical determination of DA using linear sweep voltammetry at Au/PEDOT-Aunano modified electrode resulted in determining DA at very low concentrations. The DA concentration could be measured in the linear range of 0.5–20 μmol L^{-1} and 25–140 μmol L^{-1} with correlation coefficients of 0.9978, and 0.9987, and detection limits of 0.39 nmol L^{-1} and 1.55 nmol L^{-1}, respectively. The validity of using this method in the determination of DA in human urine was also demonstrated. It has been shown that modified electrode can be used as a sensor with high reproducibility, sensitivity, selectivity, and long term stability. Promising voltammetric sensors based on the modification of Pt and poly(3-methylthiophene) (PMT) electrodes with Pd nanoparticles were achieved for the determination of catecholamine neurotransmitters, ascorbic acid and acetaminophen [234]. Electrochemistry of the indicated compoundswas studied at these electrodes and interesting electrocatalytic effects were found. Furthermore, simple, easily prepared one electrochemical step Pd-modified Pt electrode (Pt/Pd) is reported for the first time. Cyclic voltammetry (CV) and chronocoulometry (CC) were used for the determination of the apparent diffusion coefficients in different electrolytes at these electrodes and the values are in the range from 10^{-4} to 10^{-5} cm^2 s^{-1}. Furthermore, it was found that the method of polymer formation had a substantial effect on the synergism between the polymer film and the loaded metal particles towards the oxidation of dopamine (DA) in different supporting electrolytes. This was confirmed by the CV, CC and EIS (electrochemical impedance spectroscopy) as well as SEM (Scanning Electron Microscopy) results. Pt and PMT electrodes modified with Pd nanoparticles showed excellent results for the simultaneous determination of tertiary and quaternary mixtures of the studied compounds. A promising modified electrode was fabricated by distributing Pt or Pd nanoparticles into conductive polymer matrix of poly(3-methylthiophene) (PMT) [235]. Electrochemical investigation of the resulting films was achieved using cyclic voltammetry and differential pulse voltammetry. Several factors affecting the electrocatalytic activity of the hybrid material were studied. Some are related to the polymer such as film thickness, method of its formation and dedoping the polymer film before loading metal particles, other factors are related to the metal particles such as type of metal deposited, method of its deposition, its amount and deposition voltage. EDX analysis was employed to confirm the loading of metal particles to the polymer film. The results suggest that the hybrid film modified electrode combining the advantages of PMT and metal nanoparticles exhibits dramatic electrocatalytic effect on the oxidation of dopamine (DA) and results in a marked enhancement of the current response. The proposed method was applied to the simultaneous determination of ascorbic acid (AA) and dopamine (DA) in physiological pH 7.4 PBS. It was observed that in the presence of AA at millimolar level (0.1 mM), the Pd nanoparticle-modified PMT electrode can sense the increase of DA at micromolar concentration (0.05–1 μM) which is typical of the physiological conditions. The interference study shows that the modified electrode exhibits excellent selectivity in the presence of AA and uric acid (UA) and glucose. It has been shown that this modified electrode can be used as a sensor with high reproducibility, sensitivity, and stability. The method was applied to urine and healthy human blood serum samples and excellent results were obtained.

An interesting electrochemical sensor has been constructed by the electrodeposition of palladium nanoclusters (Pd_{nano}) on poly(N-methylpyrrole) (PMPy) film-coated platinum (Pt) electrode (Fig. 5.7) [236]. It was demonstrated that the electroactivity of the modified electrode depends strongly on the electrosynthesis conditions of the PMPy film and Pd_{nano}. Moreover, the modified electrode exhibits strong electrocatalytic activity toward the oxidation of a mixture of dopamine (DA), ascorbic acid (AA), and uric acid (UA) with obvious reduction of overpotentials. The simultaneous analysis of this mixture at conventional (Pt, gold [Au], and glassy carbon) electrodes usually struggles. However, three well-resolved oxidation peaks for AA, DA, and UA with large peak separations allow this modified electrode to individually or simultaneously analyze AA, DA, and UA by using differential pulse voltammetry (DPV) with good stability, sensitivity, and selectivity. This sensor is also ideal for the simultaneous analysis of AA, UA and either of epinephrine (E), norepinephrine (NE) or L-DOPA. Additionally, the sensor shows strong electrocatalytic activity towards acetaminophen (ACOP) and other organic compounds. The calibration curves for AA, DA, and UA were obtained in the ranges of 0.05 to 1 mM, 0.1 to 10 μM, and 0.5 to 20 μM, respectively. The detection limits (signal/noise [S/N] = 3) were 7 μM, 12 nM, and 27 nM for AA, DA, and UA, respectively. The practical application of the modified electrode was demonstrated by measuring the concentrations of AA, DA, and UA in injection sample, human serum, and human urine samples, respectively, with satisfactory results. The reliability and stability of the modified electrode gave a good possibility for applying the technique to routine analysis of AA, DA, and UA in clinical tests.

A promising electrochemical biosensor was developed by electrodeposition of palladium nanoclusters on polyfuran film modified platinum electrode [237]. This biosensor electrode was used to determine some catecholamines, namely dopamine, epinephrine and norepinephrine, ascorbic acid and paracetamol. The method of formation of the polymer film and deposition of Pd particles plays a key role in the electroactivity of the resulting hybrid material. This sensor effectively resolved the overlapping anodic peaks of ascorbic acid (AA), dopamine (DA) and paracetamol (ACOP) into three well-defined voltammetric peaks in differential pulse voltammetry analysis. The detection limit of DA in the absence and presence of AA and ACOP are eventually the same which indicates that the oxidation processes of DA, AA and ACOP are independent and that the simultaneous measurements of the three analytes are possible without interference. The electrodeposition of Pd on polyfuran improved exceptionally the detection limit about four decades. Moreover, diffusion coefficient measurements confirmed the fast electron transfer kinetics of the electrochemical oxidation of the analyte molecules at the sensor/solution interface. It is very interesting to note that the electrocatalytic effect of PF/Pd composite has been increased to be sometimes 21 times that of the pristine PF which has been considered for a long time to be of low conductivity and attracted low attention as a result of the difficulty of its formation and poor conductivity.

The development of a microneedle-based multiplexed drug delivery actuator that enables the controlled delivery of multiple therapeutic agents was reported by Wang et al. [238]. Two individually addressable channels on a single microneedle array, each paired with its own reservoir and conducting polymer nanoactuator, are used to deliver various permutations of two unique chemical species (Fig. 5.8). Upon application of suitable redox potentials to the selected actuator, the conducting polymer is able to undergo reversible volume changes, thereby serving to release a model chemical agent in a controlled fashion through the corresponding microneedle channels.

248

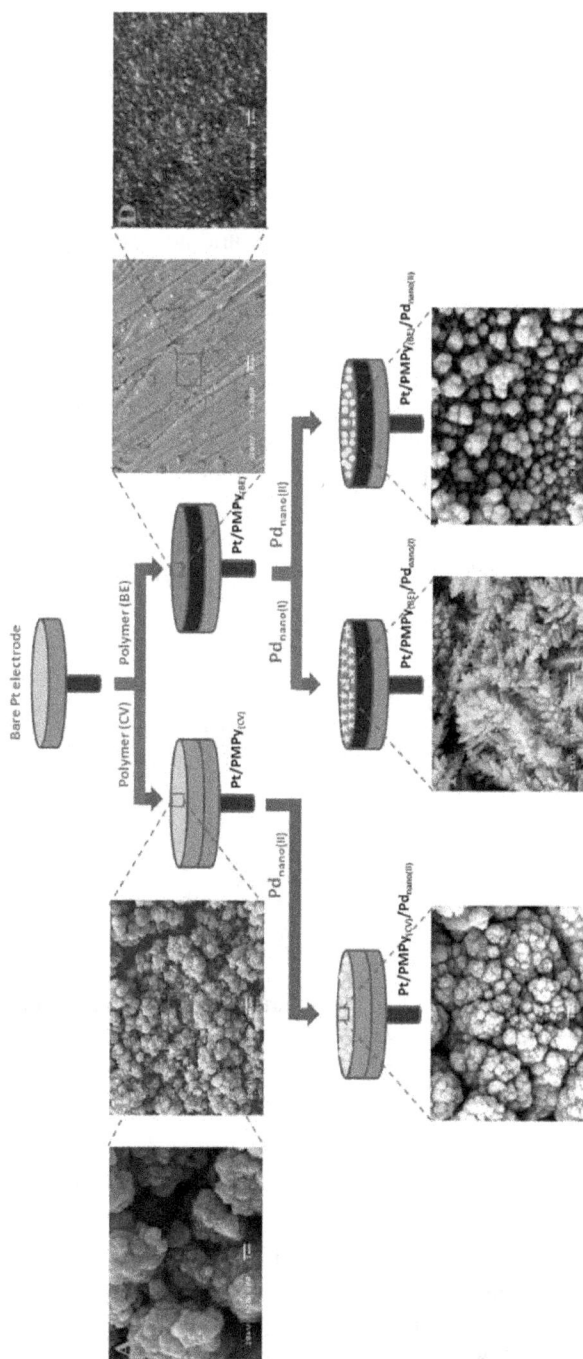

Fig. 5.7. Schematic showing two-step process for the modification of Pt electrode. In the first step, a polymer film is electrodeposited on Pt electrode using two electrochemical methods (BE and CV). In the second step, Pdnano are deposited on the polymer film using two approaches (double potential step [I] and CV program [II]). The notation of every electrode is described. SEM micrographs for all cases are shown. The SEM image of Pt/PMPy(CV)/Pdnano(I) is omitted because of its low electroactivity [236].

Time-lapse videos offer direct visualization and characterization of the membrane switching capability and, along with calibration investigations, confirm the ability of the device to alternate the delivery of multiple reagents from individual microneedles of the array with higher precision and temporal resolution than conventional drug delivery actuators.Analytical modeling offers prediction of the volumetric flow rate through a single microneedle and accordingly can be used to assist in the design of subsequent microneedle arrays. The robust solid-state design and lack of mechanical components circumvent reliability issues that challenge fragile conventional microelectromechanical drug delivery devices. This proof-of-concept study demonstrates the potential of the drug delivery actuator system to aid in the rapid administration of multiple therapeutic agents and indicates the potential to counteract diverse biomedical conditions.

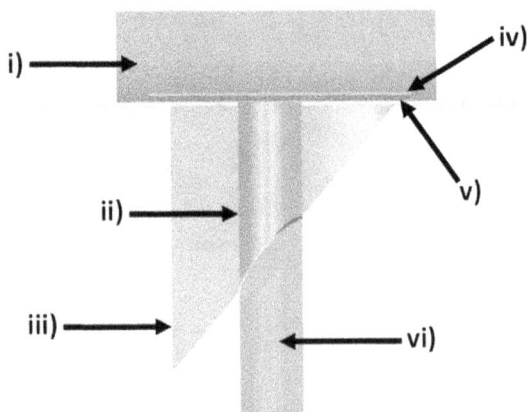

Fig. 5.8. Schematic of a single microneedle during drug delivery. Microneedle components include the following: (i) reservoir, (ii) lumen (342 _m diameter), (iii) hollow microneedle, (iv) Au/PPy/DBS nanoporous membrane, (v) PC membrane, and (vi) the released drug [238].

Amperometric detection of tolazoline (TL) was carried out on a gold nanoparticles (AuNPs)/poly-oaminothiophenol (PoAT)-modified electrode by a molecular imprinting technique and electropolymerization method (Fig. 5.9) [239]. The recognition between the imprinted sensor and target molecule was observed by measuring the variation of amperometric response of the oxidation-reduction probe, $K_3Fe(CN)_6$ on electrode. Under the optimal experimental conditions, the peak currents were proportional to the concentrations of tolazoline in two ranges of 0.05–5.0 µg mL^{-1} and 5.0–240 µg mL^{-1} with the detection limit of 0.016 µg mL^{-1}. Meanwhile the prepared sensor showed sensitive and selective binding sites for tolazoline. The enhancement of sensitivity was attributed to the presence of AuNPs which decreased the electron-transfer impedance.

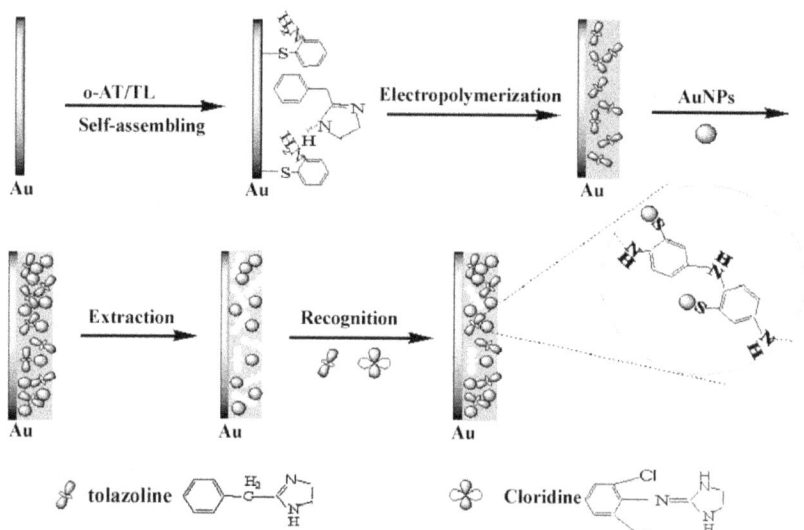

Fig. 5.9. Diagram of construction of the tolazoline imprinted sensor for recognition of tolazoline [239].

5.4.5. Humidity Sensor

Humidity sensors are useful for the detection of the relative humidity in various environments. These sensors attracted a lot of attention in the medical and industrial fields. The measurement and control of humidity are important in many areas, including industry (paper, food, electronic), domestic environment (air conditioner), medical (respiratory equipment), etc. Polymer, polymer composites and modified polymers with hydrophilic properties have been used in humidity sensor devices. Ion conducting polymeric systems has been used in humidity sensor devices based on variation of the electrical conductivity with water vapor. Polymer electrolytes containing polymer cation/polymer anion with its counter ions and mixtures or complexes of inorganic salts with polymer are the major materials for fabrication of humidity sensor. For example, lithium chloride dispersed in hydrophobic polyvinyl acetate held in the pores of polyvinyl alcohol film [240] and $LiClO_4$ doped polyethylene oxide [241] are reported for humidity sensors.

Layer-by-layer nano-assembly for deposition of ultrathin poly(anilinesulfonic acid) (SPANI) films for fabrication of highly sensitive and rapid response humidity sensors was used [242]. Spin coating was also used for fabrication of SPANI based humidity sensors for comparison. The change in electrical sheet resistance of the sensing film was monitored as the device was exposed to humidity. For 5 % change in relative humidity, the sensitivity was measured to be 11 and 6 % from layer-by-layer based and the spin coated humidity sensors, respectively. An intended application for these layer-by-layer assembled devices is in disposable handheld instruments to monitor the presence of humidity in humidity sensitive environments. Humidity sensor based on polyaniline nanofibres was fabricated and its response to humidity was investigated by Liu et al., [243]. It was found that the sensor behaved differently compared to commonly known conducting polymer based sensors. The sensor responded to low relative humidity (<50 % RH) normally by decreasing electrical resistance with increasing humidity. However, at higher RH the sensor reversed its responses by increasing the electrical resistance with humidity. It is believed that the unique structure of nanofibres, which is more susceptible to distortion during polymer swelling, is responsible for the 'reversed behaviour'. IR data confirmed that excess water absorption occurred and that a change in polymer oxidation state might also have taken place.

Resistive-type humidity sensors were fabricated through in situ photopolymerization of pure polypyrrole (PPy) and TiO_2 nanoparticles/polypyrrole (TiO_2 NPs/PPy) composite thin films on an alumina substrate [244]. The electronic properties of PPy and TiO_2 NPs/PPy composite thin films were investigated based on the components $AgNO_3$ and TiO_2 used as electron acceptor and photoinitiator, respectively. The humidity sensing mechanism of TiO_2 NPs/PPy composite thin films was investigated via the results of activation energy and impedance spectroscopy. The sensor made of TiO_2 NPs/PPy composite thin films, using the added amount of TiO_2 NPs as 0.0012 g showed the highest sensitivity, smaller hysteresis and best linearity. A novel flexible resistive-type humidity sensor was fabricated by the *in situ* photopolymerization of TiO_2 nanoparticles/polypyrrole (TiO_2 NPs/PPy) and TiO_2

nanoparticles/polypyrrole/poly-[3-(methacrylamino)propyl] trimethyl ammonium chloride (TiO$_2$ NPs/PPy/PMAPTAC) composite thin films on a polyester (PET) substrate [245]. The effect of the TiO$_2$ NPs concentration on the electrical and humidity sensing properties of the TiO$_2$ NPs/PPy composite thin films on a PET substrate was investigated. PMAPTAC was incorporated into the TiO2 NPs/PPy composite thin films to increase the flexibility and sensitivity of the composite films for practical use. Novel low-humidity sensors were fabricated by the *in situ* photopolymerization of polypyrrole/Ag/TiO$_2$ nanoparticles (PPy/Ag/TiO$_2$ NPs) composite thin films on a quartz-crystal microbalance (QCM) [246]. The sensitivity increased with the doping amount of TiO$_2$ NPs. The PPy/Ag/50 wt% of TiO$_2$ NPs composite thin films exhibited excellent sensitivity (0.0246 Hz/ppmv at 171.1 ppmv), linearity and a short response time (12 s at 55.0 ppmv). The low-humidity sensing mechanism was elucidated in terms of surface texture and the nanostructural morphology of the composite materials. Additionally, based on the dynamic analysis of adsorption, the association constants of water vapor molecules with PPy/Ag and PPy/Ag/50 wt% TiO$_2$ NPs composite thin films were estimated to be 81.6 and 227.9 M^{-1}, respectively, explaining the effect of adding 50 wt% TiO$_2$ NPs to PPy/Ag; the sensitivity to low humidity increased as the association constant increased.

5.4.6. DNA Biosensor

There has been considerable interest in the development of DNA sensors for the analysis of unknown or mutant genes, the diagnosis of infectious agents in various environments and the detection of drugs, pollutants, etc. which interact with the structure of the double stranded DNA. Single strand DNA probes are immobilized by techniques of adsorption, direct covalent binding, entrapment in a polymer matrix, etc. The feasibility of fabricating single polypyrrole and polyaniline nanowires and their application as DNA sensors (1 nm) was also studied [247]. Such an enzyme based glucose sensor has been fabricated and characterized, based on co-electrodeposition of redox polymer poly (vinylimidazole) and glucose oxidase onto a low-noise carbon fibre nanoelectrode [248]. This nanosensor offers a highly sensitive and rapid response to glucose at an operating potential of 0.22 V, with a linear range of 0.01–15 mM and a detection limit of 0.004 mM. The use of conducting polymer substrates and the amplification afforded by semiconductor nanoparticles can be combined to construct a novel DNA sensor as illustrated in Fig. 5.10 [249].

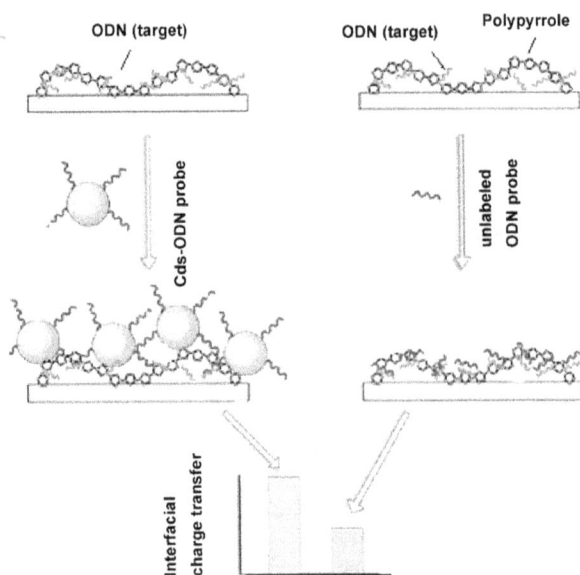

Fig. 5.10. Mechanism underlying the sensor response. (Top) formation of a layer of CdS-ODN nanoparticles on the PPy-ODN film after hybridization. (Bottom) binding of unlabelled ODN probes to the PPy-ODN film (from Ref. [249]), CdS: cadmium sulfide; ODN: oligonucleotides.

A prerequisite for exploiting sensing devices based on semiconductor nanowires is ultra-sensitive and selective direct electrical detection of biological and chemical species. A transducer was constructed based on copolymer of poly(3,4,-ethylenedioxythiophene) (PEDOT) and carboxylic group functionalized PEDOT single nanowire in between gold electrodes, followed by covalent attachment of amino-modified probe oligonucleotide [250]. The target ODNs specific to Homo sapiens Breast and ovarian cancer cells were detected at femtomolar concentration and incorporation of negative controls (non-complementary ODN) were clearly discriminated by the sensor. The ex situ measurements were performed by using two terminal device setup and the changes in the interface of the nanowire associated with the association or dissociation of ODNs were measured as change in resistance. In addition, in situ measurements were performed by utilizing scanning ion conductance microscopy to measure the change in resistance of probe modified nanowire upon addition of different concentration of target ODNs in presence of relevant buffer. The constructed, nano sensor showed highly sensitive concentration dependent resistance change.

A novel DNA sensor based on a polypyrrole substrate and nanoparticle labeled ODN probes was prepared and characterized by Travas-Sejdic et al. [251]. In this sensor type the DNA sample to be analyzed was physically entrapped into the polypyrrole film during electropolymerization. The resulting sensor film was then exposed to ODN probes in solution. A further amplification in sensor response (as compared to using unlabeled ODN probes) was achieved by tagging ODN probes with CdS nanoparticles. It was suggested that the sensor response is caused by an increase in charge transfer resistance upon binding of complementary CdS–ODN nanoparticle probes. The selectivity of the sensor was tested with various match and mismatch sequences. By appropriate choice of hybridization conditions the sensor was able to robustly discriminate between the exact match and a two-point mismatch sequence.

5.5. Biomolecular Immobilization on CPs for Biosensing Applications

With the advance in the development of biosensors, several problems surfaced relating to functioning of enzyme system like loss of enzyme (especially expensive enzymes), maintenance of enzyme stability and shelf life of the biosensors. In addition to this, there grew a need to reduce the time of enzymatic response and offer disposable devices, which can easily be used in stationary or in, flow systems [252]. In order to overcome these problems, several immobilization procedures have been developed. The procedure of biomolecule immobilization on conductive surfaces remains a key step for the performance of the resulting electrochemical devices. For efficient deposition of a biomolecule, it must satisfy a few pre-requisites, that:

- There must be an efficient and stable immobilization of the biological macromolecules on transducer surfaces;

- It must retain its biological properties completely;

- It should be compatible and chemically inert towards host structure;

- It should be accessible when immobilized [253].

Immobilized enzyme has many operational advantages over free enzyme such as reusability, enhanced stability, continuous operational mode, rapid termination of reaction, easy separation of biocatalyst from product, and reduced cost of operation [254]. Immobilization of biomolecules can be carried out using many different procedures, like physical adsorption, covalent linking and electroimmobilization, while retaining the biological recognition properties of the biomolecules.

5.5.1. Physical Adsorption

In physical adsorption process, the biomolecule gets adsorbed in the polymer/solution interface due to static interactions between the polycationic matrix of the oxidated polymer and the total negative enzyme charge, provided pH is higher than the isoelectric point (pI) of the enzyme [255]. But this process has some known disadvantages like binding forces can change with pH, adsorption is limited to one monolayer on the polymer surface hence amount of enzyme incorporated is very small [256]. Also, since biomaterial is immobilized on the outer layer of conducting polymer, it gets leached out into sample solution during measurements. This decreases the lifetime stability of enzyme electrode [257]. This method has been used for the preparation of many biosensors [258–265]. Immobilization by physical adsorption also involves cross-linking. Cross-linking gives sensors with short response times but poor stability because enzyme is directly exposed to the bulk solution and partly denaturized by the cross-linking. Immobilization of acetylcholine esterase (AChE) by cross-linking involving glutaraldehyde significantly increased the attachment of enzyme to the transducer and hence led to a more direct electron exchange. Here AChE was co immobilized with choline oxidase (ChO) on to a Pt surface using a solution of gluteraldehyde [252]. Immobilization of AChE in nylon net using gluteraldehyde has been reported by Gulla et al. [266]. The method involves immobilization by sandwiching the enzyme layer between two cellophane membranes kept in close contact with a gold electrode. Milagres et al. [267] immobilized salicylate hydroxylase by cross-linking. Similarly, immobilization of tyrosinase [268] and lactate dehydroganase [269] has also been reported in literature. A dual electrode amperometric biosensor for simultaneous determination of acetylcholine and choline was fabricated by Guerrieri et al. by electrosynthesising poly (pyrrole)/poly (2-napthol) bilayer membrane as an effective anti interference layer. This device is potentially useful for the analysis of real matrices, (e.g. brain homogenates & CSF) without the use of a chromatographic step typically required to separate acetylcholine and choline [270]. Rigid conducting carbon polymer composites were reported to be ideal for construction of electrochemical sensors by Cespedes et al. A great number of biological materials can be immobilized by blending them with these composites. This technique is very attractive for the mass fabrication of amperometric biosensors [271]. A cholinesterase potentiometric biosensor based on a glassy electrode modified with processible PANI has been developed and explored for the detection of organophosphorous and carbamic pesticides by Ivanov et al. Acetyl and butyryl-cholinesterase from various sources were immobilized on the surface of PANI modified electrode by cross linking with gluteraldehyde. The detection limits of the pesticides investigated were found lower than those obtained with other similar cholinesterase [272]. Brahim et al. utilized PPy-hydrogel composites for the construction of clinically important biosensors. They reported the use of poly (2-hydroxyethyl methacrylate) in which PPy and various oxidoreductase enzymes were physically entrapped to function as a viable matrix. The optimized glucose oxidase (GOx) biosensor displayed a wide linear glucose response range (5.0×10^{-5}–2.0×10^{-2} M) and a response time of 35–40 s [273]. As per the reported literature, configuration of a biosensor plays a pivotal role in performance of a biosensor device. Shan et al. studied the immobilization of enzymes with clay. It was reported that if the enzyme/clay ratio was decreased from 1 to 0.125, a sharp decrease in sensitivity of biosensor was observed. They also studied the influence of thickness of the clay film on the biosensor performance and observed that higher sensitivity to cyanide and lower detection limits with thinner coatings [274]. Another group used a lower amount of immobilized acetylcholine-esterase to measure lower concentration of paraoxan [275].

5.5.2. Covalent Immobilization

In order to achieve an increased lifetime stability of enzyme electrode, it is necessary that there should be a strong and an efficient bonding between the enzyme and the immobilizing material. Hence, covalent linking of biomolecules on transducer is an efficient method of immobilization. It comes with an advantage of low diffusional resistance and such a sensor shows good stability under adverse conditions. Covalent linking of a biomolecule to polymer matrix is accomplished in a twostep process i.e. synthesis of functionalized polymer followed by covalent immobilization. Since, immobilization takes place only on the outer surface of the polymer, it permits selection of optimum reaction conditions for each step [255]. Guilbault bound GOx via acyl azide derivatives of polyacrylamide for glucose detection [276]. Ramanathan et al. used GOx bound to poly (o-amino benzoic acid) as glucose biosensor [277]. Rahman et al. recently fabricated a biosensor by covalent immobilization of pyruvate oxidase on the nanoparticle-comprised poly (terthiophene carboxylic acid), poly-TTCA on glassy carbon electrode for amperometric detection of phosphate ions [278]. A linear response for the detection of phosphate ions was observed between 1.0 and 100 mM with a detection limit of about 0.3 mM. Rajesh et al. have developed an amperometric phenol biosensor by covalent immobilization of tyrosinase onto an electrochemically prepared novel copolymer poly (N-3-aminopropyl pyrrole-co-pyrrole) film. The covalent linkage of enzyme and porous morphology of the polymer film lead to high enzyme loading and an increased lifetime stability of the enzyme electrode. An amperometric biosensor based on the use of pH sensitive, redox active dissolved hematin molecule was also developed by the same group and has been used for the quantitative determination of urea in aqueous solution. This biosensor had a 95 % steady-state current value and 80 % of the enzyme activity was retained for 2 months [257]. Similarly, they have also reported other techniques for the fabrication of biosensors and enhancement of related properties [279–282]. Oennerfjord et al. immobilized tyrosinase covalently on the surface of carbodiimide-activated graphite and used in a flow injection system [283]. Another technique of covalent linking involves integrating enzymes in monomers, which can be electrochemically oxidized to generate polymers. By virtue of this process greater amount of enzymes get immobilized [255]. Immobilization of GOx by covalent bonding has been reported by many research groups [284–290].

5.5.3. Electrochemical Immobilization

Although, the conventional procedures for deposition of biomolecules on to the transducers like, physical adsorption, covalent linkage, entrapment and cross-linking have been used extensively, they suffer from a low reproducibility and have a poor spatially controlled deposition. In recent years, the focus of immobilization has been shifted towards the entrapment of biomolecules in the layers of electrochemically synthesized polymers. Aizawa and Yabuki had pioneered the application of PPy and PANI to the immobilization of enzymes. Since, then hundreds of papers have been published and patents have been granted on this subject [291,292]. Serge Cosnier reviewed different electrochemical procedures for biomolecule immobilization based on electro polymerized films and describes the adsorption of amphiphilic monomers and biomolecule before the polymerization step. Examples of organic phase enzyme electrode and electrical wiring of immobilized enzymes are presented. Furthermore, the construction of controlled architectures based on spatially segregated multilayers, exhibiting complimentary biological activities is described. The electrochemical formation of polymer layers of controlled thickness and enzyme activity in galvanostatic or potentiostatic conditions offers numerous advantages over conventional procedures for the design of biosensors, such as complete coverage of the active surface, greater control over film thickness and greater reproducibility. The major advantages of the electrochemical immobilization techniques are: it is a one step, fast than in all immobilization procedures [293]; the distribution of the immobilized enzyme is spatially controlled irrespective of geometry, shape and dimension of the electrode [294]; the film thickness can be precisely controlled through, e.g. the charge involved in the deposition step and it can be used to build up multiplayer/multienzyme structures; it allows the reproducible and precise formation of a polymer coating over surfaces whatever their size and geometry may be [295, 296]. The fabrication of biosensor by entrapment of biomolecules in electro-polymerized films is the most popular approach. Since, the pioneering work by Umana and Waller and Bartlett et al. [297] numerous papers have been published on this approach [298–305]. Immobilization by enzyme entrapment involves electrochemical oxidation of the monomer in the presence of the enzyme to form a polymer, which incorporates the homogenously distributed enzyme molecules during its growth process. This method is the most straightforward method of immobilization with some inherent advantages like simplicity and reproducibility of the one step preparation procedure and the possibility of immobilizing of mediators, co-enzymes or the other enzymes simultaneously simply by adding them to the polymerization solution. The mechanism of incorporation of the enzyme into electro synthesized polymer films has been an object of various speculations. It has been suggested that electrostatic interaction plays the major role in enzyme incorporation and this assumption has been adequately supported by indirect experimental evidence [306]. This is not the case for non-ionic polymer films where pre-adsorption of enzyme onto the electrode surface seems to be the important step in determining the enzyme loading [307]. In this respect, enzyme/polymer interaction is of paramount importance to improve the fundamental knowledge about the biological interface of the biosensor. Incorporation of the enzyme results in changes in the apparent optical properties and density of the polymeric films, suggesting a mutual stabilization of polymer and the enzyme. Griffith et al. have reported the amount of GOx entrapped in conducting and non-conducting electro polymerized film (on golf surface) at different adsorption time and enzyme concentrations in the growing solution [308]. Although, biomolecules are only physically confined near transducer, it has a high diffusion barrier [276]. Entrapment of GOx during the growth of PANI films at low pH due to its high tolerance has been extensively reported [305, 309–317]. Compagnone et al. used thiophene derivative polymers synthesized in organic solvents for entrapment [318], whileGambhir et al. entrapped urease and glutamate dehydrogenase in PPy/polyvinyl sulfonate films for urea detection. Similarly, incorporation of a number of other such enzymes has been studied and reported in literature [319–328]. This process involves application of an appropriate potential to the working electrode soaked in aqueous solution containing both biomolecule and electro polymerizable monomer. Biomolecules present in the immediate vicinity of the electrode surface are thus physically incorporated into the growing polymer. This, reagentless electrochemical approach is easily applicable to a wide variety of biological macromolecules. Ramanathan [329] has immobilized GOx on different conducting matrices such as PPy, PANI, polyammino benzoic acid (PAB) and studied the response characteristics, lifetime detail, etc.. Amongst all CP PPy has been extensively used for immobilization, as it has relatively stable electrical conductivity and can be electro synthesized under biocompatible conditions [330–333]. But there are few problems with sensors based on this polymer, like the film is degraded by hydrogen peroxide. Covalent attachment of mediating components onto pyrrole monomers can avoid these problems but the process is technically complex and there is no guarantee that the modified monomer will polymerize. Recently, many reports have been published on the immobilization of enzymes into the PANI film [334, 335]. Metheby et al. developed an amperometric biosensor by insito deposition of horseradish peroxide (HRP) enzyme on PANI doped Pt disk electrode. HRP was electrostatically immobilized onto the surface of the PANI film and voltametry was used to monitor the electro catalytic reduction of H_2O_2 under diffusioncontrolled conditions. The linear range of this biosensor was from 2.5×10^{-4} to 5×10^{-3} M [336]. Chun-Xiang Li et al. reported a sensitive H_2O_2 amperometric sensor based on HRP-labelled nano-Au colloids. Nano-Au colloids were immobilized by the thiol group of cysteamine, which was associated with the carboxyl groups of poly (2,6-pyridinedicarboxylic acid). With the aid of hydroquinone the sensor displayed excellent electrocatalytical response to the reduction of H_2O_2 in the range of $3.0 \times 10^{-7} - 2 \times 10^{-3}$ M. The detection limit was 1.0×10^{-7} M [337]. Earlier Nakabayashi et al. reported an amperometric biosensor for sensing H_2O_2 based on electron transfer between HRP and ferrocene as a mediator [338]. Similarly, Thanachasai et al. developed novel H_2O_2 sensors based on peroxide carrying poly {pyrrole-co-[4-(3-pyrrolyl) butanesulfonate]} copolymer [339]. A novel method of electro polymerization of GOx in PPy was reported by Wolowacz et al. [340]. GO_x was chemically derivatized by coupling N-(2-carboxyethyl) pyrrole to surface lysyl residues with a carbodiimidepromoted reaction. The modified enzyme displays almost identical properties to the native enzyme except that the pH was slightly more acidic and stability was 6 fold higher at pH 7.0 and 60 °C. This new procedure incorporated more enzyme activity into the conducting polymer than the previously reported physical entrapment technique at equivalent enzyme concentrations in the polymerization media. The covalently entrapped enzyme was 171 fold more stable than the soluble native enzyme. Yasuzawa et al. discussed the properties of glucose sensors based on the immobilization of GOx in N substituted PPy film [341]. Sensors were prepared by electro polymerization of 3-(1-pyrrolyl)propionic acid in the presence of the enzyme following the treatment with water soluble carbodiimide to provide covalent bonding between GOx and PPy derivatives. The treatment with water-soluble carbodiimide improves the response stability of the enzyme immobilized electrodes. The entrapment of enzymes in CP provides a facile means for ensuring proximity between the active site of the enzyme and conducting surface of the electrode with considerable potential for biosensor construction. This method provides a facile and controllable method for the deposition of biologically active molecules to defined areas on the electrodes. This well-controlled procedure of enzyme immobilization by electro polymerization is of great significance in fabrication of micro-sensors in the preparation of multilayer devices [342, 343] and multienzyme biosensor [343–347]. Vedrine et al. constructed an amperometric enzyme biosensor using tyrosinase for the determination of phenolic compounds and herbicides. The enzyme was entrapped in a conducting polymer, poly 3,4-ethylene dioxythiophene (PEDT) electrochemically generated on a glassy carbon electrode. Detection of herbicides was obtained from the inhibition of tyrosinase electrode responses [348]. Mulchandani et al. have reported an electrochemically deposited ferrocene modified phenylene diamine film on a glassy carbon electrode and used it for the entrapment of peroxide enzyme, thereby utilizing the assembly as a biosensor [349]. The preparation of a cholesterol amperometric biosensor using a platinized Pt electrode as a support for the electropolymerization of a PPy film, in which cholesterol oxidase and ferrocene monocarboxylic acid were co-entrapped has been described by Vidal et al. All the biosensor preparation steps (platinized and electro polymerization) and the cholesterol determination take place in the same flow system. The presence of the mediator enhances the sensitivity and selectivity of the platinized biosensor without modifying the dynamic parameters of the response and the platinized layer improves the operational lifetime of the mediated sensor [350]. Arslan et al. constructed and characterized poly acrylic acid and poly (1-vinyl imidazole) based biosensor and entrapped invertase in the interpenetrating polymer network [351]. Retama et al. developed a glucose biosensor by synthesizing semi-conducting PPy/polyacrylamide micro particles and immobilized GOx into the aqueous phase of the concentrated emulsion before starting polymerization. The micro particles with immobilized enzymes were used as biological component of amperometric glucose biosensor. Bartlett et al. electrochemically polymerized phenol and its derivatives and used them to immobilize GOx at a platinum electrode surface. The resulting films were thin and blocked the electrochemistry of ferro/ferricyanide to a greater or lesser degree. The films are active towards the oxidation of glucose and thus can be used to detect hydrogen peroxide generated by reaction between glucose and oxygen catalyzed by GOx trapped within the film. The response times for different phenolic films were all similar, typically o4 s [352]. Boyukbayram et al. grafted copolymers of thiophene capped polytetrahydrofuran and pyrrole using SDS as supporting electrolyte to produce biosensor for determination of phenolic compound in red wine. Nunes et al. presented a comparison of several AChE immobilization procedures carried out on tetracyanoquinone diamino methane (TCNQ) modified graphite working electrode. The biosensor produced by this procedure showed good results, fast response, good reproducibility, wide working range and excellent sensitivity [353]. Although, this method of immobilization is mainly used for entrapment of enzymes, a few examples are reported wherein immobilization of other biomolecules has been done by electrochemical entrapment [354, 355].

Kaotit et al. developed a biosensor for evaluation of atrazine which provides rapid and technically simple system at concentration below the ppm level. This technique involved immobilization of enzyme polyphenol oxidase (PPO) during the anodic electro polymerization of PPy [356]. With the emergence of nanotechnology, many researchers tried to incorporate it in biosensor development. Qu et al. prepared conducting PANI films and assembled these films with multiwalled carbon nano tubes layer by layer on glassy carbon electrodes. The assembly was used as an amperometric biosensor for choline detection. This biosensor displayed a rapid response and an expanded linear response range as well as excellent reproducibility and stability [357]. Ramanavicius et al. have discussed self-encapsulation of redox enzyme GOx within conducting polymer PPy recently. GOx based nanoparticles coated by pyrrole were developed, which exhibited detectable catalytic activity with increased Km value. GOx/PPY nanoparticles were applied as novel transducer in the design of amperometric biosensors [358]. Recently, Zhu et al. applied PANI nanowire modified electrode as DNA biosensor. The nanowires of CP were directly synthesized through a three-step electrochemical deposition procedure in Aniline containing electrolyte solution by using the glassy carbon electrode as working electrode. Such wires have been found to be of great promise for sensitive electrochemical biosensor applications [359]. Similarly, nanocomposites of poly (o-anisidine) containing TiO_2 nanoparticles, carbon black and multiwalled carbon nanotubes (MWCNT) were prepared by Carrora et al. The synthesized materials were deposited in thin films and analysis of such films demonstrated that materials based on polymeric nanocomposites are best suited for biosensor applications because they provide the best electron transfer and assure the faster ion mass transfer [360]. Sotiropoulou and Chaniotakis stabilized AChE using nano porous carbon matrix. By the use of this matrix, the enzyme was significantly stabilized and its detection unit was decreased [361]. However, it should be noted that electropolymerization process requires high concentrations of monomer and biomolecules. On the basis of the numerous studies reported before, the enzymatic immobilization is one of the most important steps involved in the biosensor design. The choice of the technique used for connecting the biological component (enzyme) to the transducer is crucial, since the stability, the longevity and the sensitivity largely depends on enzymatic layer configuration.

References

[1]. J. D. Strenger-Smith, Intrinsically electrically conducting polymers. Synthesis, characterization, and their applications, *Prog. Polym. Sci.*, Vol. 23, 1998, pp. 57–79.

[2]. E. T. Kang, K. G. Neoh, K. L. Tan, Polyaniline: a polymer with many interesting redox states, *Prog. Polym. Sci.*, Vol. 23, 1998, pp. 277–324.

[3]. N. Gospodinova, L. Terlemezyan, Conducting polymers prepared by oxidative polymerization: polyaniline, *Prog. Polym. Sci.*, Vol. 23, 1998, pp. 1443–1484.

[4]. S. Palaniappan, A. John, Polyaniline materials by emulsion polymerization pathway, *Prog. Polym. Sci.*, Vol. 33, 2008, pp. 732–758.

[5]. S. Bhadra, D. Khastgir, N. K. Singha, J. H. Lee, Progress in preparation, processing and applications of polyaniline, *Prog. Polym. Sci.*, Vol. 34, 2009, pp. 783–810.

[6]. A. Pud, N. Ogurtsov, A. Korzhenko, G. Shapoval, Some aspects of preparation methods and properties of polyaniline blends and composites with organic polymers, *Prog. Polym. Sci.*, Vol. 28, 2003, pp. 1701–1753.

[7]. J. Huang, R. B. Kaner, The intrinsic nanofibrillar morphology of polyaniline, *Chem. Commun.*, 2006, pp. 367–376.

[8]. D. Zhang, Y. Wang, Synthesis and applications of one-dimensional nano-structured polyaniline, *Mater. Sci. Eng. B*, Vol. 134, 2006, pp. 9–19.

[9]. M. Wan, A template-free method towards conducting polymer nanostructures, *Adv. Mater.*, Vol. 20, 2008, pp. 2926–2932.

[10]. M. Wan, Some issues related to polyaniline micro-/nanostructures, *Macromol. Rapid Commun.*, Vol. 30, 2009, pp. 963–975.

[11]. D. Li, J. Huang, R. B. Kaner, Polyaniline nanofibers: a unique polymer nanostructure for versatile applications, *Acc. Chem. Res.*, Vol. 42, 2009, pp. 135–145.

[12]. H. D. Tran, D. Li, R. B. Kaner, One-dimensional conducting polymer nanostructures: bulk synthesis and applications, *Adv. Mater.*, Vol. 21, 2009, pp. 1487–1499.

[13]. C. Li, H. Bai, G. Q. Shi, Conducting polymer nanomaterials: electrosynthesis and applications, *Chem. Soc. Rev.*, Vol. 38, 2009, pp. 2397–2409.

[14]. C. R. Martin, Template synthesis of electronically conductive polymer nanostructures, *Acc. Chem. Res.*, Vol. 28, 1995, pp. 61–68.

[15]. J. Huang, S. Virji, B. H. Weiller, R. B. Kaner, Polyaniline nanofibers: facile synthesis and chemical sensors, *J. Am. Chem. Soc.*, Vol. 125, 2003, pp. 314–315.

[16]. X. Zhang, W. J. Goux, S. K. Manohar, Synthesis of polyaniline nanofibers by "nanofiber seeding", *J. Am. Chem. Soc.*, Vol. 126, 2004, pp. 4502–4503.

264

[17]. J. Huang, R. B Kaner, Nanofiber formation in the chemical polymerization of aniline: a mechanistic study, *Angew. Chem. Int. Ed.*, Vol. 43, 2004, pp. 5817–5821.

[18]. S. K. Pillalamarri, F. D. Blum, A. T. Tokuhiro, J. G. Story, M. F. Bertino, Radiolytic synthesis of polyaniline nanofibers: a new templateless pathway, *Chem. Mater.*, Vol. 17, 2005, pp. 227–229.

[19]. X. Jing, Y. Wang, D. Wu, L. She, Y. Guo Y, Polyaniline nanofibers prepared with ultrasonic irradiation, *J. Polym. Sci. Part A Polym. Chem.*, Vol. 44, 2006, pp. 1014–1019.

[20]. X. Lu, H. Mao, D. Chao, W. Zhang, Y. Wei, Fabrication of polyaniline nanostructures under ultrasonic irradiation: from nanotubes to nanofibers, *Macromol. Chem. Phys.*, Vol. 207, 2006, pp. 2142–2152.

[21]. D. W. Hatchett, M. Josowicz, Composites of intrinsically conducting polymers as sensing nanomaterials, *Chem. Rev.*, Vol. 108, 2008, pp. 746–769.

[22]. X. Xu, Q. Wang, H. C. Choi, Encapsulation of iron nanoparticles with PVP nanofibrous membranes to maintain their catalytic activity, *J. Membr. Sci.*, Vol. 348, 2010, pp. 231–237.

[23]. M. Tong, S. Yuan, H. Long, M. Zheng, Reduction of nitrobenzene in groundwater by iron nanoparticles immobilized in PEG/nylon membrane, *J. Contam. Hydrol.*, 2010.

[24]. Z. Xiong, D. Zhao, G. Pan, Rapid and complete destruction of perchlorate in water and ion-exchange brine using stabilized zero-valent iron nanoparticles, *Water Res.*, Vol. 41, 2007. pp. 3497–3505.

[25]. L. Guo, Q. Huang, X. Li, S. Yang, PVP-coated iron nanocrystals: Anhydrous synthesis, characterization, and electrocatalysis for two species, *Langmuir*, Vol. 22, 2006. pp. 7867–7872.

[26]. Z Liu, F. Zhang, Nano-zerovalent iron contained porous carbons developed from waste biomass for the adsorption and dechlorination of PCBs, *Bioresour. Technol.*, Vol. 101, 2010, pp. 2562–2564.

[27]. S. M. Ponder, J. G. Darab, T. E. Mallouk, Remediation of Cr(VI) and Pb(II) aqueous solutions using supported, nanoscale zero-valent iron, *Environ. Sci. Technol.*, Vol. 34, 2000, pp. 2564–2569.

[28]. C. J. Lin, S. L. Liou, Y. H. Lo, Degradation of aqueous carbon tetrachloride by nanoscale zerovalent copper on a cation resin, *Chemosphere*, Vol. 59, 2005, pp. 1299–1307.

[29]. S. J. Wu, T. H. Liou, F. L. Mi, Synthesis of zero-valent copper-chitosan nanocomposites and their application for treatment of hexavalent chromium, *Bioresour. Technol.*, Vol. 100, 2009, pp. 4348–4353.

[30]. L. M. Blaney, S. Cinar, A. K. SenGupta, Hybrid anion exchanger for trace phosphate removal from water and wastewater, *Water Res.*, Vol. 41, 2007, pp. 1603–1613.

[31]. L. Cumbal, A. K. Sengupta, Arsenic removal using polymer-supported hydrated iron(III) oxide nanoparticles: role of Donnan membrane effect, *Environ. Sci. Technol.*, Vol. 39, 2005, pp. 6508–6515.

[32]. Q. J. Zhang, B. C. Pan, X. Q. Chen, W. M. Zhang, B. J. Pan, Q. X. Zhang, L. Lv, X. S. Zhao, Preparation of polymer-supported hydrated

ferric oxide based on Donnan membrane effect and its application for arsenic removal, *Sci. China B*, Vol. 51, 2008, pp. 379–385.

[33]. A. I. Zouboulis, I. A. Katsoyiannis, Arsenic removal using iron oxide loaded alginate beads, *Ind. Eng. Chem. Res.*, Vol. 41, 2002, pp. 6149–6155.

[34]. K. L. Chen, S. E. Mylon, M. Elimelech, Enhanced aggregation of alginate-coated iron oxide (hematite) nanoparticles in the presence of calcium, strontium, and barium cations, *Langmuir*, Vol. 23, 2007, pp. 5920–5928.

[35]. P Sylvester, P. Westerhoff, T. Möller, M. Badruzzaman, O. Boyd, A hybrid sorbent utilizing nanoparticles of hydrous iron oxide for arsenic removal from drinking water, *Environ. Eng. Sci.*, Vol. 24, 2007, pp. 104–112.

[36]. G. N. Manju, K. A. Krishnan, V. P. Vinod, T. S. Anirudhan, An investigation into the sorption of heavy metals from wastewaters by polyacrylamide-grafted iron(III) oxide, *J. Hazard. Mater. B*, Vol. 91, 2002, pp. 221–238.

[37]. M. J. DeMarco, A. K. SenGupta, J. E. Greenleaf, Arsenic removal using a polymeric/inorganic hybrid sorbent, *Water Res.*, Vol. 37, 2003, pp. 164–176.

[38]. T. Moller, P. Sylvester, Effect of silica and pH on arsenic uptake by resin/iron oxide hybrid media, *Water Res.*, Vol. 42, 2008, pp. 1760–1766.

[39]. I. A. Katsoyiannis, A. I. Zouboulis, Removal of arsenic from contaminated water sources by sorption onto iron-oxide-coated polymeric materials, *Water Res.*, Vol. 36, 2002, pp. 5141–5155.

[40]. O. M. Vatutsina, V. S. Soldatov, V. I. Sokolova, J. Johann, M. Bissen, A. Weissenbacher, A new hybrid (polymer/inorganic) fibrous sorbent for arsenic removal from drinking water, *React. Funct. Polym.*, Vol. 67, 2007, pp. 184–201.

[41]. L. R. D. da Silva, Y. Gushikem, L. T. Kubota, Horseradish peroxidase enzyme immobilized on titanium(IV) oxide coated cellulose microfibers: study of the enzymatic activity by flow injection system, *Colloid. Surf. B Biointerf.*, Vol. 6, 1996, pp. 309–315.

[42]. B. J. Jordan, R. Hong, B. Gider, J. Hill, T. Emrick, V. M. Rotello, Stabilization of chymotrypsin at air–water interface through surface binding to gold nanoparticle scaffolds, *Soft Matter*, Vol. 2, 2006, pp. 558–560.

[43]. B. Chico, C. Camacho, P. Marilín, M. A. Longo, M. A. Sanromán, J. M. Pingarrón, R. Villalonga, Polyelectrostatic immobilization of gold nanoparticles-modified peroxidase on alginate-coated gold electrode for mediatorless biosensor construction, *J. Electroanal. Chem.*, Vol. 629, 2009, pp. 126–132.

[44]. G. B. Shan, J. M. Xing, H. Y. Zhang, H. Z. Liu, Biodesulfurization of dibenzothiophene by microbial cells coated with magnetite nanoparticles, *Appl. Environ. Microbiol.*, Vol. 71, 2005, pp. 4497–4502.

[45]. Z. Tang, J. Qian, L. Shi, Characterizations of immobilized neutral proteinase on chitosan nano-particles, *Proc. Biochem.*, Vol. 41, 2006, pp. 1193–1197.

[46]. V. C. F. da Silva, F. J. Contesini, Enantioselective behavior of lipases from Aspergillus niger immobilized in different supports, *J. Ind. Microbiol. Biotechnol.*, Vol. 36, 2009, pp. 949–954.

[47]. P. Tripathi, A. Kumari, P. Rath, A. M. Kayastha, Immobilization of a-amylase from mung beans (Vigna radiata) on Amberlite MB 150 and chitosan beads:A comparative study, *J. Mol. Catal. B: Enzym.*, Vol. 49, 2007, pp. 69–74.

[48]. J. Kim, J. W. Grate, Single-Enzyme Nanoparticles armored by a nanometerscale organic/inorganic network, *Nano Lett.*, Vol. 3, 2003, pp. 1219–1222.

[49]. J. Kim, J. W. Grate, P. Wang, Nanostructures for enzyme stabilization, *Chem. Eng. Sci.*, Vol. 61, 2006, pp. 1017–1026.

[50]. P. Wang, Nanoscale biocatalyst systems, *Curr. Opin. Biotechnol.*, Vol. 17, 2006, pp. 574–579.

[51]. M. Darder, P. Aranda, E. Ruiz-Hitzky, Bionanocomposites: A new concept of ecological, bioinspired, and functional hybrid materials, *Adv. Mater.*, Vol. 19, 2007, pp. 1309–1319.

[52]. J. Kim, J. W. Grate, P. Wang, Nanobiocatalysis and its potential applications, *Trends Biotechnol.*, Vol. 26, 2008, pp. 639–646.

[53]. M. Yan, J. Ge, Z. Liu, P. Ouyang, Encapsulation of single enzyme in nanogel with enhanced biocatalytic activity and stability, *J. Am. Chem. Soc.*, Vol. 128, 2006, pp. 11008–11009.

[54]. Z. Yang, S. Si, C. Zhang, Magnetic single-enzyme nanoparticles with high activity and stability, *Biochem. Biophys. Res. Commun.*, Vol. 367, 2008, pp. 169–175.

[55]. A. Sinsawat, K. L. Anderson, R. A. Vaia, B. L. Farmer, Influence of polymer matrix composition and architecture on polymer nanocomposite formation: Coarsegrained molecular dynamics simulation, *J. Polym. Sci. B: Polym. Phys.*, Vol. 41, 2003, pp. 3272–3284.

[56]. C. R. Martin, Nanomaterials: a membrane-based synthetic approach, *Science,* Vol. 266, 1994, pp. 1961–1966.

[57]. C. R. Martin, Template synthesis of electronically conductive polymer nanostructures, *Acc. Chem. Res.*, Vol. 28, 1995, pp. 61–68.

[58]. M. Granstrom, O. Inganas, Electrically conductive polymer fibres with mesoscopic diameters: 1. Studies of structure and electrical properties, *Polymer,* Vol. 36, 1995, pp. 2867–2872.

[59]. Z. H. Cai, J. T. Lei, W. B. Liang, V. Menon, C. R. Martin, Molecular and super-molecular origins of enhanced electronic conductivity in template-synthesized polyheterocyclic fibrils. 1. Supermolecular effects, *Chem. Mater.*, Vol. 3, 1991, pp. 960–967.

[60]. J. Duchet, R. Legras, S. Demoustier-Champagne, Chemical synthesis of polypyrrole: structure-properties relationship, *Synth. Met.*, Vol. 98, 1998, pp. 113–122.

[61]. J. L. Duvail, P. Retho, V. Fernandez, G. Louarn, P. Molinie, O. Chauvet, Effects of the confined synthesis on conjugated polymer transport properties, *J. Phys. Chem. B*, Vol. 108, 2004, pp. 18552–18556.

[62]. J. M. Mativetsky, W. R. Datars, Morphology and electrical properties of template-synthesized polypyrrole nanocylinders, *Physica B*, Vol. 324, 2002, pp. 191–204.

[63]. B. H. Kim, D. H. Park, J. Joo, S. G. Yu, S. H. Lee, Synthesis, characteristics, and field emission of doped and de-doped polypyrrole, polyaniline, poly(3, 4-ethylenedioxy-thiophene) nanotubes and nanowires, *Synth. Met.*, Vol. 150, 2005, pp. 279–284.

[64]. F. Massuyeau, J. L. Duvail, H. Athalin, J. M. Lorcy, S. Lefrant, J. Wery, E. Faulques, Elaboration of conjugated polymer nanowires and nanotubes for tunable photoluminescence properties, *Nanotechnology*, Vol. 20, 2009, pp. 1-8.

[65]. Y. B. Guo, Q. X. Tang, H. B. Liu, Y. J. Zhang, Y. L. Li, W. P. Hu, S. Wang, D. B. Zhu, Light-controlled organic/inorganic P-N junction nanowires, *J. Am. Chem. Soc.*, Vol. 130, 2008, pp. 9198–9199.

[66]. Y. B. Guo, Y. J. Zhang, H. B. Liu, S. W. Lai, Y. L. Li, Y. J. Li, W. P. Hu, S. Wang, C. M. Che, D. B. Zhu, Assembled organic/inorganic p-n junction interface and photovoltaic cell on a single nanowire, *J. Phys. Chem. Lett.*, Vol. 1, 2010, pp. 327–330.

[67]. L. Gence, S. Faniel, C. Gustin, S. Melinte, V. Bayot, V. Callegari, O. Reynes, S. Demoustier-Champagne, Structural and electrical characterization of hybrid metal-polypyrrole nanowires, *Phys. Rev. B*, Vol. 76, 2007, pp. 1-8.

[68]. V. Callegari, L. Gence, S. Melinte, S. Demoustier-Champagne, Electrochemically template-grown multi-segmented gold-conducting polymer nanowires with tunable electronic behavior, *Chem. Mater.*, Vol. 21, 2009, pp. 4241–4247.

[69]. R. Liu, S. B. Lee, MnO$_2$/poly(3, 4-ethylenedioxythiophene) coaxial nanowires by one-step coelectrodeposition for electrochemical energy storage, *J. Am. Chem. Soc.*, Vol. 130, 2008, pp. 2942–2943.

[70]. J. M. Lorcy, F. Massuyeau, P. Moreau, O. Chauvet, E. Faulques, J. Wery, J. L. Duvail, Coaxial nickel/poly(p-phenylene vinylene) nanowires as luminescent building blocks manipulated magnetically, *Nanotechnology*, Vol. 20, 2009, pp. 1-7.

[71]. X. Y. Zhang, W. J. Goux, S. K. Manohar, Synthesis of polyaniline nanofibers by "nanofiber seeding", *J. Am. Chem. Soc.*, Vol. 126, 2004, pp. 4502–4503.

[72]. X. Y. Zhang, S. K. Manohar, Bulk synthesis of polypyrrole nanofibers by a seeding approach, *J. Am. Chem. Soc.*, Vol. 126, 2004, pp. 12714–12715.

[73]. X. Y. Zhang, S. K. Manohar, Narrow pore-diameter polypyrrole nanotubes, *J. Am. Chem. Soc.*, Vol. 127, 2005, pp. 14156–14157.

[74]. L. J. Pan, L. Pu, Y. Shi, S. Y. Song, Z. Xu, R. Zhang, Y. D. Zheng, Synthesis of polyaniline nanotubes with a reactive template of manganese oxide, *Adv. Mater.*, Vol. 19, 2007, pp. 461–464.

[75]. X. Li, S. J. Tian, Y. Ping, D. H. Kim, W. Knoll, One-step route to the fabrication of highly porous polyaniline nanofiber films by using PS-b-PVP diblock copolymers as templates, *Langmuir*, Vol. 21, 2005, pp. 9393–9397.

[76]. P. Nickels, W. U. Dittmer, S. Beyer, J. P. Kotthaus, F. C. Simmel, Polyaniline nanowires synthesis template by DNA, *Nanotechnology*, Vol. 15, 2004, pp. 1524–1529.

[77]. A. Houlton, A. R. Pike, M. A. Galindo, B. R. Horrocks, DNA-based routes to semiconducting nanomaterials, *Chem. Commun.*, 2009, pp. 1797–1806.

[78]. X. Li, M. X. Wan, X. N. Li, G. L. Zhao, The role of DNA in PANI-DNA hybrid: template and dopant, *Polymer*, Vol. 50, 2009, pp. 4529–4534.

[79]. J. H. Rong, F. Oberbeck, X. N. Wang, X. D. Li, J. Oxsher, Z. W. Niu, Q. Wang, Tobacco mosaic virus templated synthesis of one dimensional inorganic-polymer hybrid fibres, *J. Mater. Chem.*, Vol. 19, 2009, pp. 2841–2845.

[80]. M. R. Abidian, D. H. Kim, D. C. Martin, Conducting-polymer nanotubes for controlled drug release, *Adv. Mater.*, Vol. 18, 2006, pp. 405–409.

[81]. M. R. Abidian, D. C. Martin, Experimental and theoretical characterization of implantable neural microelectrodes modified with conducting polymer nanotubes *Biomaterials*, Vol. 29, 2008, pp. 1273–1283.

[82]. M. R. Abidian, K. A. Ludwig, T. C. Marzullo, D. C. Martin, D. R. Kipke, Interfacing conducting polymer nanotubes with the central nervous system: chronic neural recording using poly (3, 4-ethylenedioxythiophene) nanotubes, *Adv. Mater.*, Vol. 21, 2009, pp. 3764–3770.

[83]. M. R. Abidian, J. M. Corey, D. R. Kipke, D. C. Martin, Conducting-polymer nanotubes improve electrical properties, mechanical adhesion, neural attachment, and neurite outgrowth of neural electrodes, *Small*, Vol. 6, 2010, pp. 421–429.

[84]. J. X. Huang, S. Virji, B. H. Weiller, R. B. Kaner, Polyaniline nanofibers: facile synthesis and chemical sensors, *J. Am. Chem. Soc.*, Vol. 125, pp. 314–315.

[85]. J. X. Huang, R. B. Kaner, A general chemical route to polyaniline nanofibers, *J. Am. Chem. Soc.*, Vol. 126, 2004, pp. 851–855.

[86]. N. R. Chiou, A. J. Epstein, Polyaniline nanofibers prepared by dilute polymerization, *Adv. Mater.*, Vol. 17, 2005, pp. 1679–1683.

[87]. M. X. Wan, A template-free method towards conducting polymer nanostructures, *Adv. Mater.*, Vol. 20, 2008, pp. 2926–2932.

[88]. M. X. Wan, J. Huang, Y. Q. Shen, Microtubes of conducting polymers, *Synth. Met.*, Vol. 101, 1999, pp. 708–711.

[89]. J. X. Huang, R. B. Kaner, Nanofiber formation in the chemical polymerization of aniline: a mechanistic study, *Angew. Chem. Int. Ed.*, Vol. 43, 2004, pp. 5817–5821.

[90]. J. Jang, H. Yoon, Facile fabrication of polypyrrole nanotubes using reverse microemulsion polymerization, *Chem. Commun.*, 2003, pp. 720–721.

[91]. H. Liu, X. B. Hu, J. Y. Wang, R. I. Boughton, Structure, conductivity, and thermopower of crystalline polyaniline synthesized by the ultrasonic irradiation polymerization method, *Macromolecules,* Vol. 35, 2002, pp. 9414–9419.

[92]. S. K. Pillalamarri, F. D. Blum, A. T. Tokuhiro, J. G. Story, M. F. Bertino, Radiolytic synthesis of polyaniline nanofibers: a new templateless pathway, *Chem. Mater.*, Vol. 17, 2005, pp. 227–229.

[93]. S. K. Pillalamarri, F. D. Blum, A. T. Tokuhiro, M. F. Bertino, One-pot synthesis of polyaniline- metal nanocomposites, *Chem. Mater.*, Vol. 17, 2005, pp. 5941–5944.

[94]. L. K. Werake, J. G. Story, M. F. Bertino, S. K. Pillalamarri, F. D. Blum, Photolithographic synthesis of polyaniline nanofibers, *Nanotechnology,* Vol. 16, 2005, pp. 2833–2837.

[95]. L. J. Zhang, Y. Z. Long, Z. J. Chen, M. X. Wan, The effect of hydrogen bonding on self-assembled polyaniline nanostructures, *Adv. Funct. Mater.*, Vol. 14, 2004, pp. 693–698.

[96]. H. J. Ding, J. Y. Shen, M. X. Wan, Z. J. Chen, Formation mechanism of polyaniline nanotubes by a simplified template-free method, *Macromol. Chem. Phys.*, Vol. 209, 2008, pp. 864–871.

[97]. H. J. Ding, M. X. Wan, Y. Wei, Controlling the diameter of polyaniline nanofibers by adjusting the oxidant redox potential, *Adv. Mater.*, Vol. 19, 2007, pp. 465–469.

[98]. H. J. Ding, Y. Z. Long, J. Y. Shen, M. X. Wan, $Fe_2(SO_4)_3$ as a binary oxidant and dopant to thin polyaniline nanowires with high conductivity, *J. Phys. Chem. B*, Vol. 114, 2010, pp. 115–119.

[99]. Y. Yan, K. Deng, Z. Yu, Z. X. Wei, Tuning the supramolecular chirality of polyaniline by methyl substitution, *Angew. Chem. Int. Edit.*, Vol. 48, 2009, pp. 2003–2006.

[100]. L. J. Zhang, M. X. Wan, Self-assembly of polyaniline—from nanotubes to hollow microspheres, *Adv. Funct. Mater.*, Vol. 13, 2003, pp. 815–820.

[101]. L. J. Zhang, M. X. Wan, Y. Wei, Hollow polyaniline microspheres with conductive and fluorescent function, *Macromol. Rapid. Commun.*, Vol. 27, 2006, pp. 888–893.

[102]. Z. M. Zhang, Z. X. Wei, L. J. Zhang, M. X. Wan, Polyaniline nanotubes and their dendrites doped with different naphthalene sulfonic acids, *Acta Mater.*, Vol. 53, 2005, pp. 1373–1379.

[103]. Q. W. Tang, J. H. Wu, X. M. Sun, Q. H. Li, J. M. Lin, Shape and size control of oriented polyaniline microstructure by a self-assembly method, *Langmuir*, Vol. 25, 2009, pp. 5253–5257.

[104]. Z. M. Huang, Y. Z. Zhang, M. Kotaki, S. Ramakrishna, A review on polymer nanofibers by electrospinning and their applications in nanocomposites, *Compos. Sci. Technol.*, Vol. 63, 2003, pp. 2223–2253.

[105]. A. Greiner, J. H. Wendorff, Functional self-assembled nanofibers by electrospinning, *Adv. Polym. Sci.*, Vol. 219, 2008, pp. 107–171.

270

[106].Y. Long, M. Li, C. Gu, M. Wan, J. Duvail, Z. Liu, Z. Fan, Recent advances in synthesis, physical properties and applications of conducting polymer nanotubes and nanofibers, *Prog. Polym. Sci.*, Vol. 36, 2011, pp. 1415–1442.

[107].I. D. Norris, M. M. Shaker, F. K. Ko, A. G. Macdiarmid, Electrostatic fabrication of ultrafine conducting fibers: polyaniline/polyethylene oxide blends, *Synth. Met.*, Vol. 114, 2000, pp. 109–114.

[108].A. G. Macdiarmid, W. E. Jones, I. D. Norris, J. Gao, A. T. Johnson, N. J. Pinto, J. Hone, B. Han, F. K. Ko, H. Okuzaki, M. Llaguno, Electrostatically-generated nanofibers of electronic polymers, *Synth. Met.*, Vol. 119, 2001, 27–30.

[109].J. R. Cardenas, M. G. O. de Franc, E. A. de Vasconcelos, W. M. de Azevedo, E. F. da Silva, Growth of sub-micron fibres of pure polyaniline using the electrospinning technique, *J. Phys. D: Appl. Phys.*, Vol. 40, 2007, pp. 1068–1071.

[110].T. S. Kang, S. W. Lee, J. Joo, J. Y. Lee, Electrically conducting polypyrrole fibers spun by electrospinning, *Synth. Met.*, Vol. 153, 2005, pp. 61–64.

[111].I. S. Chronakis, S. Grapenson, A. Jakob, Conductive polypyrrole nanofibers via electrospinning: electrical and morphological properties, *Polymer,* Vol. 47, 2006, pp. 1597–1603.

[112].A. Laforgue, L. Robitaille, Fabrication of poly-3-hexylthiophene/ polyethylene oxide nanofibers using electrospinning, *Synth. Met.*, Vol. 158, 2008, pp. 577–584.

[113].M. K. Shin, Y. J. Kim, S. I. Kim, S. K. Kim, H. Lee, G. M. Spinks, S. J. Kim, Enhanced conductivity of aligned PANi/PEO/MWNT nanofibers by electrospinning, *Sensor Actuat B-Chem*, Vol. 134, 2008, pp. 122–126.

[114].N. J. Pinto, A. T. Johnson, C. H. Mueller, N. Theofylaktos, D. C. Robinson, F. A. Miranda, Electrospun polyaniline/polyethylene oxide nanofiber field-effect transistor, *Appl. Phys. Lett.*, Vol. 83, 2003, pp. 4244–4246.

[115].H. Q. Liu, C. H. Reccius, H. G. Craighead, Single electrospun regioregular poly(3- hexylthiophene) nanofiber field-effect transistor, *Appl. Phys. Lett.*, Vol. 87, 2005, pp. 1-3.

[116].H. Q. Liu, J. Kameoka, D. A. Czaplewski, H. G. Craighead, Polymeric nanowire chemical sensor, *Nanolett.*, Vol. 4, 2004, pp. 671–675.

[117].F. L. Zhang, T. Nyberg, O. Inganas, Conducting polymer nanowires and nanodots made with soft lighography, *Nanolett.*, Vol. 2, 2002, pp. 1373–1377.

[118].Z. J. Hu, B. Muls, L. Gence, D. A. Serban, J. Hofkens, S. Melinte, B. Nysten, S. Demoustier-Champagne, A. M. Jonas, High-throughput fabrication of organic nanowire devices with preferential internal alignment and improved performance, *Nanolett.*, Vol. 7, 2007, pp. 3639–3644.

[119].C. Y. Huang, B. Dong, N. Lu, N. J. Yang, L. G. Gao, L. Tian, D. P. Qi, Q. Wu, L. F. Chi, A strategy for patterning conducting polymers using

nanoimprint lithography and isotropic plasma etching, *Small,* Vol. 5, 2009, pp. 583–586.

[120]. X. Z. Niu, S. L. Peng, L. Y. Liu, W. J. Wen, P. Sheng, Characterizing and patterning of PDMS-based conducting composites, *Adv. Mater.,* Vol. 19, 2007, pp. 2682–2686.

[121]. B. Ozturk, C. Blackledge, B. N. Flanders, D. R. Grischkowsky, Reproducible interconnects assembled from gold nanorods, *Appl. Phys. Lett.,* Vol. 88, 2006, pp. 1-3.

[122]. M. Yun, N. V. Myung, R. P. Vasquez, C. Lee, E. Menke, R. M. Penner, Electrochemically grown wires for individual addressable sensor arrays, *Nanolett.,* Vol. 4, 2004, pp. 419–422.

[123]. K. Ramanathan, M. A. Bangar, M. H. Yun, W. Chen, A. Mulchandani, N. V. Myung, Individually addressable conducting polymer nanowires array, *Nanolett.,* Vol. 4, 2004, pp. 1237–1239.

[124]. A. Das, C. H. Lei, M. Elliott, J. E. Macdonald, M. L. Turner, Non-lithographic fabrication of PEDOT nano-wires between fixed Au electrodes, *Org. Electron.,* Vol. 7, 2006, pp. 181–187.

[125]. I. Lee, H. I. Park, S. Park, M. J. Kim, M. H. Yun, Highly reproducible single polyaniline nanowire using electrophoresis method, *Nano,* Vol. 3, 2008, pp. 75–82.

[126]. P. S. Thapa, D. J. Yu, J. P. Wicksted, J. A. Hadwiger, J. N. Barisci, R. H. Baughman, B. N. Flanders, Directional growth of polypyrrole and polythiophene wires, *Appl. Phys. Lett.,* Vol. 94, 2009, pp. 1-3.

[127]. A. Noy, A. E. Miller, J. E. Klare, B. L. Weeks, B. W. Woods, J. J. DeYoreo, Fabrication of luminescent nanostructures and polymer nanowires using dip-pen nanolithography, *Nanolett.,* Vol. 2, 2002, pp. 109–112.

[128]. S. Samitsu, T. Shimomura, K. Ito, M. Fujimori, S. Heike, T. Hashizume, Conductivity measurements of individual poly(3,4-ethylenedioxythiophene)/ poly(styrenesulfonate) nanowires on nanoelectrodes using manipulation with an atomic force microscope, *Appl. Phys. Lett.,* Vol. 86, 2005, pp. 1-3.

[129]. S. Samitsu, Y. Takanishi, J. Yamamota, Self-assembly and one dimensional alignment of a conducting polymer nanofiber in a nematic liquid crystal, *Macromolecules,* Vol. 42, 2009, pp. 4366–4368.

[130]. H. Yonemura, K. Yuno, Y. Yamamoto, S. Yamada, Y. Fujiwara, Y. Tanimoto, Orientation of nanowires consisting poly(3-hexylthiophene) using strong magnetic field, *Synth. Met.,* Vol. 159, 2009, pp. 955–960.

[131]. Y. Xia, Y. Xiong, B. Lim, S. E. Skrabalak, Shape-controlled synthesis of metal nanocrystals: simple chemistry meets complex physics?, *Angew. Chem. Int. Ed.,* Vol. 48, 2008, pp. 60–103.

[132]. J. Huang, S. Virji, B. H. Weiller, R. B. Kaner, Nanostructured polyaniline sensors, *Chem. A Eur. J.,* Vol. 10, 2004, pp. 1314–1319.

[133]. R. J. Tseng, J. Huang, J. Ouyang, R. B. Kaner, Y. Yang, Polyaniline nanofiber/gold nanoparticle nonvolatile memory, *Nanolett.,* Vol. 5, 2005, pp. 1077–1080.

[134].B. J. Gallon, R. W. Kojima, R. B. Kaner, P. L. Diaconescu, Palladium nanoparticles supported on polyaniline nanofibers as a semiheterogeneous catalyst in water, *Angew. Chem. Int. Ed.*, Vol. 46, 2007, pp. 7251–7254.

[135].X. Zhang, S. K. Manohar, Narrow pore-diameter polypyrrole nanotubes, *J. Am. Chem. Soc.*, Vol. 127, 2005, pp. 14156–14157.

[136].J. Xu, J. Hu, B. Quan, Z. Wei, Decorating polypyrrole nanotubes with Au nanoparticles by an in situ reduction process, *Macromol. Rapid Commun.*, Vol. 30, 2009, pp. 936–940.

[137].Z. Chen, L. Xu, W. Li, M. Waje, Y. Yan, Polyaniline nanofibre supported platinum nanoelectrocatalysts for direct methanol fuel cells, *Nanotechnology*, Vol. 17, 2006, pp. 5254–5259.

[138].Y. Ma, S. Jiang, G. Jian, H. Tao, L. Yu, X. Wang, X. Wang, J. Zhu, Z. Hu, Y. Chen, CNx nanofibers converted from polypyrrole nanowires as platinum support for methanol oxidation, *Energy Environ. Sci.*, Vol. 2, 2009, pp. 224–229.

[139].S. Guo, S. Dong, E. Wang, Polyaniline/Pt hybrid nanofibers: high efficiency nanoelectrocatalysts for electrochemical devices, *Small,* Vol. 5, 2009, pp. 1869–1876.

[140].H. Cao, Z. Xu, H. Sang, D. Sheng, C. Tie, Template synthesis and magnetic behavior of an array of cobalt nanowires encapsulated in polyaniline nanotubules, *Adv. Mater.*, Vol. 13, pp. 121–123.

[141].H. Cao, Z. Xu, D. Sheng, J. Hong, H. Sang, Y. Du, An array of iron nanowires encapsulated in polyaniline nanotubules and its magnetic behavior, *J. Mater. Chem.*, Vol. 11, 2001, pp. 958–960.

[142].H. Cao, C. Tie, Z. Xu, J. Hong, H. Sang, Array of nickel nanowires enveloped in polyaniline nanotubules and its magnetic behavior, *Appl. Phys. Lett.*, Vol. 78, 2001, pp. 1592–1594.

[143].J. Zhang, G. Shi, C. Liu, L. Qu, M. Fu, F. Chen, Electrochemical fabrication of polythiophene film coated metallic nanowire arrays, *J. Mater. Sci.*, Vol. 38, 2003, pp. 2423–2427.

[144].A. Drury, S. Chaure, M. Kroll, V. Nicolosi, N. Chaure, W. J. Blau, Fabrication and characterization of silver/polyaniline composite nanowires in porous anodic alumina, *Chem. Mater.*, Vol. 19, 2007, pp. 4252–4258.

[145].S. Park, J. H. Lim, S. W. Chung, C. A. Mirkin, Self-assembly of mesoscopic metal-polymer amphiphiles, *Science,* Vol. 303, 2004, pp. 348–351.

[146].M. Lahav, E. A. Weiss, Q. Xu, G. M. Whitesides, Core-shell and segmented polymer-metal composite nanostructures, *NanoLett.*, Vol. 6, 2006, pp. 2166–2171.

[147].R. M. Hernandez, L. Richter, S. Semancik, S. Stranick, T. E. Mallouk, Template fabrication of protein-functionalized gold-polypyrrole-gold segmented nanowires, *Chem. Mater.*, Vol. 16, 2004, pp. 3431–3438.

[148].O. Reynes, S. Demoustier-Champagne, Template electrochemical growth of polypyrrole and gold-polypyrrole-gold nanowire arrays, *J. Electrochem. Soc.*, Vol. 152, 2005, pp. D130–135.

[149].L. Kong, X. Lu, E. Jin, S. Jiang, C. Wang, W. Zhang, Templated synthesis of polyaniline nanotubes with Pd nanoparticles attached onto their inner walls and its catalytic activity on the reduction of p-nitroanilinum, *Composites Sci. Technol.*, Vol. 69, 2009, pp. 561–566.

[150].X. Feng, Z. Sun, W. Hou, J. Zhu, Synthesis of functional polypyrrole/ prussian blue and polypyrrole/Ag composite microtubes by using a reactive template, *Nanotechnology,* Vol. 18, 2007, pp. 1–7.

[151].A. Chen, H. Wang, X. Li, One-step process to fabricate Ag–polypyrrole coaxial nanocables, *Chem. Commun.*, 2005, pp. 1863–1864.

[152].D. Munoz-Rojas, J. Oro-Sole, O. Ayyad O, P. Gomez-Romero P, Facile one-pot synthesis of self-assembled silver@polypyrrole core/shell nanosnakes, *Small,* Vol. 4, 2008, pp. 1301–1306.

[153].K. Huang, Y. Zhang, Y. Long, J. Yuan, D. Han, Z. Wang, L. Niu, Z. Chen, Preparation of highly conductive, self-assembled gold/polyaniline nanocables and polyaniline nanotubes, *Chem. Eur. J.,* Vol. 12, 2006, pp. 5314–5319.

[154].G. Lu, C. Li, J. Shen, Z. Chen, G. Shi, Preparation of highly conductive gold-poly(3, 4-ethylenedioxythiophene) nanocables and their conversion to poly(3, 4-ethylenedioxythiophene) nanotubes, *J. Phys. Chem. C,* Vol. 111, 2007, pp. 5926–5931.

[155].K. Mallick, M. J. Witcomb, A. Dinsmore, M. S. Scurrell, Fabrication of a metal nanoparticles and polymer nanofibers composite material by an in situ chemical synthetic route, *Langmuir*, Vol. 21, 2005, pp. 7964–7967.

[156].A. Chen, H. Xie, H. Wang, H. Li, X. Li, Fabrication of Ag/polypyrrole coaxial nanocables through common ions adsorption effect, *Synth. Met.*, Vol. 156, 2006, pp. 346–350.

[157].S. K. Pillalamarri, F. D. Blum, A. T. Tokuhiro, M. F. Bertino, One-pot synthesis of polyaniline-metal nanocomposites, *Chem. Mater.*, Vol. 17, 2005, pp. 5941–5944.

[158].Z. Zhang, M. Wan, Y. Wei, Electromagnetic functionalized polyaniline nanostructures, *Nanotechnology,* Vol. 16, 2005, pp. 2827–2832.

[159].L. Zhang, M. Wan, Polyaniline/TiO$_2$ composite nanotubes, *J. Phys. Chem. B*, Vol. 107, 2003, pp. 6748–6753.

[160].X. Lu, H. Mao, D. Chao, W. Zhang, Y. Wei, Ultrasonic synthesis of polyaniline nanotubes containing Fe$_3$O$_4$ nanoparticles, *J. Solid State Chem.*, Vol. 179, 2006, pp. 2609–2615.

[161].X. Wang, N. Liu, X. Yan, W. Zhang, Y. Wei, Alkali-guided synthesis of polyaniline hollow microspheres, *Chem. Lett.*, Vol. 34, 2005, pp. 42–43.

[162].H. Ding, M. Wan, Y. Wei, Controlling the diameter of polyaniline nanofibers by adjusting the oxidant redox potential, *Adv. Mater.*, Vol. 19, 2007, pp. 465–469.

[163].H. Ding, J. Shen, M. Wan, Z. Chen, Formation mechanism of polyaniline nanotubes by a simplified template-free method, *Macromol. Chem. Phys.*, Vol. 209, 2008, pp. 864–871.

[164].Z. Zhang, J. Deng, J. Shen, M. Wan, Z. Chen, Z. Zhang, M. Wan, Y. Wei, Chemical one step method to prepare polyaniline nanofibers

with electromagnetic function, *Macromol. Rapid Commun.*, Vol. 28, 2007, pp. 585–590.

[165]. J. Han, G. Song, R. Guo, Synthesis of rectangular tubes of polyaniline/NiO composites, *J. Polym. Sci. Part A Polym. Chem.*, Vol. 44, 2006, pp. 4229–4234.

[166]. G. Song, J. Han, R. Guo, Synthesis of polyaniline/NiO nanobelts by a self-assembly process, *Synth. Met.*, Vol. 157, 2007, pp. 170–175.

[167]. G. Ciric -Marjanovic, L. Dragicevic, M. Milojevic, M. Mojovic, S. Mentus, B. Dojcinovic, B. Marjanovic, J. Stejskal, Synthesis and characterization of self-Assembled polyaniline nanotubes/silica nanocomposites, *J. Phys. Chem. B*, Vol. 113, 2009, pp. 7116–7127.

[168]. H. Mao, X. Lu, D. Chao, L. Cui, Y. Li, W. Zhang, Preparation and characterization of PEDOT/Fe^{3+}O(OH, Cl) nanospindles with controllable sizes in aqueous solution, *J. Phys. Chem. C*, Vol. 112, 2008, pp. 20469–20480.

[169]. G. Li, C. Zhang, H. Peng, K. Chen, One-dimensional V$_2$O$_5$@polyaniline core/shell nanobelts synthesized by an in situ polymerization method *Macromol. Rapid Commun.*, Vol. 30, 2009, pp. 1841–1845.

[170]. X. Lu, Q. Zhao, X. Liu, D. Wang, W. Zhang, C. Wang, Y. Wei, Preparation and characterization of polypyrrole/TiO$_2$ coaxial nanocables, *Macromol. Rapid Commun.*, Vol. 27, 2006, pp. 430–434.

[171]. X. Lu, H. Mao, W. Zhang, Surfactant directed synthesis of polypyrrole/TiO2 coaxial nanocables with a controllable sheath size, *Nanotechnology*, Vol. 18, 2007, pp. 1–5.

[172]. Q. Cheng, V. Pavlinek, Y. He, C. Li, A. Lengalova, P. Saha, Facile fabrication and characterization of novel polyaniline/titanate composite nanotubes directed by block copolymer, *Eur. Polym. J.*, Vol. 43, 2007, pp. 3780–3786.

[173]. J. Xu, X. Li, J. Liu, X. Wang, Q. Peng, Y. Li, Solution route to inorganic nanobelt-conducting organic polymer core-shell nanocomposites, *J. Polym. Sci. Part A Polym. Chem.*, Vol. 43, 2005, pp. 2892–2900.

[174]. U. Sree, Y. Yamamoto, B. Deore, H. Shugi, T. Nagaoka, Characterization of polypyrrole nanofilms for membrane based sensors, *Synth. Met.*, Vol. 131, 2002, pp. 161–165.

[175]. D. T. McQuade, A. E. Pullen, T. M. Swager, Conjugated polymer based chemical sensors, *Chem. Rev.*, Vol. 100, 2000, pp. 2537–2574.

[176]. S. Geeta, C. R. K. Roa, M. Vijayan, D. C. Trivedi, Biosensing and drug delivery of polypyrrole, *Anal. Chim. Acta.*, Vol. 568, 2006, pp. 119–125.

[177]. M. K. Ram, O. Yavuz, V. Lahsangah, M. Aldissi, CO gas sensing from ultrathin nanocomposite conducting polymer film, *Sens. Actuators, B*, Vol. 106, 2005, pp. 750–757.

[178]. C. K. Chiang, Y. W. Park, A. J. Heeger, H. Shirakawa, E. J. Louis, A. G. MacDiarmid, Conducting polymers: halogen-doped polyacetylene, *J. Chem. Phys.*, Vol. 69, 1978, pp. 5098–5104.

[179]. H. Shirakawa, E. J. Louis, A. G. MacDiarmid, C. K. Chiang, A. J. Heeger, Synthesis of electrically conducting organic polymers: halogen

derivatives of polyacetylene, (CH)x, *J. Chem. Soc. Chem. Comms.*, 1977, pp. 578–580.

[180].A. F. Diaz, K. K. Kanazawa, G. P. Gardini, Electrochemical polymerization of pyrrole, *J. Chem. Soc. Chem. Comms.*, 1979, pp. 635–636.

[181].J. Roncali, Conjugated poly(thiophenes): synthesis, functionalization, and applications, *Chem. Rev.*, Vol. 92, 1992, pp. 711–738.

[182].C. Nylander, M. Armgarth, I. Lundstrom, An ammonia detector based on a conducting polymer, *Anal. Chem. Symp. Ser.*, Vol. 17, 1983, pp. 203–207.

[183].K. C. Persaud, P. Pelosi, An approach to an artificial nose, *Trans. Am. Soc. Artif. Int. Organs.*, Vol. 31, 1985, pp. 297–300.

[184].E. H. Amrani, S. Ibrahim, K. C. Persaud, Synthesis, chemical characterisation and multifrequency measurements of poly N-(2-pyridyl) pyrrole for sensing volatile chemicals, *Mat. Sci. Engng.*, Vol. C1, 1993, pp. 17–22.

[185].J. M. Slater, E. J. Watt, Examination of ammonia-poly(pyrrole) interactions by piezoelectric and conductivity measurements, *Analyst*, Vol. 116, 1991, pp. 1125–1130.

[186].P. N. Bartlett, S. K. Ling-Chung, Conducting polymer gas sensors Part III: Results for four different polymers and five different vapours, *Sensors Actuators*, Vol. 20, 1989, pp. 287–292.

[187].M. J. Marsella, P. J. Carroll, T. M. Swager, Design of chemoresistive sensory materials: polythiophene-based pseudopolyrotaxanes, *J. Am. Chem. Soc.*, Vol. 117, 1995, pp. 9832–9841.

[188].P. Bruschi, F. Cacialli, A. Nannini, B. Neri, Gas and vapour effects on the resistance fluctuation spectra of conducting polymer thin-film resistors, *Sensors Actuators B*, Vol. 18-19, 1994, pp. 421–425.

[189].L. Torsi, M. Pezzuto, P. Siciliano, R. Rella, L. Sabbatini, L. Valli, P. G. Zambonin, Conducting polymers doped with metallic inclusions: new materials for gas sensors, *Sensors Actuators B*, Vol. 48, 1998, pp. 362–367.

[190].M. Hirata, L. Sun, Characteristics of an organic semiconductor polyaniline film as a sensor for NH_3 gas, *Sensors Actuators A*, Vol. 40, 1994, pp. 159–163.

[191].S. Unde, J. Ganu, S. Radhakrishnan, Conducting polymerbased chemical sensor: characteristics and evaluation of polyaniline composite films, *Adv. Mater. Optics Electr.*, Vol. 6, 1996, pp. 151–157.

[192].K. Ogura, T. Saino, M. Nakayama, H. Shiigi, The humidity dependence of the electrical conductivity of a soluble polyaniline-poly(vinyl alcohol) composite film, *J. Mater. Chem.*, Vol. 7, 1997, pp. 2363–2366.

[193].D. Li, Y. Jiang, Z. Wu, X. Chen, Y. Li, Self-assembly of polyaniline ultrathin films based on doping-induced deposition effect and applications for chemical sensors, *Sensors Actuators B*, Vol. 66, 2000, pp. 125–127.

[194]. M. E. Nicho, M. Trejo, A. García-Valenzuela, J. M. Saniger, J. Palacios, H. Hu, Polyaniline composite coatings interrogated by nulling optical-transmittance bridge for sensing low concentrations of ammonia gas, *Sensors Actuators B*, Vol. 76, 2001, pp. 18–24.

[195]. D. S. Sutar, N. Padma, D. K. Aswal, S. K. Deshpande, S. K. Gupta, J. V. Yakhmi, Preparation of nanofibrous polyaniline films and their application as ammonia gas sensor, *Sensors Actuators B*, Vol. 128, 2007, pp. 286–292.

[196]. J. Yadong, W. Tao, W. Zhiming, L. Dan, C. Xiangdong, X. Dan, Study on the NH_3-gas sensitive properties and sensitive mechanism of polypyrrole, *Sensors Actuators B*, Vol. 66, 2000, pp. 280–282.

[197]. R. Gangopadhyay, A. De, Conducting polymer composites: noble materials for gas sensing, *Sensors Actuators B*, Vol. 77, 2001, pp. 326–329.

[198]. I. Lahdesmaki, W. W. Kubiak, A. Lewenstam, A. Ivaska, Interferences in a polypyrrole-based amperometric ammonia sensor, *Talanta*, Vol. 52, 2000, pp. 269–275.

[199]. D. Kim, B. Yoo, A novel electropolymerization method for Ppy nanowire-based NH_3 gas sensor with low contact resistance, *Sensors Actuators B*, Vol. 160, 2011, pp. 1168–1173.

[200]. G. Shi, M. Luo, J. Xue, Y. Xian, L. Jin, J. Y. Jin, The study of $PVP/Pd/IrO_2$ modified sensor for amperometric determination of sulfur dioxide, *Talanta*, Vol. 55, 2001, pp. 241–247.

[201]. W. H. Christensen, D. N. Sinha, S. F. Conductivity of polystyrene film upon exposure to nitrogen dioxide: a novel NO_2 sensor, Agnew, *Sensors Actuators B*, Vol. 10, 1993, pp. 149–153.

[202]. D. Xie, Y. Jiang, W. Pan, D. Li, Z. Wu, Y. Li, Fabrication and characterization of polyaniline-based gas sensor by ultra-thin film technology, *Sensors Actuators B*, Vol. 81, 2002, pp. 158–164.

[203]. S. Prakash, S. Rajesh, S. K. Singh, K. Bhargava, G. Ilavazhagan, V. Vasu, C. Karunakaran, Copper nanoparticles entrapped in SWCNT-PPy nanocomposite on Pt electrode as NOx electrochemical sensor, *Talanta*, Vol. 85, 2011, pp. 964–969.

[204]. L. Torsi, M. Pezzuto, P. Siciliano, R. Rella, L. Sabbatini, L. Valli, P. G. Zambonin, Conducting polymers doped with metallic inclusions: new materials for gas sensors, *Sensors Actuators B*, Vol. 48, 1998, pp. 362–367.

[205]. S. Roy, S. Sana, B. Adhikari, S. Basu, Preparation of doped polyaniline and polypyrrole films and applications for hydrogen gas sensors, *J. Polym. Mater.*, Vol. 20, 2003, pp. 173–180.

[206]. H. M. Fog, B. Rietz, Piezoelectric crystal detector for the monitoring of ozone in working environments, *Anal. Chem.*, Vol. 57, 1985, pp. 2634–2638.

[207]. P. G. Su, C. C. Shiu, Flexible H_2 sensor fabricated by layer-by-layer self-assembly of thin films of polypyrrole and modified in situ with Pt nanoparticles, *Sensors Actuators B*, Vol. 157, 2011, pp. 275–281.

[208].S. Virji, J. Huang, R. B. Kaner, B. H. Weiller, Polyaniline nanofibres gas sensor: examination of response mechanism, *NanoLett.*, Vol. 4, 2004, pp. 491–496.

[209].H. X. He, C. Z. Li, N. J. Tao, Conductance of polymer nanowires fabricated by a combined electro deposition and mechanical break junction method, *Appl. Phys. Lett.*, Vol. 78, 2001, pp. 811–813.

[210].J. Huang, S. Virji, B. H. Weiller, R. B. Kaner, Nanostructured polyaniline sensors, *Chem. Eur. J.*, Vol. 10, 2004, pp. 1314–1319.

[211].Q. J. Wang, X. J. Zhang, C. G. Zhang, X. Y. Zhou, Study on carbon fiber pH ultramicrosensor modified by polyaniline film and its applications to the in vivo detection on Brassica stigmata, *Gaodeng Xuexiao Huaxue Xuebao*, Vol. 18, 1997, pp. 226–228.

[212].C. A. Lindino, L. O. S. Bulhoes, The potentiometric response of chemically modified electrodes, *Anal. Chim. Acta.*, Vol. 334, 1996, pp. 317–322.

[213].D. T. McQuade, A. E. Pullen, T. M. Swager, Conjugated polymerbased chemical sensors, *Chem. Rev.*, Vol. 100, 2000, pp. 2537–2574.

[214].A. A. Karyakin, M. Vuki, L. V. Lukachova, E. E. Karyakina, A. V. Orlov, G. P. Karpachova, J. Wang, Processible polyaniline as an advanced potentiometric pH transducer. Application to biosensors, *Anal. Chem.*, Vol. 71, 1999, pp. 2534–2540.

[215].S. Demarcos, O. S. Wolfbeis, Optical sensing of pH based on polypyrrole films, *Anal. Chim. Acta*, Vol. 334, 1996, pp. 149–153.

[216].Z. Ge, C. W. Brown, L. Sun, S. C. Yang, Fiber-optic pH sensor based on evanescent wave absorption spectroscopy, *Anal. Chem.*, Vol. 65, 1993, pp. 2335–2338.

[217].E. Pringsheim, E. Terpetschnig, O. S. Wolfbeis, Optical sensing of pH using thin films of substituted polyanilines, *Anal. Chim. Acta*, Vol. 357, 1997, pp. 247–252.

[218].U. W. Grummt, A. Pron, M. Zagorska, S. Lefrant, Polyaniline based optical pH sensor, *Anal. Chim. Acta*, Vol. 357, 1997, pp. 253–259.

[219].Z. Jin, Y. Su, Y. Duan, An improved optical pH sensor based on polyaniline, *Sensors Actuators B*, Vol. 71, 2000, pp. 118–122.

[220].J. C. Chiang, A. G. MacDiarmid, Polyaniline: protonic acid doping of the emeraldine form to the metallic regime, *Synth. Met.*, Vol. 13, 1986, pp. 193–205.

[221].S. Sukeerthi, A. Q. Contractor, Applications of conducting polymers as sensors, *Indian J. Chem. Sect. A*, Vol. 33, 1994, pp. 565–571.

[222].V. Hatfield, P. Neaves, P. J. Hicks, K. Persaud, P. Travers, Towards an integrated electronic nose using conducting polymer sensors, *Sensors Actuators B*, Vol. 18-19, 1994, pp. 221–228.

[223].Y. Xia, J. M. Wiesinger, A. G. MacDiarmid, A. J. Epstein, Camphorsulfonic acid fully doped polyaniline emeraldine salt: conformations in different solvents studied by an ultraviolet/visible/near-infrared spectroscopic method, *Chem. Mater.*, Vol. 7, 1995, pp. 443–445.

[224].A. G. MacDiarmid, A. J. Epstein, The concept of secondary doping as applied to polyaniline, *Synth. Met.*, Vol. 65, 1994, pp. 103–116.

[225]. A. G. MacDiarmid, A. J. Epstein, Secondary doping in polyaniline, *Synth. Met.*, Vol. 69, 1995, pp. 85–92.

[226]. V. Svetlicic, A. J. Schmidt, L. L. Miller, Conductometric sensors based on the hypersensitive response of plasticized polyaniline films to organic vapors, *Chem. Mater.*, Vol. 10, 1998, pp. 3305–3307.

[227]. A. A. Athawale, M. V. Kulkarni, Polyaniline and its substituted derivatives as sensor for aliphatic alcohols, *Sensors Actuators B*, Vol. 67, 2000, pp. 173–177.

[228]. H. K. Jun, Y. S. Hoh, B. S. Lee, S. T. Lee, J. O. Lim, D. D. Lee, J. S. Huh, Electrical properties of polypyrrole gas sensors fabricated under various pretreatment conditions, *Sensors Actuators B*, Vol. 96, 2003, pp. 576–581.

[229]. P. N. Bartlett, S. K. Ling-Chung, Conducting polymer gas sensors part II: response of polypyrrole to methanol vapour, *Sensors Actuators*, Vol. 19, 1989, pp. 141–150.

[230]. S. S. Barkade, J. B. Naik, S. H. Sonawane, Ultrasound assisted miniemulsion synthesis of polyaniline/Ag nanocomposite and its application for ethanol vapor sensing, *Colloid. Surf. A Physicochem. Eng. Aspects*, Vol. 378, 2011, pp. 94–98.

[231]. A. Ambrosi, A. Morrin, M. R. Smyth, A. J. Killard, The application of conducting polymer nanoparticle electrodes to the sensing of ascorbic acid, *Anal. Chim. Acta*, Vol. 609, 2008, pp. 37–43.

[232]. C. Zanardi, F. Terzi, R. Seeber, Composite electrode coatings in amperometric sensors. Effects of differently encapsulated gold nanoparticles in poly(3, 4-ethylendioxythiophene) system, *Sensors Actuators B*, Vol. 148, 2010, pp. 277–282.

[233]. N. F. Atta, A. Galal, E. H. El-Ads, Gold nanoparticles-coated poly(3, 4-ethylene-dioxythiophene) for the selective determination of sub-nano concentrations of dopamine in presence of sodium dodecyl sulfate, *Electrochim. Acta*, Vol. 69, 2012, pp. 102–111.

[234]. N. F. Atta, M. F. El-Kady, Poly(3-methylthiophene)/palladium sub-micro-modified sensor electrode. Part II: Voltammetric and EIS studies, and analysis of catecholamine neurotransmitters, ascorbic acid and acetaminophen, *Talanta*, Vol. 79, 2009, pp. 639–647.

[235]. N. F. Atta, M. F. El-Kady, Novel poly(3-methylthiophene)/Pd, Pt nanoparticle sensor: Synthesis, characterization and its application to the simultaneous analysis of dopamine and ascorbic acid in biological fluids, *Sensors Actuators B*, Vol. 145, 2010, pp. 299–310.

[236]. N. F. Atta, M. F. El-Kady, A. Galal, Simultaneous determination of catecholamines, uric acid and ascorbic acid at physiological levels using poly(N-methylpyrrole)/Pd-nanoclusters sensor, *Anal. Biochem.*, Vol. 400, 2010, pp. 78–88.

[237]. N. F. Atta, M. F. El-Kady, A. Galal, Palladium nanoclusters-coated polyfuran as a novel sensor for catecholamine neurotransmitters and paracetamol, *Sensors Actuators B*, Vol. 141, 2009, pp. 566–574.

[238]. G. V. Ramírez, J. R. Windmiller, J. C. Claussen, A. G. Martinez, F. Kuralay, M. Zhou, N. Zhou, R. Polsky, P. R. Miller, R. Narayan,

J. Wang, Multiplexed and switchable release of distinct fluids from microneedle platforms via conducting polymer nanoactuators for potential drug delivery, *Sensors Actuators B*, Vol. 161, 2012, pp. 1018– 1024.

[239].J. Zhang, Y. Wang, R. Lv, L. Xu, Electrochemical tolazoline sensor based on gold nanoparticles and imprinted poly-o-aminothiophenol film, *Electrochim. Acta*, Vol. 55, 2010, pp. 4039–4044.

[240].Y. Sakai, M. Matsuguchi, H. Hara, S. Suzuki, N. Honda, Moisture sensitive component and method for its manufacture, *Eur. Pat.* 875752, 1998.

[241].Y. Sadaoka, Y. Sakai, Humidity sensor using lithium doped poly(ethylene oxide) thin film, *Denki Kagaku*, Vol. 52, 1984, pp. 132–133.

[242].R. Nohria, R. K. Khillan, Y. Su, R. Dikshit, Y. Lvov, K. Varahramyan, layer nano-assembly, *Sensors Actuators B*, Vol. 114, 2006, pp. 218–222.

[243].F. W. Zeng, X. X. Liu, D. Diamond, K. T. Lau, Humidity sensors based on polyaniline nanofibres, *Sensors Actuators B*, Vol. 143, 2010, pp. 530–534.

[244].P. G. Su, L. N. Huang, Humidity sensors based on TiO_2 nanoparticles/polypyrrole composite thin films, *Sensors Actuators B*, Vol. 123, 2007, pp. 501–507.

[245].P. G. Su, C. P. Wang, Flexible humidity sensor based on TiO_2 nanoparticles-polypyrrole-poly-[3-(methacrylamino)propyl] trimethyl ammonium chloride composite materials, *Sensors Actuators B*, Vol. 129, 2008, pp. 538–543.

[246].P. G. Su, Y. P. Chang, Low-humidity sensor based on a quartz-crystal microbalance coated with polypyrrole/Ag/TiO_2 nanoparticles composite thin films, *Sensors Actuators B*, Vol. 129, 2008, pp. 915–920.

[247].Y. Im, R. P. Vasquez, C. Lee, N. Myung, R. Penner, M. Yun, Single metal and conducting polymer nanowire sensors for chemical & DNA detections, *J. Phys. Conf. Ser.*, Vol. 38, 2006, pp. 61–64.

[248].J. Fei, K. Wu, F. Wang, S. Hu, Glucose nanosensor based on redox polymer/glucose oxidase modified carbon fibre nanoelectrodes, *Talanta*, Vol. 65, 2005, pp. 918–924.

[249].H. Peng, C. Soeller, M. B. Cannell, G. A. Bowmaker, R. P. Cooney, J. Travas-Sejdic, Electrochemical detection of DNA hybridization amplified by nanoparticles, *Biosens. Bioelectron.*, Vol. 21, 2006, pp. 1727–1736.

[250].B. Kannan, D. E. Williams, C. Laslau, J. T. Sejdic, A highly sensitive, label-free gene sensor based on a single conducting polymer nanowire, *Biosens. Bioelectron.*, Vol. 35, 2012, pp. 258– 264.

[251].J. T. Sejdic, H. Peng, R. P. Cooney, G. A. Bowmaker, M. B. Cannell, C. Soeller, Amplification of a conducting polymer-based DNA sensor signal by CdS nanoparticles, *Cur. Appl. Phys.*, Vol. 6, 2006, pp. 562–566.

[252]. A. Amine, H. Mohammadi, I. Bourais, G. Palleschi, tal monitoring. Enzyme inhibition-based biosensors for food safety and environmental monitoring, *Biosens Bioelectron*, Vol. 21, 2006, pp. 1405–1423.

[253]. S. Cosnier, Biosensors based on electropolymerised films: new trends, *Anal. Bioanal. Chem.*, Vol. 377, 2003, pp. 507–520.

[254]. A. E. Boyukbayram, S. Kiralp, L. Toppare, Y. Yagci, Preparation of biosensors by immobilization of polyphenol oxidase in conducting co-polymers and their use in determination of phenolic compounds in red wine, *Bioelectrochem.*, Vol. 69, 2006, pp. 164–171.

[255]. J. C. Vidal, E. G. Ruiz, J. R. Castillo, Recent advances in electro polymerized conducting polymers in amperometric biosensors, *Microchim. Acta*, Vol. 143, 2003, pp. 93–111.

[256]. J. M. Dicks, M. F. Cardosi, A. P. F. Turner, I. Karube, The application of ferrocene-modified n type silicon in glucose biosensors, *Electroanal.*, Vol. 5, 1993, pp. 1–9.

[257]. Rajesh, V. Bisht, W. Takashima, K. Kaneto, An amperometric urea biosensor based on covalent immobilization of urease onto an electrochemically prepared co polymer poly (N-3-amino propyl pyrrole-co-pyrrole) film, *Biomater.*, Vol. 26, 2005, pp. 3683–3690.

[258]. S. Mu, H. Xue, Bioelectrochemical characteristics of glucose oxidase immobilized in a polyaniline film, *Sensors Actuators B*, Vol. 31, 1996, pp. 155–160.

[259]. A. Kumar, Rajesh, A. Chaubey, S. K. Grover, B. D. Malhotra BD, Immobilization of cholesterol oxidase and potassium ferricyanide on dodecyl-benzene sulfonate ion-doped polypyrrole film, *J. Appl. Polym. Sci.*, Vol. 82, 2001, pp. 3486–3491.

[260]. N. Gajovic, K. Habermuller, A. Warsinke, W. Schuhmann, F. W. Scheller, A pyruvate oxidase electrode based on an electrochemically deposited redox polymer, *Electroanal.*, Vol. 11, 1999, pp. 1377–1383.

[261]. K. Ramanathan, M. K. Ram, B. D. Malhotra, A. S. N. Murthy, Application of polyaniline Langmuir–Blodgett films as a glucose biosensor, *Mater. Sci. Eng. C*, Vol. 3, 1995, pp. 159–163.

[262]. K. Ramanathan, R. Mehrotra, B. Jayaram, A. S. N. Murthy, B. D. Malhotra, Simulation of electrochemical process for glucose oxidase immobilized conducting polymer, *Anal. Lett.*, Vol. 29, 1996, pp. 1477–1484.

[263]. K. Ramanathan, M. K. Ram, M. M. Verghese, B. D. Malhotra, Dielectric spectroscopic studies on polypyrrole glucose oxidase films, *J. Appl. Polym. Sci.*, Vol. 60, 1996, pp. 2309–2316.

[264]. A. Gambhir, M. Gerard, A. Mulchandani, B. D. Malhotra, Coimmobilization of urease & glutamate dehydrogenase in electrochemically prepared polypyrrole-polyvinyl sulphonate films, *Appl. Biochem. Biotechnol.*, Vol. 96, 2001, pp. 249–257.

[265]. J. L. Romette, J. S. Yang, H. Kusakabe, D. Thomas, Enzyme electrode for specific determination of L-lysine, *Biotechnol. Bioeng.*, Vol. 25, 1983, pp. 2557–2566.

[266]. K. C. Gulla, M. D. Gouda, M. S. Thakur, N. G. Karanth, Reactivation of immobilized acetyl cholinesterase in an amperometric biosensor for organophosphorous pesticide, *Biochem. Biophys. Acta*, Vol. 1597, 2002, pp. 133–139.

[267]. B. G. Milagres, G. D. Neto, L. T. Kubota, H. Yamanka, A new amperometric biosensor for salicylate based on salicylate hydroxylase immobilized on polypyrrole film doped with hexacyanoferrate, *Anal. Chim. Acta*, Vol. 347, 1997, pp. 35–41.

[268]. A. Chaubey, M. Gerard, R. Singhal, V. S. Singh, B. D. Malhotra, Immobilization of lactate dehydrogenase on electrochemically prepared polypyrrole-polyvinylsulphonate composite films for application to lactate biosensor, *Electrochim. Acta*, Vol. 46, 2000, pp. 723–729.

[269]. F. A. Mc Ardle, K. C. Persaud, Development of an enzyme-based biosensor for atrazine detection, *Analyst*, Vol. 4, 1993, pp. 419–423.

[270]. A. Guerrieri, V. Lattanzio, F. Palmisano, P. Zambonin, Electro synthesized poly(pyrrole)/poly(2-napthol) bilayer membrane as an effective anti interference layer for simultaneous determination of acetylcholine and choline by a dual electrode amperometric biosensor, *Biosens. Bioelectron.*, Vol. 21, 2006, pp. 1710–1718.

[271]. F. Cespedes, S. Alegret, New materials for electrochemical sensing II. Rigid carbon-polymer biocomposites, TrAC Trends *Anal. Chem.*, Vol. 9, 2000, pp. 276–285.

[272]. A. N. Ivanov, G. A. Evtugyn, L. V. Lukachova, E. E. Karyakina, H. C. Budnikov, S. G. Kiseleva, New polyaniline-based potentiometric biosensor for pesticides detection, *IEEE Sensors J.*, Vol. 3, 2003, pp. 333–340.

[273]. S. Brahim, D. Narinesingh, A. Guiseppi-Elie, Polypyrrole-hydrogel composites for the construction of clinically important biosensors, *Biosens. Bioelectron.*, Vol. 17, 2002, pp. 53–59.

[274]. P. Shan, C. Mousty, S. Cosnier, Subnanomolar cyanide detection at polyphenol oxidase/clay biosensors, *Anal. Chem.*, Vol. 76, 2004, pp. 178–183.

[275]. A. Ciucu, C. Negulescu, R. P. Baldwin, Detection of pesticides using an amperometric biosensor based on ferophthalocyanine chemically modified carbon paste electrode and immobilized bienzymatic system, *Biosens. Bioelectron.*, Vol. 18, 2003, pp. 303–310.

[276]. G. G. Guilbault, Enzyme electrodes probes, *Meth. Enzymol.*, Vol. 137, 1988, pp. 14–29.

[277]. K. Ramanathan, S. S. Pandey, R. Kumar, A. Gulati, A. S. N. Murthy, B. D. Malhotra, Covalent immobilization of glucose oxidase to poly (o-amino benzoic acid) for application to glucose biosensor, *J. Appl. Polym. Sci.*, Vol. 78, 2000, pp. 662–667.

[278]. M. A. Rahman, D. Park, S. Chang, C. J. McNeil, Y. Shim, The biosensor based on the pyruvate oxidase modified conducting polymer for phosphate ion determinations, *Biosens. Bioelectron.*, Vol. 21, 2006, pp. 1116–1124.

[279]. Rajesh, V. Bisht, W. Takashima, K. Kaneto, A novel thin film urea biosensor based on copolymer poly (N-3-aminopropylpyrrole-copyrrole) film, *Surf. Coat. Technol.*, Vol. 198, 2005, pp. 231–236.

[280]. Rajesh, K. Kaneto, A new tyrosinase biosensor based on covalent immobilization of enzyme on (N-3-aminopropyl) pyrrole polymer film, *Curr. Appl. Phys.*, Vol. 5, 2005, pp. 178–183.

[281]. Rajesh, V. Bisht, W. Takashima, K. Kaneto, Development of a pltentiometric urea biosensor based on copolymer poly(N-3-aminopropyl pyrrole-co-pyrrole) film, *React. Funct. Polym.*, Vol. 62, 2005, pp. 51–59.

[282]. Rajesh, W. Takashima, K. Kaneto, Electrochemical biosensor based on thin conducting polymer film for phenol, *Trans. Mater. Res. Soc. Japan*, Vol. 29, 2004, pp. 771–774.

[283]. P. Oennerfjord, J. Emneus, G. Marko-Varga, L. Gorton, F. Ortega, E. Dominguez, Tyrosinase graphite-epoxy based composite electrodes for detection of phenols, *Biosens. Bioelectron.*, Vol. 10, 1995, pp. 607–619.

[284]. W. Schuhmann, Functionalized polypyrrole covalently attached with glucose oxidase & its application to glucose sensing, *Synth. Met.*, Vol. 429, 1991, pp. 41–43.

[285]. B. F. Y. Yon-Hin, M. Smolander, T. Crompton, C. R. Lowe, Covalent electro polymerization of glucose oxidase in polypyrrole. Evaluation of methods of pyrrole attachment to glucose oxidase on the performance of electropolymerised glucose sensors, *Anal. Chem.*, Vol. 65, 1993, pp. 2067–2071.

[286]. B. F. Y. Yon-Hin, C. R. Lowe, An investigation of 3 functionalized pyrrole-modified glucose oxidase for the covalent electro polymerization of enzymes films, *J. Electroanal. Chem.*, Vol. 374, 1994, pp. 167–172.

[287]. K. Kojima, T. Unuma, T. Yamauchi, M. Shimomura, S. Miyauchi, Preparation of polypyrrole covalently attached with glucose oxidase & its application to glucose sensing, *Synth. Met.*, Vol. 85, 1997, pp. 1417–1418.

[288]. K. Kojima, T. Yamauchi, M. Shimomura, S. Miyauchi, Covalent immobilization of glucose oxidase on poly [1-(2-carboxyethyl) pyrrole] film for glucose sensing, *Polymer*, Vol. 39, 1998, pp. 2079–2082.

[289]. T. Yamauchi, K. Kojima, K. Oshima, M. Shimomura, S. Miyauchi, Glucose-sensing characteristics of conducting polymer bound with glucose oxidase, *Synth. Met.*, Vol. 102, 1992, pp. 1320.

[290]. K. Ramanathan, S. S. Pandey, R. Kumar, A. Gulati, A. S. N. Murthy, B. D. Malhotra, Covalent immobilization of glucose oxidase to poly (o-aminobenzoic acid) for application to glucose biosensor, *J. Appl. Polym. Sci.*, Vol. 78, 2000, pp. 662–667.

[291].M. Aizawa, S. Yabuki, Electrochemical characteristics of an enzyme immobilized conducting polymer membrane, in *Proceedings of the 51st Annual Meeting Japan Chemical Society*, 1985, pp. 6.

[292].P. P. Srangopol, Sandulovipium: M. Trojanowicz, M. L. Hitchman, K. Cammann, (Eds.), Conducting polymers in design of biosensors, *Curr. Top. Biophys.*, Vol. 5, 1996.

[293].F. Palmisano, D. Centonze, P. G. Zambonin, An in situ electro synthesized amperometric biosensor based on lactate oxidase immobilized in a poly-o-phenylenediamine film: determination of lactate in serum by flow injection analysis, *Biosens. Bioelectron.*, Vol. 9, 1994, pp. 471–479.

[294].G. E. De Benedetto, F. Palmisano, P. G. Zambonin, Flow-through tyrosinase enzyme reactor based on reticulated vitreous carbon functionalized by an electrochemically synthesized film, *Anal. Chim. Acta*, Vol. 326, 1996, pp. 149–154.

[295].J. Davis, D. H. Vaughan, M. F. Cardosi, Elements of biosensor construction, *Enzyme Microb. Technol.*, Vol. 17, 1995, pp. 1030–1035.

[296].M. Trojanowicz, O. Geschke, V. Krawozynski, T. Krawczyk, K. Cammann, Biosensors based on oxidase immobilized in various conducting polymers, *Sens. Actuators B*, Vol. 28, 1995, pp. 191–199.

[297].P. N. Bartlett, R. G. Whitaker, M. J. Green, J. Frew, Covalent binding of electron relays to glucose oxidase, *J. Chem. Soc. Chem. Commun.*, 1987, pp. 1603–1604.

[298].D. C. Trivedi, H. S. Nalwa, (Ed.), Handbook of organic conductive molecules and polymers, *Wiley,*New York, 1997, pp. 506.

[299].M. Trojanowicz, V. Krawozynski, T. Krawczyk, Electrochemical biosensor based on enzyme immobilized in electropolymerised films, *Mikrochim. Acta*, Vol. 121, 1995, pp. 167–181.

[300].W. Schuhmann, Conducting polymer based amperometric enzyme electrode, *Mikrochim. Acta*, Vol. 121, 1995, pp. 1–29.

[301].F. Palmisano, P. G. Zamborin, D. Contonzo, Amperometric biosensor based on electro synthesized polymeric films, *Fresenius J. Anal. Chem.*, Vol. 366, 2000, pp. 586–601.

[302].E. Bakker, M. Telting-Diaz, Electrochemical sensors, *Anal. Chem.*, Vol. 74, 2002, pp. 2781–2800.

[303].W. Schuhmann, Amperometric enzyme biosensor based on optimizedelectron transfer pathways and non-manual immobilization procedure, *Rev. Mol. Biotechnol.*, Vol. 82, 2002, pp. 425–441.

[304].P. N. Bartlett, R. G. Whitaker, Strategies for the development of amperometric enzyme electrodes, *Biosensors*, Vol. 3, 1988, pp. 359–379.

[305].J. C. Cooper, E. A. H. Hall, EAH, Electrochemical response of an enzymeloaded polyaniline film, *Biosens. Bioelectron.*, Vol. 7, 1992, pp. 473–485.

[306].Y. Kajiya, H. Sugai, C. Iwakura, H. Yoneyama, Glucose sensitivity of polypyrrole films containing immobilized glucose oxidase and hydroquinone sulphonate ions, *Anal. Chem.*, Vol. 63, 1991, pp. 49–54.

[307]. P. N. Bartlett, Z. Ali, V. Eastwick-Field V, Electrochemical immobilization of enzymes, *J. Chem. Soc. Faraday Trans.,* Vol. 88, 1992, pp. 2677–2683.

[308]. A. Griffith, A. Glidle, J. M. Cooper, Probing enzyme polymer biosensors using X-ray photoelectron spectroscopy: determination of glucose oxidase in electropolymerised films, *Biosens. Bioelectron.,* Vol. 11, 1996, pp. 625–631.

[309]. J. C. Vidal, E. Garcia, J. R. Castillo, Electro polymerization of pyrrole & immobilization of glucose oxidase in a flow system: influence of the operating conditions on analytical performance, *Biosens. Bioelectron.,* Vol. 13, 1998, pp. 371–382.

[310]. W. J. Sung, Y. H. Bae, A glucose oxidase electrode based on electropolymerised conducting polymer with polyanion-enzyme conjugated dopant, *Anal. Chem.,* Vol. 72, 2000, pp. 2177–2181.

[311]. K. Habermuller, W. Schuhmann, A low-volume electrochemical cell for the deposition of conducting polymers and entrapment of enzymes, *Electroanalysis,* Vol. 10, 1998, pp. 1281–1284.

[312]. S. Reiter, K. Habermuller, W. Schuhmann, A reagentless glucose biosensor based on glucose oxidase entrapped into osmium-complex modified polypyrrole, *Sens. Actuators B,* Vol. 79, 2001, pp. 150–156.

[313]. F. Tian, G. Zhu, Bienzymatic amperometric biosensor for glucose based on polypyrrole/ceramic carbon electrode material, *Anal. Chim. Acta,* Vol. 451, 2000, pp. 251–258.

[314]. S. Arjsiriwat, M. Tanticharoen, K. Kirtikara, K. Aoki, M. Somasundrum, Metal-dispersed conducting polymer-coated electrode used for oxidase-based biosensors, *Electrochem. Commun.,* Vol. 2, 2000, pp. 441–444.

[315]. R. Garjonyte, A. Malianauskas, Glucose biosensor based on glucose oxidase immobilized in electropolymerised polypyrrole and poly (ophenylenediamine) films on Prussian Blue-modified electrode, *Sens. Actuators B,* Vol. 63, 2000, pp. 122–128.

[316]. N. G. Skinner, E. A. H. Hall, Investigation of the origin of the glucose response in a glucose oxidase/polyaniline system, *J. Electroanal. Chem.,* Vol. 420, 1997, pp. 179–188.

[317]. H. Xue, Z. Shen, Y. Li, Polyaniline-polyisoprene composite based glucose biosensor with high perm selectivity, *Synth. Met.,* Vol. 124, 2001, pp. 345–349.

[318]. D. Compagnone, G. Federici, J. V. Bannister, A new conducting polymer glucose sensor based polythianaphthene, *Electroanalysis,* Vol. 7, 1995, pp. 1151–1155.

[319]. T. Tatsuma, M. Gondiara, T. Watanable, Peroxidase-incorporated polypyrrole membrane electrodes, *Anal. Chem.,* Vol. 64, 1992, pp. 1183–1187.

[320]. W. Tretinak, I. Lionti, M. Mascini, Cholesterol biosensors prepared by electro polymerization of pyrrole, *Electroanalysis,* Vol. 5, 1993, pp. 753–763.

[321]. J. C. Vidal, E. Garcia, J. R. Castillo, In situ preparation of a cholesterol biosensor, entrapment of cholesterol oxidase in an over oxidized polypyrrole film electrodeposited in a flow system: determination of total cholesterol in serum, *Anal. Chim. Acta*, Vol. 385, 1999, pp. 213–222.

[322]. S. B. Adeloju, J. N. Barisci, G. G. Wallace, Electro immobilization of sulphite oxidase into a polypyrrole film and its utilization for flow amperometric detection of sulphite, *Anal. Chim. Acta*, Vol. 332, 1996, pp. 145–153.

[323]. S. Uchiyama, Y. Hasebe, M. Tanaka, L-ascorbate sensor with polypyrrole-coated carbon felt membrane electropolymerised in a cucumber juice solution, *Electroanalysis*, Vol. 9, 1997, pp. 176–178.

[324]. H. G. Xue, Z. Q. Shen, A highly stable biosensor for phenols prepared by immobilizing polyphenol oxidase into polyaniline-polyacrylonitrile composite matrix, *Talanta*, Vol. 57, 2002, pp. 289–295.

[325]. N. Kizilayar, V. Akbulut, L. Toppare, M. Y. Ozden, Y. Yagci, Immobilization of invertase in conducting polypyrrole/polytetrahydrofuran graft polymer matrices, *Synth. Met.*, Vol. 104, 1999, pp. 45–50.

[326]. R. Dobay, G. Harsanyi, C. Visy, Detection of uric acid with a new type of conducting polymer-based enzymatic sensor by bipotentiostatic technique, *Anal. Chim. Acta*, Vol. 385, 1999, pp. 187–194.

[327]. T. Haruyana, H. Shinohara, Y. I. Kariyama, M. Aizawa, Modulation of the function of an enzyme immobilized in a conductive polymer by electrochemical changing of the substrate concentration, *J. Electroanal. Chem.*, Vol. 347, 1993, pp. 293–301.

[328]. S. Yabuki, F. Mitzutani, M. Asai, Preparation & characterization of an electroconductive membrane containing glutamate dehydrogenase, NADP and mediator, *Biosens. Bioelectron.*, Vol. 6, 1991, pp. 311–315.

[329]. K. Ramanathan, Application of some conducting polymers to biosensors, PhD Thesis, *IIT Delhi*, India, 1995.

[330]. Y. M. Uang, T. C. Chon, Criteria for designing a polypyrrole glucose biosensor by galvanostatic electro polymerization, *Electroanalysis*, Vol. 14, 2002, pp. 1564–1570.

[331]. J. L. Besombes, S. Cosnier, P. Labbe, G. Reverdy, Improvement of the analytical characterstics of an enzyme electrode for free and total cholesterol via laponite clay additives, *Anal. Chim. Acta*, Vol. 317, 1995, pp. 275–280.

[332]. S. Cosnier, M. Fontccave, C. Innocent, V. Niviere, An original electro enzymatic system: flavin reductase-riboflavin for the improvement of dehydrogenase-based biosensors. Application to the amperometric detection of lactate, *Electroanalysis*, Vol. 9, 1997, pp. 685–688.

[333]. C. Mousty, B. Galland, S. Cosnier, Electro generation of a hydrophilic cross-linked polypyrrole film for enzyme electrode fabrication. Application to the amperometric detection of glucose, *Electroanalysis*, Vol. 13, 2001, pp. 186–190.

[334]. Y. Yang, M. Shaolin, Bioelectrochemical responses of polyaniline horseradish peroxidase electrodes, *J. Electroanal. Chem.*, Vol. 432, 1997, pp. 71–78.

[335]. A. Chaubey, K. K. Pande, V. S. Singh, B. D. Malhotra, Co-immobilization of lactate oxidase and lactate dehydrogenase on conducting polyaniline films, *Anal. Chim. Acta*, Vol. 407, 2000, pp. 97–103.

[336]. N. G. R. Mathebe, A. Morrin, E. I. Iwuoha, Electrochemistry and scanning electron microscopy of polyaniline/peroxidase-based biosensor, *Talanta*, Vol. 64, 2004, pp. 115–120.

[337]. A. Callegavi, S. Cosnier, M. Marcaccio, D. Paolucci, F. Paolucci, V. Creorgakilas, Functionalized single wall carbon nanotubes/polypyrrole composites for the preparation of amperometric glucose biosensors, *J. Mater. Chem.*, Vol. 5, 2004, pp. 807–810.

[338]. Y. Nakabayashi, H. Yoshikawa, Amperometric biosensors for sensing of hydrogen peroxide based on electron transfer between horseradish peroxide and ferrocene as a mediator, *Anal. Sci.,* Vol. 16, 2000, pp. 609–613.

[339]. S. Thanachasai, S. Rokutanzono, S. Yoshida, T. Watanabe, Novel hydrogen peroxide sensors based on peroxidase-carrying poly{pyrrole-co-[4-(3-pyrrolyl)butanesulfonate]} copolymer films, *Anal. Sci.,* Vol. 18, 2002, pp. 773–777.

[340]. S. E. Wolowacz, B. F. Y. Yon Hin, C. R. Lowe, Covalent electro polymerization of glucose oxidase in polypyrrole, *Anal. Chem.*, Vol. 64, 1992 pp. 1541–1545.

[341]. M. Yasuzava, T. Nieda, T. Hirano, A. Kunugi, Properties of glucose sensors based on the immobilization of glucose oxidase in Nsubstituted polypyrrole film, *Sens. Actuators B*, Vol. 66, 2000, pp. 77–79.

[342]. Z. Sun, H. Tachikawa, Enzyme based bilayer conducting polymer electrode consisting of poly-metallopthalocyanines and polypyrroleglucose oxidase thin films, *Anal. Chem.*, Vol. 64, 1992, pp. 1112–1117.

[343]. D. J. Strike, N. F. De-Rooij, M. Koudelka-Hep, Electro deposition of GOx for the fabrication of miniature sensors, *Sens. Actuators B*, Vol. 13, 1993, pp. 61–64.

[344]. B. F. Y. Yon-Hin, C. R. Lowe, Amperometric response of polypyrrole entrapped bienzyme films, *Sens. Actuators B,* Vol. 7, 1992, pp. 339–342.

[345]. T. Tatsuma, T. Watanabe, Polypyrrole bioenzyme electrodes with GOx and peroxidase, *Sens. Actuators B,* Vol. 14, 1993, pp. 752–753.

[346]. T. Tatsuma, T. Watanabe, Electrochemical characterization of polypyrrole bioenzyme electrodes with glucose oxidase and peroxidase, *J. Electroanal. Chem.*, Vol. 356, 1993, pp. 245–253.

[347]. M. G. Garguilo, N. Huynh, A. Proctor, A. C. Michael, Amperometric sensors for peroxide, choline and acetylcholine based on electron transfer between horseradish peroxidase and a redox *polymer, Anal. Chem.*, Vol. 65, 1993, pp. 523–528.

[348]. C. Vedrine, S. Fabiano, C. Tran-Minh, Amperometric tyrosinase based biosensor using an electro generated polythiophene film as an entrapment support, *Talanta*, Vol. 59, 2003, pp. 535–544.

[349]. A. Mulchandani, S. Pan, Ferrocene-conjugated m-phenylenediamine conducting polymer-incorporated peroxidase biosensors, *Anal. Biochem.*, Vol. 267, 1999, pp. 141–147.

[350]. J. C. Vidal, E. Garcia, J. R. Castillo, Development of a platinized and ferrocene-mediated cholesterol amperometric biosensor based on electro polymerization of polypyrrole in a flow system, *Anal. Sci.*, Vol. 18, 2002, pp. 537–542.

[351]. A. Arslan, S. Kiralp, L. Toppare, A. Bozkurt, Novel conducting polymer electrolyte biosensor based on poly(1-vinyl imidazole) &poly(acrylic acid) networks, *Langmuir*, Vol. 22, 2006, pp. 2912–2915.

[352]. P. N. Bartlett, P. Tebbutt, C. H. Tyrrell, Electrochemical immobilization of enzymes. 3. Immobilization of glucose oxidase in thin films of electrochemically polymerized phenols, *Anal. Chem.*, Vol. 64, 1992, pp. 138–142.

[353]. G. S. Nunes, G. Jeanty, J. L. Marty, Enzyme immobilization procedures on screen printed electrodes used for detection of anticholine esterase pesticide, comparative study, *Anal. Chim. Acta*, Vol. 523, 2004, pp. 107–115.

[354]. S. Cosnier, C. Innocent, Immobilization of flavin coenzyme in poly (pyrrole-alkylammonium) and characterization of the resulting bielectrode, *J. Electroanal. Chem.*, Vol. 398, 1992, pp. 339–345.

[355]. T. E. Campbell, A. J. Hodgson, G. G. Wallace, Incorporation of erythrocytes into polypyrrole to form the basis of a biosensor to screen for rhesus (D) blood groups and rhesus (D) antibodies, *Electroanalysis*, Vol. 11, 1999, pp. 215–222.

[356]. M. El Kaotit, B. Bouchta, H. Zejli, N. Izaouman, K. R. Temsamami, A simple conducting polymer based biosensor for the determination of atrazine, *Anal. Lett.*, Vol. 37, 2004, pp. 1671–1681.

[357]. Q. Fengli, M. Yang, J. Jiang, G. Shen, R. Yu, Amperometric biosensor for choline based on layer-by-layer assembled functionalized carbon nanotube and polyaniline multilayer film, *Anal. Biochem.*, Vol. 344, 2005, pp. 108–114.

[358]. A. Ramanavicius, A. Kausaite, A. Ramanaviciene, Polypyrrolecoated glucose oxidase nanoparticles for biosensor design, *Sens. Actuators B*, Vol. 111-112, 2005, pp. 532–539.

[359]. N. Zhu, Z. Chang, P. He, Y. Fang, Electrochemically fabricated polyaniline nanowire modified electrode for voltammetric detection of DNA hybridization, *Electrochim. Acta*, Vol. 51, 2005, pp. 3758–3762.

[360]. S. Carrara, V. Bavastrello, D. Ricci, E. Stura, C. Nicolini, Improved nanocomposite materials for biosensor application investigated by electrochemical impedance spectroscopy, *Sens. Actuators B*, Vol. 104, 2005, pp. 221–226.

[361]. S. Sotiropolulou, N. A. Chaniotakis, Lowering the detection unit of acetylcholine esterase biosensor using nanoporous carbon matrix, *Anal. Chim. Acta*, Vol. 530, 2005, pp. 199–204.

Index

www.ingramcontent.com/pod-product-compliance
Lightning Source LLC
Chambersburg PA
CBHW050455190326
41458CB00005B/1291